ALLE ZEIT WACH 1842

T. Wieland, M. Bodanszky

# The World of Peptides

## A Brief History of Peptide Chemistry

With 138 Figures

Springer-Verlag
Berlin Heidelberg New York
London Paris Tokyo
Hong Kong Barcelona

Professor Dr. Theodor Wieland
Max-Planck-Institut für Medizinische Forschung
Jahnstraße 29, 6900 Heidelberg, FRG

Professor Dr. Miklos Bodanszky
One Markham Road 1E, Princeton, NJ 08540, USA

ISBN-13: 978-3-642-75852-2 e-ISBN-13: 978-3-642-75850-8
DOI: 10.1007/978-3-642-75850-8

**Library of Congress Cataloging-in-Publication Data.**
Wieland, Theodor. The world of peptides: a brief history of peptide chemistry/T. Wieland, M. Bodanszky. p. cm. Includes bibliographical references.
ISBN-13: 978-3-642-75852-2
1. Peptides. I. Bodanszky, Miklos. II. Title.
QD431.W46 1991 547.7'56—dc20 90-23821 CIP


Softcover reprint of the hardcover 1st edition 1991



Typesetting: Thomson Press (India) Ltd.;

2151/3020-543210 – Printed on acid-free paper

*To Irmgard and to the memory of Agnes*

# Preface

Almost two centuries ago proteins were recognized as *the* primary materials (proteios = primary) of life, but the significance and wide role of peptides (from pepsis = digestion) in practically all life processes has only become apparent in the last few decades. Biologically active peptides are now being discovered at rapid intervals in the brain and in other organs including the heart, in the skin of amphibians and many other tissues. Peptides and peptide-like compounds are found among toxins and antibiotics. It is unlikely that this process, an almost explosive broadening of the field, will come to a sudden halt. By now it is obvious that Nature has used the combination of a small to moderate number of amino acids to generate a great variety of agonists with specific and often highly sophisticated functions. Thus, peptide chemistry must be regarded as a discipline in its own right, a major branch of biochemistry, fairly separate from the chemistry of proteins. Because of the important role played by *synthesis* both in the study and in the practical preparation of peptides, their area can be considered as belonging to bio-organic chemistry as well.

The already overwhelming and still increasing body of knowledge renders an account of the history of peptide chemistry more and more difficult. It appears therefore timely to look back, to take stock and to recall the important stages in the development of a new discipline. Also, with the passing of time the principal contributors to peptide chemistry, become, as persons, gradually too distant and somewhat forgotten. A few of us are still around who had the good luck to be participants in the exciting early endeavors of peptide research and had also the distinct privilege of knowing some of the ground-breaking investigators in person. The stories they told us about their predecessors, their teachers and about their own work provide an invaluable link to the past. This kind of oral tradition is usually absent from scientific publications and can be found mainly in biographies and autobiographies of famous scientists, for instance in "Aus meinem Leben" by Emil Fischer. Yet, we thought that remembrance of the past of peptide chemistry, in order to reach its full value, should be presented as an integrated continuum, in the form of a book, and undertook the task of writing one.

A noteworthy break in the development of peptide chemistry can be discerned in the nineteen thirties, a geographical change, marked, at least symbolically, by the emigration in 1934 of Max Bergmann from Germany to the United States. Prior to this time contributions by German authors exceeded by far, both in number and in significance, the publications of researchers from other countries. In the following period the center of activity in peptide research shifted west, to Great Britain and to the United States and in later years also to Japan. Only in the last decades could Europe regain a part of the ground lost. The somewhat different backgrounds of the two authors of this volume led to a natural division of the material covering these two major periods. Yet, in spite of the undeniable dichotomy, the book attempts to show an uninterrupted intellectual enterprise. Its aim is to present the history of peptide chemistry as a continued human effort toward lofty goals. The authors will feel rewarded if this objective has been reached, even if only in part. Should "The World of Peptides" also provide entertaining reading for the experienced researcher and some stimulus for the uninitiated, their aim would be fully achieved.

Heidelberg, Princeton

Theodor Wieland
Miklos Bodanszky

# Table of Contents

# Acknowledgements

The authors express their thanks to many colleagues for providing photographs and biographical information. They are grateful to Ms. Barbara J. Gilson, Archivist, The Rockefeller University, for the portraits of Max Bergmann, Lymann C. Craig, Stanford Moore and William Stein and to Mrs. Edita Rudinger, to Mrs. Helene Lederer, Mrs. Jill Kenner, to Mrs. Ursula Geiger to Mrs. Agnes Henschen-Edman and to Mrs. Franca Scoffone for photos of their late husbands, to Prof. A. Zeeck for the picture of H. Brockmann, and to the Max-Planck-Gesellschaft (Minerva Ges.) for the portrait of E. Katchalski. Prof. V.T. Ivanov, Soviet Academy of Sciences kindly provided the picture of M.M. Shemyakin. Helmut Zahn and the insulin team was photographed by Forschelen (Aachen), V. du Vigneaud (with cigar) by Dr. F. Sipos.

The authors are particularly grateful to Prof. Heiner Schirmer for literature references on glutathion and to Frau Beate Isert for her valuable help in the preparation of the manuscript. Agnes Bodanszkys participation in the composition of the text is fondly remembered.

# 1 Introduction. Amino Acids and a Few Early Paradigmatic Peptides

Today, as we are approaching the end of the 20th century, research in chemistry of natural products can look back on two hundred successful years. Organic chemistry, the chemistry of living matter has dealt, since the beginning of its systematic development, with the isolation and analysis of simple natural products. Substances, that became conspicuous on account of their taste, color, odor or some biological activity stood at the focus of interest. As the skills of chemists increased, they turned to more complex problems, such as the structure and chemistry of fats, sugars, or the building components of nucleic acids. Thus began the study of natural products of greater molecular weight. Proteins, the functional molecules of all life processes, belong to this category. They initially appeared as most unsuited objectives of research: many are insoluble in water (keratin from horn, hair and hide; collagen from tendon; silk), the soluble ones, were usually obtained, according to the methods at hand, in non-homogeneous form not as crystalline materials, but as ill-defined substances with indeterminable molecular weight. Peptides, building components of proteins, while more accessible to chemical manipulations, were found in nature in concentrations too low to stimulate systematic chemical studies until the turn of the century.

Peptide chemistry, a subdiscipline of chemistry, developed, therefore, with considerable delay. Its growth is not comparable to the rapid blossoming of the chemistry of aromatic compounds that started with the isolation of benzene from coal gas by Faraday in 1825 and the proposition of its structure by A.v. Kekulé in 1865. These discoveries gave birth to the application of aromatic compounds to the production of dyes and medicines, later followed the emergence of petrochemistry.

The prehistory of peptide chemistry lies hidden in early studies of proteins, in physiological chemistry, for instance in the efforts toward understanding of the relationships between nutrients and the composition of blood.

The importance of "albuminoids", now known as proteins, in animal nutrition was recognized by the first half of the last century through the work of F. Magendie. The analytical studies of G. Mulders in 1840 on substances such as egg albumin, milk casein and blood fibrin led to the formulation of a theory on the energy rich character of protoplasmic proteins. In 1881 Oscar Loew put forward a theory that, for the dynamic properties of protoplasmic proteins, the aldehyde group should be responsible and, according to the chemical knowledge of the time, P.W. Latham wrote in 1897 that albumin "is a compound of cyan-alcohols united to a benzene nucleus, these being derived from the various aldehydes, glycols, ketones or that they may be formed in the living body by the dehydration of the amino-acids". Although several amino acids had been isolated from natural sources by 1897, only three of them, the much earlier described tyrosine, leucine and aspartic acid were considered by Neumeister to be true components of proteins.

Near the end of the 19th century it became firmly established that proteins are composed of α-amino acids, which can be liberated through hydrolytic cleavage catalyzed by enzymes of the digestive system or from microorganisms and also by alkali (P. Schützenberger in 1875 introduced a hot solution of barium hydroxide) or by acids. Major importance was attributed to albumoses and peptones produced in the partial digestion of proteins by the gastric juice (pepsis, gr. = digestion). Today these are recognized to be polypeptides with moderate molecular weight. Such protein fragments were the starting point in the studies of Franz Hofmeister (1850–1922) in the field of proteins, a work that led to the present concept of protein structure.

Even as a medical student and later as professor of pharmacology in Prague and then, from 1896 on, as professor of physiological chemistry in Strasbourg, he dedicated his efforts to this central theme, to both its physiological and chemical aspects. Thus he investigated the separation of proteins from their aqueous solutions by the addition of various salts. Through precipitation with ammonium sulfate be succeeded, in 1889, in the crystallization (!) of ovalbumin. In a by now famous plenary lecture at the 74th meeting of the Society of German Scientists and Physicians, in Karlsbad in 1902, he presented his views on protein structure according to which these are long chains of α-amino acids linked to each other through amide bonds between carboxyl and amino groups [1]. He stressed the importance of the characteristic biuret reaction, the formation of a violet color on addition of copper ions to alkaline solutions of compounds which have several amide groups in their molecule, as seen in the simplest case, biuret: $H_2NCONHCONH_2$. A suggestion of amide-like linkages in proteins had been made by the French chemist E. Grimaux as early as 1882.

At the same Karlsbad meeting, after the general lectures, Emil Fischer (1852–1919) read a paper in which he summarized the isolation of amino acids and peptides from protein hydrolysates and discussed the coupling of amino acids in proteins [2]. In his own words (translated from German): "the idea that acid-amide-like groups play the principal role comes readily to mind, as Hofmeister also assumed in his general lecture this morning". Fischer had begun experiments to link amino acids to each other by methods of organic chemistry as early as 1900. The two lectures mark the emergence of the so called Fischer-Hofmeister theory of protein structure. Emil Fischer, who in a paper with E. Fourneau [3] had just described the first prototype, glycyl-glycine, proposed in his lecture also the designations "peptide, dipeptide, tripeptide, etc." and, later "polypeptides". Proteins, accordingly were considered as polypeptides, but some doubts about protein structure persisted until 1950 when Frederick Sanger dispelled them through the determination of the amino acid sequence of insulin [4]. An excellent survey on early theories of protein structure has been given by Joseph S. Fruton [5].

Assumptions about fundamental structural differences between proteins as they exist in the living organism and as they are present in the test tube after isolation lingered on until the nineteen thirties. Also, hypotheses on the aggregation of relatively small molecules held together by secondary valences were proposed, among them the misleading theory of Emil Abderhalden about proteins as association products of diketopiperazines. The evidence offered by him was less than convincing. In the same year, 1924, Max Bergmann offered a hypothesis that also assumed the association of diketopiperazine-like structures with the difference that his presumed units were unsaturated. Diffraction patterns of X-rays on silk fibroin, obtained by R. Brill in 1923, could have been interpreted in favor of such structural regularity.

Small cyclic units of pyrrol and piperidine type were also postulated as building elements of proteins as late as 1942 by N. Troensegaard as a consequence of his experiments with proteins hydrogenated in water free solvents [6].

A regular structure that for several years attracted more attention was formulated by Dorothy Wrinch in 1937 [7]. Her "cyclol hypothesis" postulates that, in proteins, secondary cyclic structures are present, similar to the crosslinks revealed later in ergot alkaloids, by the addition of amide, —NH-groups to opposite carbonyl groups forming an —N—C(OH)— bond and so preferably 6-membered rings (Fig. 1). Later it was shown in several instances that cyclols indeed occur in small cyclic peptides, but they were never found in proteins.

As the last episode, a rule on the number and sequences of amino acids in the polypeptide chain of proteins should be mentioned, a result of questionable amino acid analysis of sometimes not quite homogeneous proteins. Max Bergmann and C. Niemann in 1936–1938 concluded from their analytical values

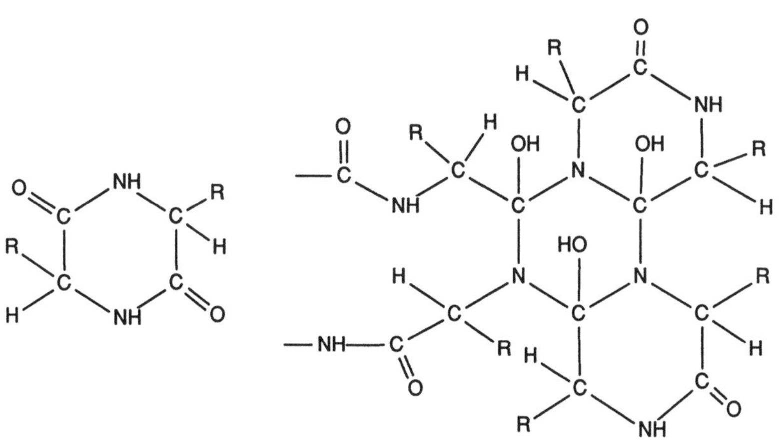

**Fig. 1.** Diketopiperazine; suggested cyclol part of a polypeptide chain

that the amino acid residues per molecule of protein fall into a series of multiples of 288, fibrin having 576 ($2^6 \times 3^2$) and silk fibroin 2592 ($9 \times 288$; $2^5 \times 3^4$) residues, respectively. Furthermore, in a protein the amino acids should occur in stoichiometric proportions, as in fibrin e.g. Glu:Trp:His like 72:18:12, corresponding to periodicities of 8, 32 and 48 respectively, meeting an arithmetical series $1 \times 2^3$, $4 \times 2^3$, $6 \times 2^3$.... From these and additional protein analyses the authors suggested that structural units are periodically arranged within the peptide chain, and "that the natural growth processes (of the protein molecules) must provide a mechanism capable of making a precise selection among the available structural units" [8].

When W. Stein, in Bergmann's laboratory, using a newly developed chromatographic method for amino acid analysis (p. 50) produced data that could not be accomodated with the periodicity rule, the Bergmann hypothesis was abandoned. The understanding of protein molecules as they appear to us at the present time was hindered by the uncertainty about their molecular weight, a major obstacle that persisted until the nineteen thirties. It was against the grain for most chemists to think of hundreds of atoms linked together through covalent, or, as it was called at that time, homoeopolar bonds. Natural products of high molecular weight were supposed to be stable aggregates of relatively small molecules. Determination of the molecular weight with the available (osmotic) methods, such as melting point depression, gave results up to 10000 (daltons) not quite worthy of belief for many scientists. Thus, at the start of his systematic experimentation with protein synthesis Emil Fischer reasoned that with the synthetic methods at hand it should be possible to combine about 20 amino acids to a peptone-like product of a molecular weight of about 3000 and to reach albeit with some difficulty, compounds twice this size, that is proteins.

Improved osmotic measurements gave higher values; for instance in 1907 S.P.L. Sørensen found the molecular weight of the crystalline albumin from eggs to be 34000 daltons. In 1926 the ultracentrifuge of The Svedberg allowed the determination of sedimentation rates and sedimentation equilibria and through these the molecular weight of numerous proteins was established, for instance 34000 was found for the above mentioned ovalbumin. Nevertheless, the concept of real macromolecules remained alien for many investigators as it was also in the case of synthetic polymers to which H. Staudinger right from the beginning assigned a structure of long chain-like molecules, but not without vigorous objections from his colleagues in the field. The contention, that in proteins an aggregation of smaller subunits is at hand, could not be completely refuted, neither through the results with the ultracentrifuge nor by the electrophoretic studies of Arne Tiselius. The last doubts about the chain structure of proteins were removed only in the last few decades by the determination of their molecular weights through electrophoresis in polyacrylamide gels, after denaturation with sodium dodecyl sulfate. In the native state, proteins are macromolecules the polypeptide chain folded in a characteristic way depending of the nature and the sequence of the amino acid constituents. The molecular architecture of proteins has been revealed by X-ray crystallography (p. 131), the first two structural studies of hemoglobin and of myoglobin by M.F. Perutz [9] and J.L. Kendrew [9] required almost twenty-five years. New structural determinations now appear at the rate of about ten or more a year. For present day information see G.E. Schulz and R.H. Schirmer: Principles of Protein Structure [10].

## 1.1 The Amino Acid Building Blocks of Proteins

One of the first natural products that readily offered itself in homogeneous form was an amino acid. In 1806 the French chemists Vauquelin and Robiquet on evaporation of an aqueous extract of *asparagus* shoots obtained a crystalline substance named asparagine. They could not realize at that time that they had a building block of an important class of compounds, proteins, in their hands. In the following years, additional amino acids were found in free form, i.e. not linked to other amino acids via peptide bonds. Some of these compounds were subsequently recognized to be hydrolysis products of proteins as well. Other amino acids were first obtained through chemical or enzymatic cleavage of proteins and two were prepared by synthesis and only later discovered to be protein constituents. Today 20 amino acids are known, as genetically encoded as protein building blocks. One of them, proline, and hydroxyproline, a post-translational hydroxylation product of proline, are cyclic imino acids, which are usually discussed together with the α-amino acids. A list of amino acids in their chronological order of discovery and the year of first synthesis is given below. The formulae are shown in the list below.

Chronology of the Discovery, Structure Elucidation and Synthesis of the Amino Acids

1806 Crystallization of L-asparagine from evaporated extract of *Asparagus* shoots by Vauquelin and Robiquet. Structure recognized only 80 years later.

1810 Cystine, as main component of bladder (gr. *Kystis*) stone, isolated by Wollaston; obtained from protein hydrolyzate only in 1899 by K.A.H. Mörner. Reduction yields cysteine: E. Baumann, 1884. Structure: E. Friedmann 1903. Synthesis: E. Erlenmeyer, 1903.

1819 L-Leucine, crystals (white, gr. *leukos*) from fermented wheat "gluten" or casein (milk curd) by Proust. Structure: E. Schulze and A. Likiernik, 1891. Synthesis of racemate: H. Limpricht, 1855.

1820 Glycine from acid hydrolyzate of gelatin by H. Braconnot. "Sucre de gelatine", "Leimsüß" (gr. *glykys*), formerly "glykokoll", present name glycine by J. Berzelius, 1848. Structure: V.D. Dessaignes, 1845. Synthesis: A. Cahours, 1858.

1846 DL-Tyrosine from alkaline hydrolysate of casein (cheese, gr. *tyros*), or L-form from acid hydrolyzate of horn by J.v. Liebig. Structure proven by synthesis. E. Erlenmeyer and A. Lipp, 1853.

1848 L-Aspartic acid by hydrolysis of its ß-amide asparagine by R. Piria. From hydrolyzate of plant protein: H. Ritthausen, 1868. Structure recognized by synthesis: A. Piutti, 1888.

1850 DL-Alanine, synthesized from acetaldehyde, ammonia, and hydrocyanic acid by A. Strecker. As constituent of a protein (silk) isolated by P. Schützenberger and A. Bourgois, 1875.

1856 L-Valine isolated from extracts of pancreas gland by E.v. Gorup-Besanez, as racemate from baryta-hydrolyzed protein by P. Schützenberger, 1879, as L-valine from HCl-hydrolyzate of casein in 1901 by E. Fischer, who also established its structure as α-amino-isovaleric acid and named it valine in 1906.

1865 L-Serine isolated from a sulfuric acid hydrolyzate of sericine, a protein component of silk (lat. *sericus* silken) by E. Cramer. Structure by the same author. Synthesis: E. Fischer and H. Leuchs, 1902.

1866 L-Glutamic acid from hyrolyzate of wheat gluten by H. Ritthausen. Structure: W. Dittmer, 1872; W. Markovnikov, 1876. Synthesis: from levulinic acid: L. Wolff, 1890. 1877: L-Glutamine from sugar beet, E. Schulze, see glutamic acid.

1879 L-Phenylalanine from alcoholic extract of seedlings of *Lupinus luteus* by E. Schulze and J. Barbieri, 1891 recognized as a component of *Lupinus* protein. Structure confirmed by synthesis: E. Erlenmeyer and A. Lipp, 1882. Resolution of racemate by E. Fischer, 1900.

1886 L-Arginine (precipitated as silver salt, lat. *argentum*) from seedlings of *Lupinus* by E. Schulze and E. Steiger. Recognized as a protein component 1895 by S.G. Hedin. Structure: E. Schulze and E. Winterstein, 1897. Synthesis: S.P.L. Sörensen, via L-ornithine, 1910.

1889 L-Lysine from acid-hydrolyzate (gr. *lysis*) of casein by E. Drechsel. Structure definitely established by synthesis by E. Fischer and F. Weigert 1902.

1896 L-Histidine (gr. *histion*, tissue) isolated by A. Kossel from acid-hydrolyzate of sturin (a fish sperm protamine), and at the same time by S.G. Hedin from hydrolyzates of other proteins. Structure established by synthesis: F.L. Pyman, 1911.

1900 L-Proline. D,L-proline was synthesized before its discovery in proteins by R. Willstätter, L-proline isolated from acid-hydrolyzate of casein by E. Fischer, 1901. Name from p(yr) roli (di) ne-carboxylic acid.

1902 L-Tryptophan. Name given as early as 1890 by R. Neumeister "appears (gr. *phainei*) after tryptic digestion". Isolated from enzymatically digested casein by F.G. Hopkins and S.W. Cole. Structure proven by synthesis by A. Ellinger and C. Flamand, 1907.

1902 L-Hydroxyproline, formerly "oxyproline" isolated from acidic hydrolyzate of gelatin by E. Fischer. Structure by synthesis by H. Leuchs, 1905. Found only in collagen.

1904 L-Isoleucine from sugarbeet molasses, a little later from protein hydrolyzates by Felix Ehrlich. Structure established by synthesis: L. Bouveault and R. Loquin, 1906.

1922 Methionine (methylthio group) from casein hydrolyzate by J.H. Mueller. Structure confirmed by synthesis: G. Barger and F.P. Coyne, 1928.

1936 Threonine from fibrin hydrolyzate by C.E. Meyer and W.C. Rose. Threose-like configuration at carbon atoms 1 and 2. Synthesis: H.D. West and H.E. Carter, 1937.

---

In the chronological listing and in Table 1 the amino acids that are protein constituents were considered without references to the extensive literature. Only a few general sources could be pointed out as the three volume work of J.P. Greenstein and M. Winitz, Chemistry of the amino acids, Wiley, New York, 1961, and Th. Wieland et al. in Houben–Weyl (E. Müller, ed.) G. Thieme, Stuttgart, Vol. XI/2, 271–509 (1958).

All amino acids can be regarded as derivatives of glycine substituted with different side chains. These side chains are depicted in Table 1 for each amino acid together with the corresponding three letter and one letter abbreviations.

The existence of some amino acids (oxyglutamic acid, α-amino-butyric acid, norvaline, norleucine) claimed as protein constituents, could not be confirmed

**Table 1.** Amino acids of proteins with three letter and one letter notation and structure of side chains

| Amino acids<br>Name | Three letter | one letter | R in $H_2N-C(CO_2H)(H)-R$ |
|---|---|---|---|
| Alanine | Ala | A | $CH_3$ |
| Arginine | Arg | R | $CH_2CH_2CH_2N(H)-C(=NH)NH_2$ |
| Asparagine | Asn | N | $CH_2CONH_2$ |
| Aspartic acid | Asp | D | $CH_2CO_2H$ |
| Cysteine | Cys | C | $CH_2SH$ |
| Glutamic acid | Glu | E | $CH_2CH_2CO_2H$ |
| Glutamine | Gln | Q | $CH_2CH_2CONH_2$ |
| Glycine | Gly | G | H |

*(Continued)*

**Table 1.** *(Continued)*

| Amino acids<br>Name | Three letter | one letter | R in $H_2N-C(CO_2H)(H)-R$ |
|---|---|---|---|
| Histidine | His | H | 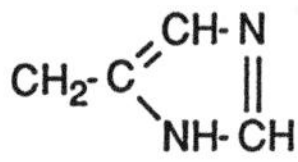 |
| Isoleucine | Ile | I | $CH(CH_3)CH_2CH_3$ (S) |
| Leucine | Leu | L | $CH_2CH(CH_3)_2$ |
| Lysine | Lys | K | $CH_2CH_2CH_2CH_2NH_2$ |
| Methionine | Met | M | $CH_2CH_2SCH_3$ |
| Phenylalanine | Phe | F | 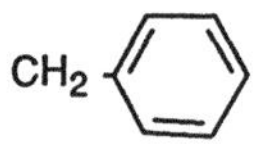 |
| Serine | Ser | S | $CH_2OH$ |
| Threonine | Thr | T | $CH(OH)CH_3$(R) |
| Tryptophan | Trp | W | 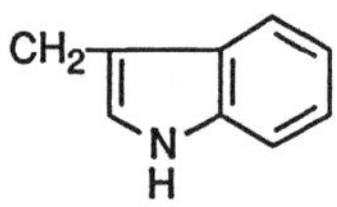 |
| Tyrosine | Tyr | Y | 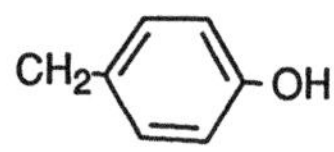 |
| Valine | Val | V | $CH(CH_3)_2$ |
| Proline | Pro | P | 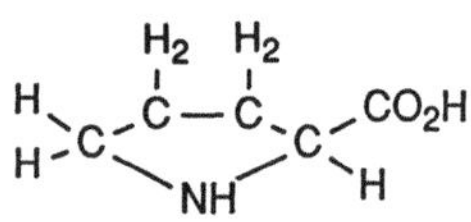 |

during the following decades [11]. Nowadays, in the era of chromatography (see pp. 49, 52), when experimental errors are unlikely to occur, hundreds of new amino acids have been detected in natural products over the past 40 years. Several of the non-protein amino acids will be mentioned in connection with interesting peptides containing such building blocks.

In proteins, there occur 20 amino acids, which take part in ribosome-mediated protein synthesis. Many non-standard amino acids, however, are found in a variety of proteins the best known of them being *trans*-4-hydroxyl-L-proline (in collagen) omitted in Table 1. These additional amino acids arise from modifications of the side chains in polypeptides after synthesis by the normal mechanism. Beside hydroxylation, *N*-methylation occurs, but also esterification of hydroxyl groups (serine, tyrosine) with phosphoric acid or sulfuric acid, and—very important—glycosylation (linking with carbohydrate) of the serine or threonine hydroxyl group or the amide nitrogen of asparagine side chain. The post-translational introduction of carboxyl groups into the 4-position of glutamic acid of clotting proteins (γ-carboxyglutamic acid) plays an important

**Table 2.** Some amino acid derivatives formed by post-translational modification in polypeptides. For references see H.B. Vickery [11]

| Name | Formula |
|---|---|
| 4-Hydroxy-L-proline | 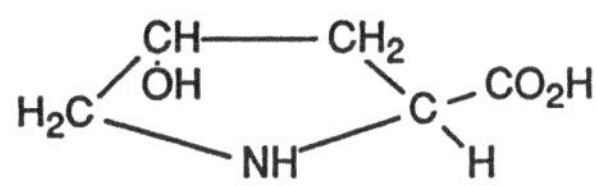 |
| 3-Hydroxy-L-proline | 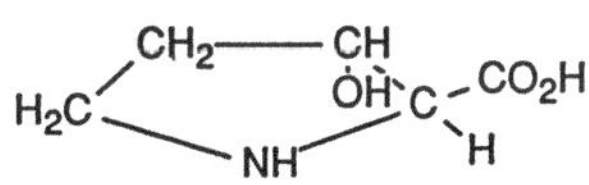 |
| 5-Hydroxy-lysine | $H_2N-CH_2-CH(OH)-(CH_2)_2-CH(NH_2)CO_2H$ |
| ε-N-Methyl-lysine | $CH_3NH-(CH_2)_3-CH(NH_2)CO_2H$ |
| ε-N-Dimethyl-lysine | $(CH_3)_2N-(CH_2)_3-CH(NH_2)CO_2H$ |
| 3-Methyl-histidine | 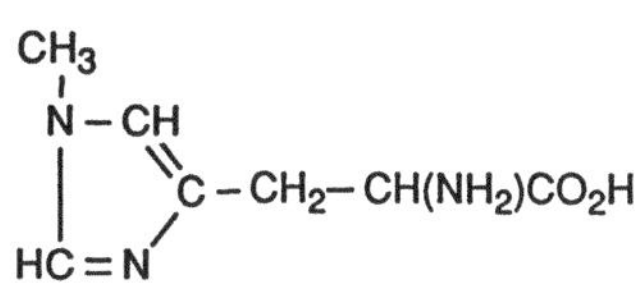 |
| 3,5-Diiodo-tyrosine | 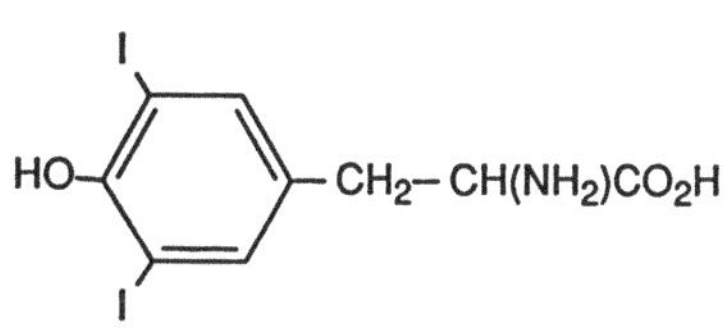 |
| Tyrosin-O-sulfate | 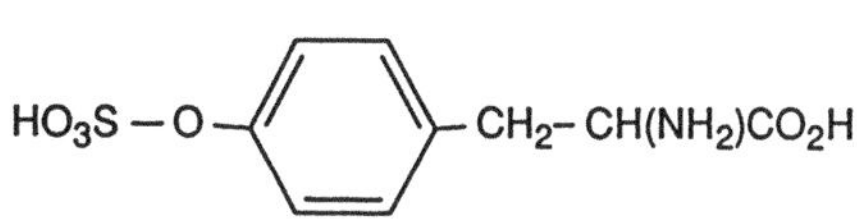 |
| γ-Carboxyglutamic acid | $HO_2C-CH(CO_2H)-CH_2CH(NH_2)CO_2H$ |

role in blood coagulation [13]. Some of enzymatically modified amino acids are listed in Table 2.

Selenocysteine, HSe—$CH_2CH(NH_2)CO_2H$, located in the active center of selenoenzymes (for example *E. coli* formate dehydrogenase, mammalian glutathione peroxidase) is incorporated in *E. coli into the polypeptide chain in an unconventional* way. There is a minor serine transfer RNA species specific for selenocysteine, recognizing UGA, one of the three stop (nonsense) codons. Its serine moiety is transformed at the hydroxyl to yield selenocysteine-t-RNA which participates in the polypeptide synthesis [12].

With the exception of glycine all amino acids which are constituents of proteins rotate the plane of polarized light. They are "optically active" because their α-carbon atom is a chiral center. All have the L, or according to a more contemporary designation the *S*-configuration, a notable exception being (*R*)-cysteine (see below). Two amino acids, threonine and isoleucine contain a second chirality center located in their side chains. As indicated in Table 1 the configurations are *R* in Thr and *S* in Ile.

From alkaline hydrolyzates the amino acids are recovered as racemates, i.e. containing equal amounts of the L- and the enantiomeric (mirror imaged) D-isomers. For the description of the spatial arrangements of the four different substituents around the "asymmetric" carbon atom Emil Fischer proposed the so called projection-formulas. In these the absolute configuration is depicted on paper by writing the carboxyl group above and the side chain below the chiral carbon atom; the amino group is drawn left (*laevus*, L) the hydrogen atom right of it, opposed to it is the D-enantiomer (*dexter*, D) (see Fig. 2).

The chiral carbon atom should be imagined lying in the plane of the paper, the amino group and the hydrogen atom above while the carboxyl group and the side chain below this plane. Thereby the four substituents are visualized in the corners of a tetrahedron with the chiral carbon atom at its center. When J.H. Bijvoet, 1951, with his X-ray diffraction method determined the absolute configuration of D-isoleucine, Fischer's originally arbitrary spatial arrangement (50% probability) turned out to be correct.

Optically active compounds were previously, according to the sign of rotation of the plane of polarized light, described as *l* (laevorotatory) or *d* (dextrorotatory) later L and D. The designations *R* and *S*, introduced in 1956 by R.S. Cahn, C.K. Ingold and V. Prelog, are based on definitions different from the above given meaning of the symbols D and L. Here, briefly, the arrangement at the chiral carbon atom is expressed in such a way that its four ligands which differ, roughly

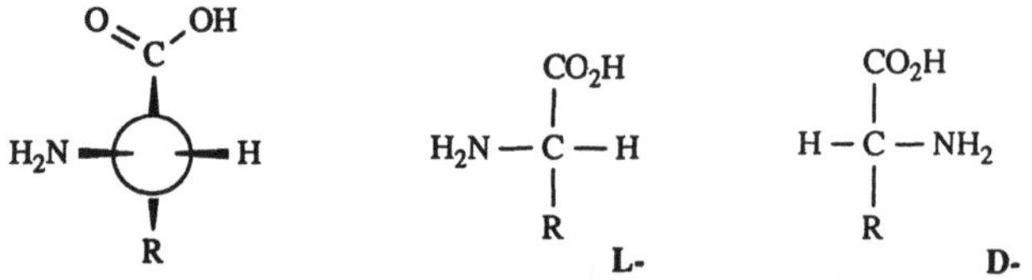


**Fig. 2.** Spatial view of an L-amino acid; Fischer-projection of L- and D-amino acids

speaking by their molecular weights are numbered according to decreasing priorities and that the direction of the sequence 1 to 4 is determined; if clockwise: *R* (lat. *rectus*, right), anticlockwise: *S* (*sinister*, left). This is the reason why L-cysteine, due to its big $CH_2SH$-residue, has to be designated (*R*)-cysteine, whereas all other natural L-amino acids fall under the (*S*)-designation.

In peptides and proteins a continued condensation links the carboxyl groups, through a peptide bond to the amino group of the next residue in the sequence: —NH—CH($R^1$)—CO—NH—CH($R^2$)—CO—. The —NH—CH(R)—CO-grouping is the amino acid residue, in which R is the side chain characteristic for the individual amino acid. Conventionally, the peptide bonds in the chain are written, as shown here, in the —CO—NH— direction. The amino acid residues are designated with a three-later symbol, the peptide bond usually with a hyphen. Thus, the segment —alanyl-glycyl— of a peptide chain appears as -Ala-Gly- or AlaGly. The same amino acids at the amino terminus of a peptide chain and at the carboxyl end, respectively, are written as HAla....GlyOH. For long polypeptide chains a single letter code can also be applied (cf. Table 1). To residues commonly used in peptide chemistry defined abbreviations like Bzl for benzyl, TFA for trifluoracetyl, Ac for acetyl, OMe, OEt for methyl- or ethylester are attributed, which are combined with the amino acid symbols by or without a hyphen: TFA-GlyOH, BzlAlaOEt.

Functional side chains of amino acids if derivatized or involved in a peptide structure can be written in parentheses e.g., *N*-tosylaspartic acid (*β*-methyl-)-benzylester (a): Tos-Asp(OMe)OBzl; *S*-benzylcysteinyl-D-valine (b): HCys-(Bzl)-D-ValOH (Fig. 3).

Assignment at the L-series (L-prefix) is omitted in most cases whereas D-amino acids are marked.

The peptide bond lends a certain rigidity to peptides. The atom participating in the —CO—NH— bond lie in the same plane: the arrangement of the atoms is stabilized through resonance. Both *cis* and *trans* configurations are possible with bond angles of about 120°.

Because of steric reasons (larger distance between R and R′) the *trans* configuration is strongly preferred. Nevertheless, peptide chains can take up many possible conformations since there is free rotation around the bonds linked to the CO and NH groups. The torsion angles $\phi$ and $\psi$ can have values ranging from 0° to +180° or to −180°. Hence, theoretically an infinite number of

a

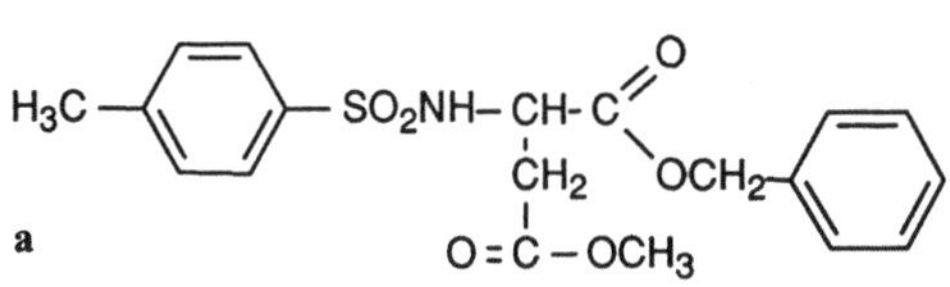


b

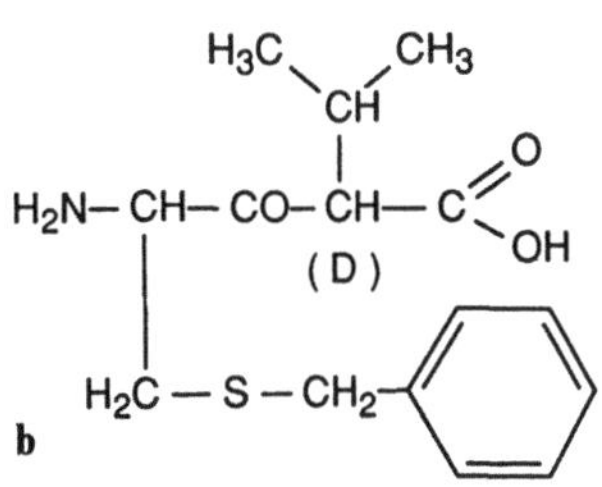


**Fig. 3 a, b.** Formulae and nomenclature

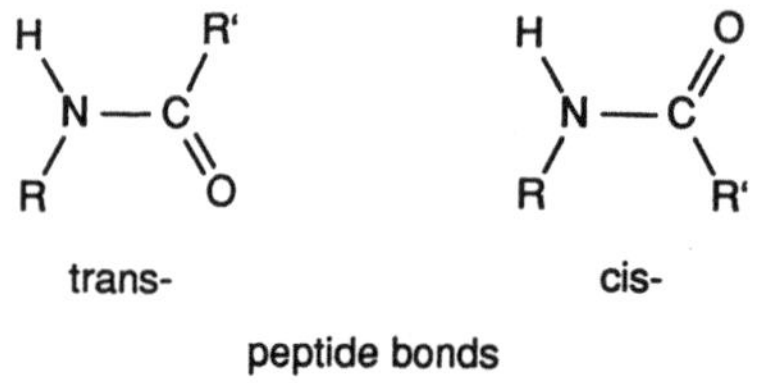


**Fig. 4.** *Trans-* and *cis*-peptide bond

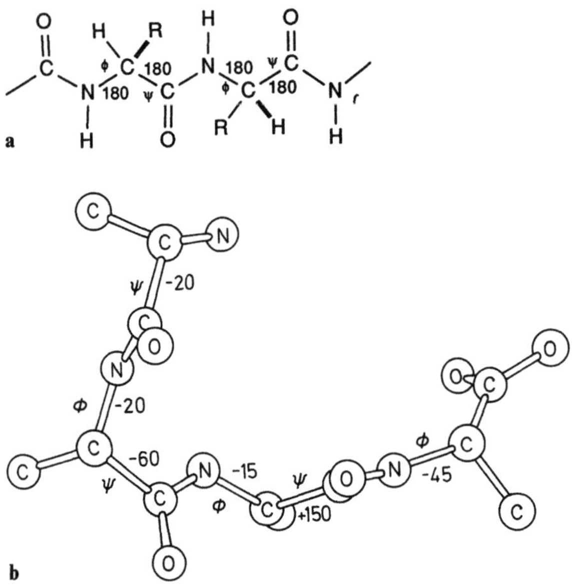


**Fig. 5 a, b.** Straight peptide chain (**a**) and random peptide chain (**b**)

conformations is possible. In reality their number is limited by interaction between the side chains (Ramachandran plot). In the here depicted linear segment (a) of a peptide chain $\phi = \pm 180°$, $\psi = \pm 180°$; in the bent, random segment (b) the values are as indicated in Fig. 5.

The conformation of peptide chains of sufficient length is stabilized by hydrogen bonds, that is by non-covalent forces between the NH group in one peptide bond and the CO group in another: $>C{=}O \cdots H{-}N<$. The most frequently occurring "secondary structures" are the $\alpha$-helix and $\beta$-pleated sheets beside random chain (Fig. 6).

Furthermore, the conformation of peptides can also be stabilized by disulfide bonds between cysteine side chains and, to a lesser extent, through ionic interactions ($-\overset{+}{N}H_3\ {}^{-}O_2C-$) and also via "hydrophobic interaction" between water-repellent side chains ($-CH_2CH(CH_3)_2$Leu; $-CH_2C_6H_5$, Phe, etc.).

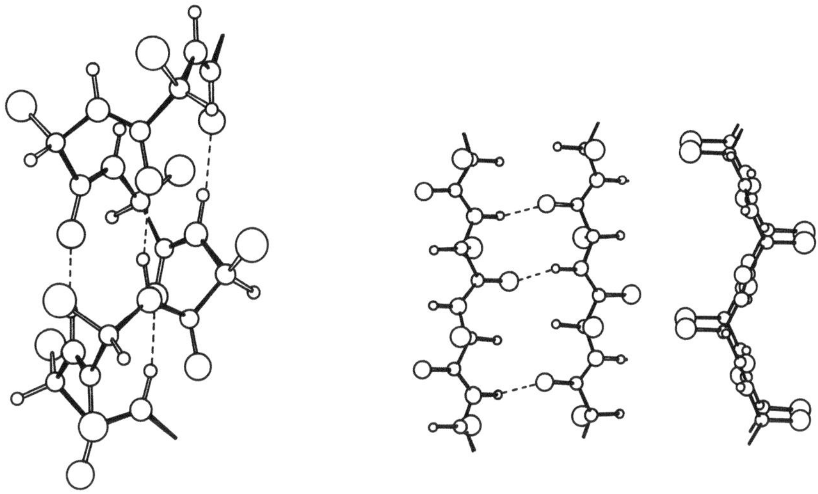

**Fig. 6.** α-Helix and β-pleated sheet, antiparallel chains

Compounds in which amino acids are linked to each other by amide bonds, according to the principle of protein structure, were called peptides by Emil Fischer. Dipeptides contain 2, tripeptides 3, etc. and polypeptides "many" amino acids. The upper limit that marks the transition to proteins, remains vague. Nowadays protein chains are described as "polypeptides" in numerous publications. It might perhaps be reasonable to consider peptides to have a mass of 10 000 daltons (10 kDa) as maximum. Some time ago B. Helferich proposed the designation "oligopeptide" for compounds up to the range of 10 amino acid residues, but this name has not been widely accepted.

In connection with the chemical structure of peptides it must be added that the requirement of normal α-peptide bonds (α-carboxyl linked to α-amino groups) was gradually relinquished. The amide bonds can involve amino groups in β, γ, etc. positions, building components other than amino acids can participate in the molecules and even other modes of binding, such as ester bonds. This kind of peptide is commonly produced by microorganisms and was designated by R. Schwyzer as "heterodetic" in contrast to "homodetic" peptides built entirely by amide bonds. In microbial peptides, D-amino-acids are also frequently found as building components.

Peptides occur in various types of structure (Fig. 7) Linear peptides obviously possess two ends, the first or N-terminal and the last or C-terminal amino acid. In *cyclic* peptides there is, of course, no N-, nor C-terminus. Partially cyclic peptides contain an amino terminus, if their carboxyl end has been bound intermolecularly to an amino group of a side chain or to a hydroxy group thus forming a smaller or larger lactam or lactone ring, respectively. Analogously a carboxyl

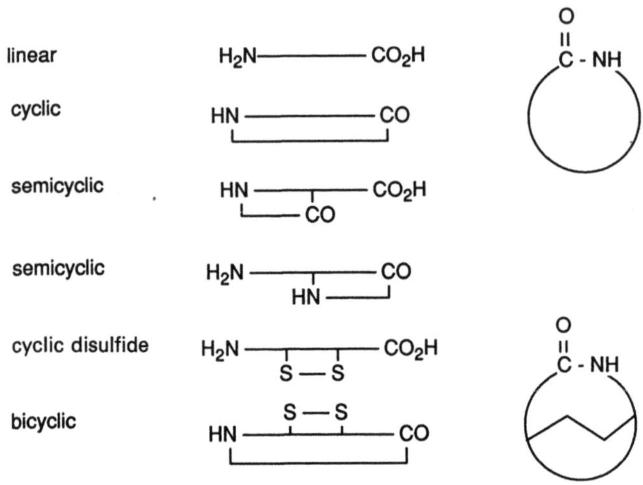


Fig. 7. Some structures of peptides

end group may be free, if the corresponding amino group has been intramolecularly linked to a side-chain carboxyl group. Cyclic peptides formed from linear peptides by an intramolecular disulfide bridge of two cysteine residues retain the two ends of a normal peptide. If these ends would be connected by an additional peptide bond a bicyclic peptide would be formed (see malformin on p. 20).

## 1.2 Some Naturally Occurring Peptides [14]

Free peptides in high concentrations are very rare in nature and were not detected in the early days of peptide research. After the introduction of chromatographic methods for separation and sensitive analysis, particularly of paper chromatography by Consden, Gordon and Martin in 1944 (see p. 52), a vast number of peptides were discovered in natural sources such as microorganisms, plants and animal cells. In the following we will only mention representative compounds which can be regarded as "classical", meaning that they had already attracted the attention and activity of organic chemists at a time when peptide chemistry was not yet so popular as it has become in the last 30 years.

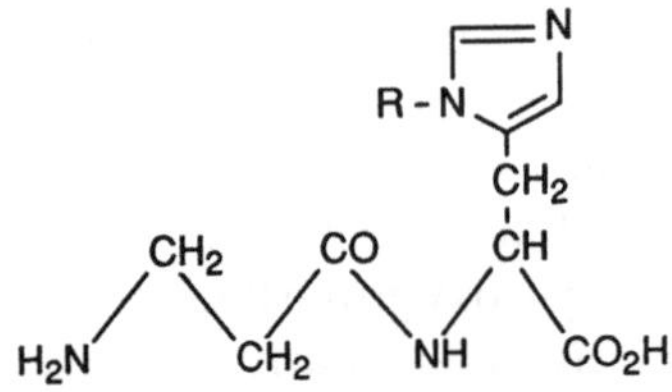


Fig. 8. Carnosine; R = H, Anserine R = $CH_3$

The first peptide, isolated from beef extract as early as 1900 by Gulevitch and Amiradzibi, *carnosine*, has the formula of β-alanyl-L-histidine; (Fig. 8) it was synthesized by Barger and Tutin in 1918. *Anserine*, the 1-methyl-histidine analog was first isolated from goose muscle in 1929 by Ackermann et al., and synthesized by Behrens and du Vigneaud in 1937. No specific physiological function could have been ascribed to these "abnormal" dipeptides at that time.

### 1.2.1 Glutathione, Structure and Biochemical Functions

Glutathione is the most widespread and probably the best studied among the naturally occurring peptides. It was observed as a reducing agent in yeast by de Rey-Pailhade as early as 1888, named "philothione", but isolated by F.G. Hopkins from yeast, liver and muscles, only in 1921 and formulated as a γ-peptide of glutamic acid, cysteine and glycine by Pirie and Pinhey, 1929 (Fig. 9). The tripeptide γ-L-gutamyl-L-cysteinyl-glycine was first synthesized by Harington and Mead in 1935 (see p. 64) and in the following years by every chemist who became engaged in a new method in preparative peptide chemistry—as a journeymen's test of his novel method. On such occasions numerous analogs have also been prepared.

Glutathione (GSH) has a multitude of biochemical functions within the cell. Due to its γ-peptide bond it is relatively stable against enzymatical hydrolysis and so can reach high concentrations (up to 0.3 g $L^{-1}$) in organs and in cells. It is the most abundant intracellular thiol in almost all aerobic biological species. Due to its SH-group it is a strong reducing compound, the redox potential $E'_0$ of the system 2 GSH$\rightleftharpoons$GSSG being −0.23 volt. This value is comparable to that of the main reducing agents NADH and NADPH ($E'_0 = -0.32$ volt). The oxidized form, GSSG, can therefore be easily reduced to GSH by the reduced nicotinamide nucleotides. The concentration of GSH inside the mammalian cell is more than a hundred times higher than that of GSSG.

Closely related to the reducing activity of the thiol group is its chemical, nucleophilic activity, which can work in the absence or the presence of enzymes (thioltransferases). Thus disulfide bonds both in homogenous ($R^1 = R^2$) as well as

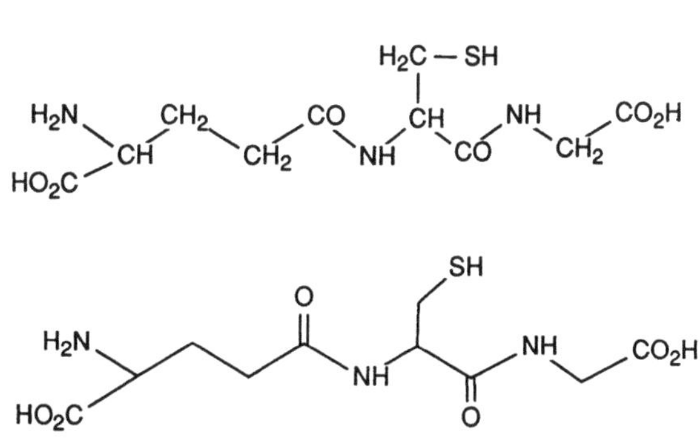

**Fig. 9.** Glutathione, ophthalmic acid

Glutathione, lower formula simplified line drawing
Ophthalmic acid: $CH_3$ instead of SH

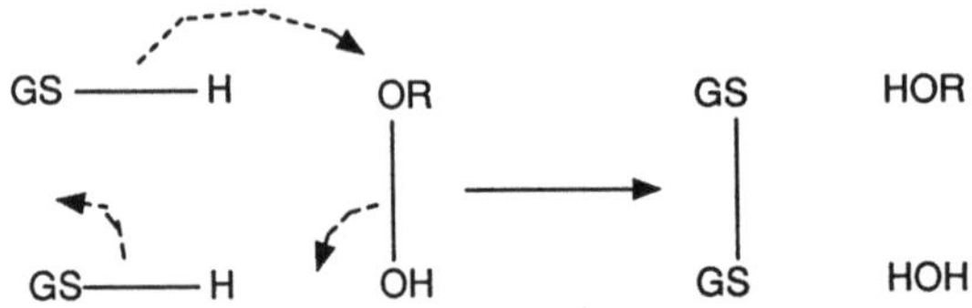


**Fig. 10.** Mechanism of glutathione peroxidase reaction

heterogenous ($R^1 \neq R^2$) compounds are exchanged in a thiol transfer reaction by GSH.

$$R^1{-}S{-}S{-}R^2 + 2GSH \longrightarrow R^1SH + R^2SH + GSSG$$

By this reaction essential SH groups of e.g. enzymes can be set free. Reduced GSH is regenerated by NADPH through the catalytic action of glutathione reductase, a flavin enzyme. Because of its high reactivity GSH also will scavenge free oxygen radical species (HO—O; HO·) or any radical generated by them, and, will generally function in protecting cells against the toxicity of oxygen in the form of non-radical products as well.

A key enzyme in protection against oxygen toxicity is glutathione peroxydase, a seleno-enzyme already mentioned on p. 10 which catalyzes the decomposition of hydrogen peroxide or organic hydroperoxides according to the reaction of Fig. 10.

In order to be recycled, GSSG has to be reduced by glutathione reductase. The mechanism of the reaction has been resolved mainly by X-ray diffraction studies [15]. From the juxtaposition of the reactive moiety, it appears that the electrons from dihydroflavin adenine dinucleotide (FADH) will reduce an intramolecular disulfide bond in the enzyme forming 2 SH-groups which in turn will reduce the GSSG disulfide.

Glutathione-*S*-transferases are the second class of enzymes protecting the cell against toxicity, also of hydroperoxides, but mainly of toxic, sulfhydryl-reactive agents generated by oxidation. The long recognized formation of "mercapturic acids" e.g. phenylmercapturic acid, $C_6H_5S{-}CH_2{-}CH(NHCOCH_3)CO_2H$, *S*-phenyl-*N*-acetyl-L-cysteine, in dogs after ingestion of aromatic compounds, here

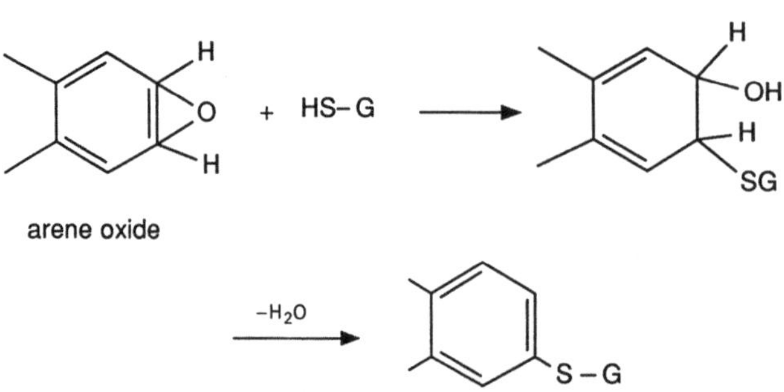


**Fig. 11.** Arylation of cysteine sulfur in glutathione (GSH)

benzene, is the consequence of the primary formation of an epoxide (arene oxide) by reaction with oxygen and subsequently of GSH with this electrophilic species (Fig. 11). *S*-Arylglutathiones are then hydrolyzed yielding *S*-aryl-cysteine which is acetylated and excreted.

If this mechanism of detoxification does not work completely as in the case of internalized carcinogenic polycyclic hydrocarbons like benzopyrene, intermediate epoxides may react with bases of DNA and cause unregulated cell growth. Many other foreign substances are, directly or after adjustment by microsomal interaction, targets for conjugation with GSH and thus detoxified and excreted by the bile. In the elimination of toxic heavy metal ions ($Hg^{++}$, $Cu^{++}$, $Ag^{+}$, $Cd^{+}$ etc.) GSH reacts with its SH group forming non-dissociating compounds.

As special cases of a detoxifying function the glyoxalase reaction and formaldehyde dehydrogenase may be considered. Oxidation processes in the cell may give rise to carbonyl compounds, among which 2-oxoaldehydes are known as strong mutagens. Glyoxalase catalyses the formation of the corresponding α-hydroxy acid, e.g. lactic acid from methylglyoxal (Fig. 12).

Similarly, GSH is a cofactor in the enzymatic oxidation of the toxic formaldehyde. Here, the *S*-methylol—GSH adduct is oxidized by $NAD^{+}$, the *S*-formyl intermediate hydrolyzed yielding GSH and formic acid.

Conjugation with GSH also functions at the formation of *leukotrienes*, mediators of inflammatory and anaphylactic reactions. Leukotriene A, an epoxide derived from arachidonic acid, reacts with GSH to form the conjugate, leukotriene $C_4$ ($LTC_4$). By the action of γ-glutamyl transpeptidase, an enzyme that transfers the γ-glutamyl moiety of GSH to other amino acids, leukotriene D responsible for slow anaphylaxis is formed. Further hydrolysis by dipeptidase

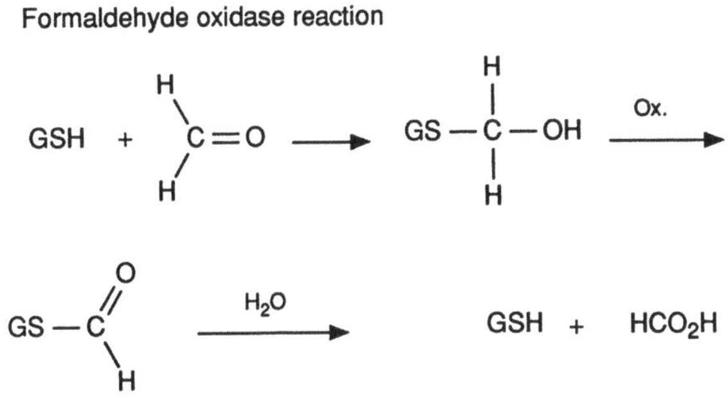


**Fig. 12.** Glyoxalase reaction, formaldehyde dehydrogenase reaction

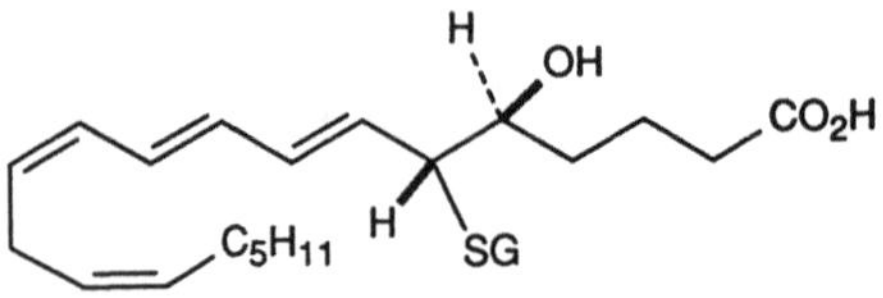


**Fig. 13.** Leucotriene $LTC_4$, a conjugate of oxidized arachidonic acid and glutathione

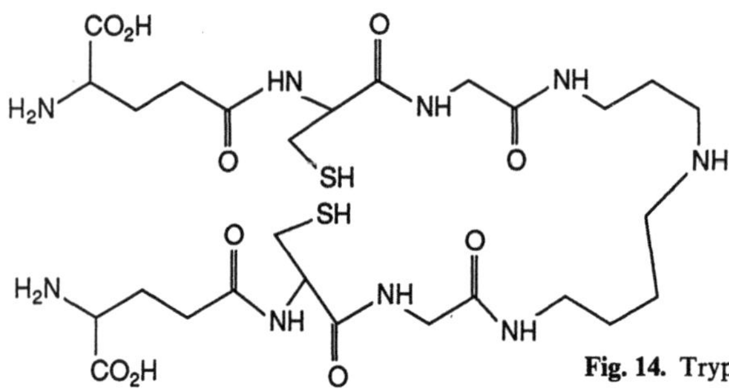


**Fig. 14.** Trypanothione

removes the glycine residue thus yielding the corresponding *S*-substituted cysteine, leukotriene E (Fig. 13).

The enzymes just mentioned also work at the inward transport of amino acids through the cytoplasmic membrane of cells, particularly of kidneys. The amino acid to be transported is converted to its γ-glutamyl conjugate by the action of γ-glutamyl transferase, permeates the membrane in this form, and is set free inside the cell (Meister [16]).

In trypanosomes a strange variant of the glutathione redox system was detected by Fairlamb in 1985 [17]. There, trypanothione, a double molecule of two glutathione moieties crosslinked at the glycine ends by a molecule of spermidine, plays the role of GSH in removing peroxides (Fig. 14).

Finally, an effect of GSH is of interest which completely differs from the chemically comprehensible reactions reported above. It is a feeding response in Hydra and possibly other organisms [18]. For intake of its prey the tentacles of the polyp contract in the direction of the mouth. This reaction is biochemically induced by $10^{-6}$ M GSH. This is also true for other Coelenterates and also for suckling motions of other animals like the tse-tse fly or the leech. The effect is quite specific in that even the closely related asparthione is inactive. That it is a biological action that does not depend on the SH group, but on the recognition of the peptide backbone by the animals, is proven by the analogous and even somewhat greater activity of *ophthalmic acid* (see in Fig. 9).

### 1.2.2 Opthalmic Acid

Ophthalmic acid is a peptide isolated from calf lens by S.G. Waley in 1956 [19]. Its structure γ-glutamyl-L-α-aminobutyryl-glycine closely resembles that of glutathione, the SH group of GSH being replaced by a $CH_3$ group. Like with most of the naturally occurring small peptides biological synthesis does not take place on ribosomes: for α-aminobutyric acid (Abu), a non-protein aminos acid, there is no codon. Two enzymes from lens catalyse the γ-coupling of Glu to Abu, and the conjugation of γ-GluAbu with Gly. Both reactions are driven by ATP hydrolysis, to form the tripeptide—in analogy to the biosynthesis of glutathione. As an antagonist to GSH, ophthalmic acid may have some regulatory functions. The scope of glutathione research since it discovery has increased to such an extent, that a two volume monograph on glutathione is being published [20].

From the multitude of microbial peptides only a few selected compounds can be discussed here. Several biologically active substances will be treated, however, in subsequent chapters of this book, for instance peptide hormones in Chap. 7, antibiotics, ionophores and toxins in Chap. 9. Here we describe three long-known compounds which attracted the interest of the, at that time still small, circle of peptide chemists.

### 1.2.3 Lycomarasmin, Malformin

Probably the longest-known microbial "peptide" is *lycomarasmin* of Plattner and Clauson-Kaas in 1945 [21]. It is a phytotoxin that causes curling and wilting

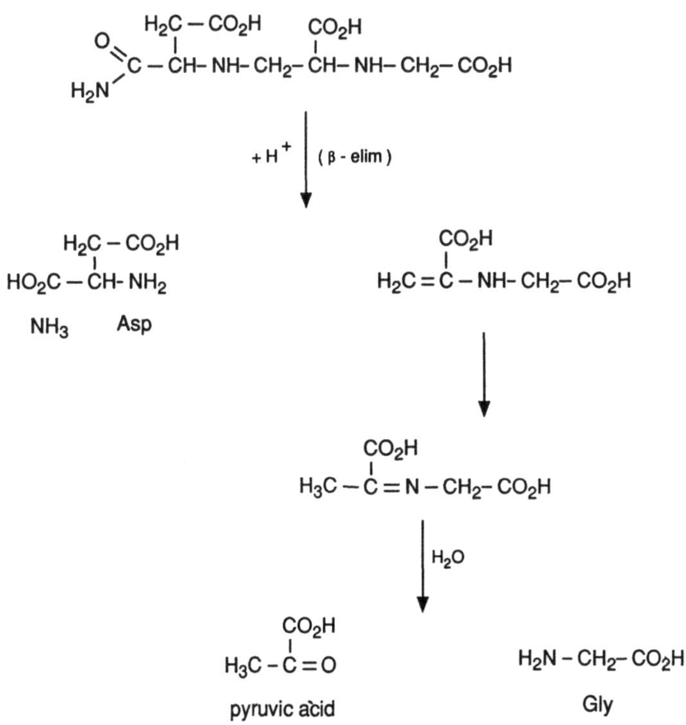


**Fig. 15.** Products of acid hydrolysis of lycomarasmine

of leaves in tomato plants. The appearance of pyruvic acid, aspartic acid, and glycine on hydrolysis with acids suggested a compound with peptide character, yet a different, rather unusual molecule seems to be at hand.

In Fig. 15 the formation of the 3 hydrolysis products is indicated.

In 1958, R.W. Curtis observed [22] that metabolic products of *Aspergillus niger* produce bizarre distortions (malformation) in growing plants, e.g. in the stem of bean plants and pronounced curvatures in the roots of germinating corn seeds. Isolation of the active principle, malformin, in pure form was followed by degradation studies which revealed a bicyclic structure consisting of a ring of five amino acid residues and a disulfide bridge:

D-Cys—Val—D-Cys—D-Leu—Ile

The synthetic compound with this structure showed, however, almost no activity. Reexamination of the amino acid sequence led to the revised structure of Fig. 16 [23].

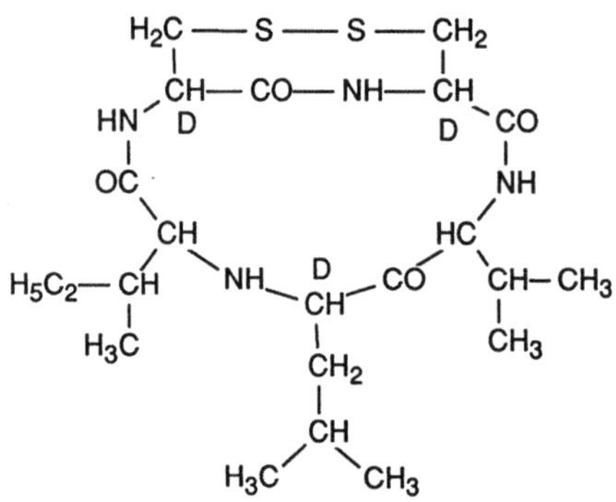


**Fig. 16.** Malformin A

### 1.2.4 Enniatins

To conclude this chapter two closely related cyclodepsipeptides, *enniatin A* and *B*, should be mentioned: they provided a beginning for the development of peptide chemistry in the Soviet Union. Enniatin A and B were isolated from culture media of a *Fusarium* by Plattner and Nager [24] in Switzerland and were found to contain equimolar amounts of D-α-hydroxisovaleric acid (Hyv) and *N*-methyl-L-isoleucine (MeIle, A) or *N*-methyl-L-valine (MeVal, B). The amino acids are linked to the hydroxy acids through ester bonds, the hydroxy acids, to the amino acids via amide bonds. The discoverers asigned cyclic structures to the enniatins, rings formed from 4 acid constituents. M.M. Shemyakin (Plate 42) and his associates in Moscow in 1962 synthesized the proposed compounds but found them to be different from the natural products [25]. They postulated rings consisting of 6 rather than 4 units and the accordingly synthesized cyclodepsipeptides were indeed identical with the enniatins [26]. The formula of enniatin B is depicted in Fig. 17.

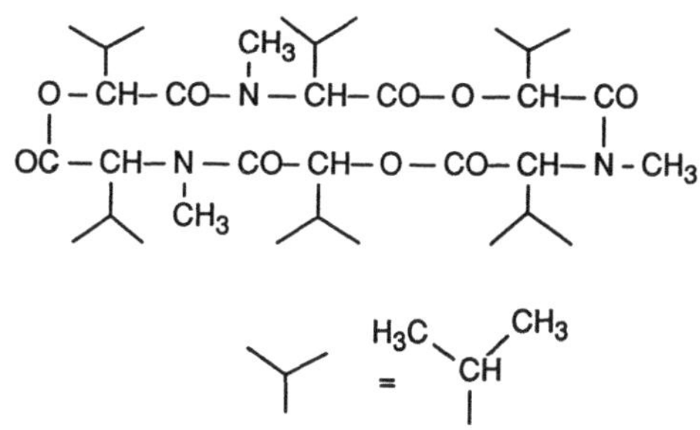

Fig. 17. Enniatin B

The enniatins belong to the class of complexones (ionophores): they transport metal ions through membranes including those in bacterial cell walls and have, therefore, antibiotic properties (see Chap. 9).

## References


1. F. Hofmeister, Naturwiss. Rundschau 17: 529–545 (1902)
2. E. Fischer, Autoreferat, Chem. Ztg. 26: 93 (1902)
3. E. Fischer, E. Fourneau, Über einige Derivate des Glykokolls Ber. dtsch. Chem. Ges. 34: 2868–2879
4. F. Sanger, The arrangement of amino acids in proteins, Advan. Protein Chem. 7: 1–67 (1952)
5. J.S. Fruton, Early theories of protein structure, Annals New York Acad. Sci. 325: 1–18 (1979)
6. N. Troensegard, Über die Struktur des Proteinmoleküls; eine chemische Untersuchung, E. Munksgaard, Kopenhagen, 1942
7. D. Wrinch, Chemical aspects of the structure of small pepidtides, Munksgaard, Copenhagen, 1960.
8. M. Bergmann, C. Niemann, On blood fibrin. A contribution to the problem of protein structure, J. Biol. Chem. 115: 77–85 (1936)
9. M.F. Perutz, M.G. Rossmann, A.F. Cullis, H. Muirhead, G. Will, A.C.T. North, Structure of haemoglobin, Nature 185: 416–422 (1960); J.C. Kendrew, R.E. Dickerson, B.E. Strandberg, R.G. Hart, D.R. Davies, D.C. Phillips, V.C. Shore, Structure of myoglobin, Nature 185: 422–427 (1960)
10. Springer Advanced Texts in Chemistry (Ch.R. Cantor, ed.) Springer Verlag New York Heidelberg Berlin, 2nd corr. printing 1979.
11. From H.B. Vickery, The history of the discovery of amino acids II. Advan. Protein Chem. 26: 82–173 (1972)
12. W. Leinfelder, E. Zehelein, M.-A. Mandra, Gene for a novel tRNA species that accepts L-serine and cotranslationally inserts selenocysteine, Nature 331: 723–725 (1988)
13. S. Magnusson, L. Sottrup-Jensen, T.E. Petersen, H.R. Morris, A. Dell, Primary structure of the vitamin K-dependent part of prothrombin, FEBS-Lett. 44: 189–193 (1974). P. Fernlund, J. Stenflo, P. Roepstorf, J. Thomsen, Carboxyglutamic acids. The vitamin K-dependent structures in prothrombin. J. Biol. Chem. 250: 6125–6133 (1975)
14. For reviews see: R.L.M. Synge, Quart. Revs. (London) 3: 245 (1949); E. Bricas, Ch. Fromageot, Advan. Prot. Chem. 8: 1–125 (1953); S.G. Waley, Advan. Prot. Chem. 21: 1–114 (1966)
15. E.F. Pai, G.E. Schulz, The catalytic mechanism of glutathione reductase as derived from X-ray diffraction analyses of reaction intermediates, J. biol. Chem. 258: 1752–1757 (1983)
16. A. Meister, M.E. Anderson, Glutathione, Ann. Rev. Biochem. 52: 711–760 (1983)
17. See G.B. Henderson, A.H. Fairlamb, Trypanothione metabolism: a chemotherapeutic target in tryptanosomatids, Parasitology Today, 3: 312–315 (1987)
18. H.M. Lenhoff, Behavior, Hormones, and Hydra, Science 161: 434–442 (1968)
19. S.G. Waley, Acidic peptides of the lens, Biochem. J. 64: 715–726 (1956)
20. Glutathione, Parts A and B, Dolphin, D., Avramovic, O., Poulson, R. eds. Wiley New York (1989)

21. P.A. Plattner, N. Clauson-Kaas, Über ein Welke-erzeugendes Stoffwechselprodukt von Fusarium lycopersici Sacc. Helv. Chim. Acta 28: 188–195 (1945)
22. R.W. Curtis, Root curvatures induced by culture filtrates of Aspergillus niger, Science 128: 661–662 (1958)
23. M. Bodanszky, G.L. Stahl, The structure and synthesis of malformin A. Proc. Natl. Acad. Sci. USA 71: 2791–2794 (1974)
24. P.A. Plattner, U. Nager, Über die Konstitution von Enniatin B, Helv. Chim. Acta 31: 665–671 (1948)
25. Yu. A. Ovchinnikov, V.T. Ivanov, A.A. Kiryushkin, M.M. Shemyakin, Synthesis of cyclic depsipeptides, Peptides Proc. 5th Europ. Symp., Oxford, Sep. 1962, 207–219 (1963)
26. M.M. Shemyakin, Yu A. Ovchinnikov, V.T. Ivanov, A.A. Kiryushkin, The structure of enniatins and related antibiotics, Tetrahedron 19: 581–591 (1963)

# 2 Syntheses of Peptides. The First Epoch

In 1902 Emil Fischer was awarded the Nobel Prize in Chemistry, the second one after the inauguration of this foundation, but not for his contributions to peptide chemistry. Fischer was then renowned as one of the most successful chemists, an authority in the field of natural product research and in various sections of organic chemistry. In 1880, only 28 years old, he entered the chemistry of uric acid, xanthine, caffeine and theobromine, until then treated but not completed by Liebig, Wöhler, Strecker and other famous scientists; he coined the term "purines" for this group which he widely expanded, particularly by syntheses of all relevant members. In 1884 he turned also to the carbohydrates the knowledge of which he, with his tool phenylhydrazine, developed to the fundamental state valid up to the present. The Fischer convention for two dimensional representation of optically active, l- and d-stereoisomers was introduced during this period (p. 10). The selective action of enzymes on $\alpha$- and $\beta$-methylglycosides led him to the famous lock-and-key metaphor. In 1899, after having finished this task, Fischer embarked on the then fascinating field of protein chemistry that had been taboo for chemists due to the lack of adequate methods and to the absence of an organized team of researchers. As a chemist who had mastered the difficult structures of sugars and polysaccharides, Fischer in a famous lecture [1] at the Deutsche Chemische Gesellschaft in Berlin, in 1906 stated: "Während vorsichtige Fachgenossen befürchten, daß eine rationelle Bearbeitung dieser Körperklasse durch ihre verwickelte Zusammensetzung und ihre höchst unbequemen physikalischen Eigenschaften heute noch auf unüberwindliche Schwierigkeiten stoßen werde, neigen andere, optimistisch veranlagte Beobachter, zu denen ich mich zählen will, zu der Ansicht, daß man wenigstens den Versuch machen soll, mit allen Hilfsmitteln der Gegenwart, die jungfräuliche Feste zu belagern; denn nur durch das Wagnis selbst kann die Grenze für die Leistungsfähigkeit unserer Methoden ermittelt werden"[1].

Five years before his summarizing lecture, Fischer had published (with E. Fourneau [2]) the preparation of the first dipeptide, glycylglycine, by partial HCl-

---

[1] "While some cautious colleagues are concerned that a rational study of this class of compounds, because of their complex composition and most unpleasant physical properties, will encounter unsurmountable difficulties at this time, other, more optimistic investigators—and I would like to be counted among them—are inclined to believe that at least an attempt should be made to besiege the impregnable fortress with all the weapons available today. Only through daring can the scope and limitations of our methodology be determined."

**Plate 1.** Emil Fischer (1852–1919), München, 1880

hydrolysis of the diketopiperazine of glycine. In the introductory paragraph of this paper which is generally regarded as the beginning of the era of peptide synthesis, the authors inspected the literature published hitherto in this field (translated): "The idea of re-uniting by anhydride-formation, the amino acids generated by hydrolysis of protein compounds, has been treated experimentally by various researchers for a long time. We would remind you of the anhydrides of aspartic acid of Schaal (1871), their transformation to the colloidal polyasparaginyl urea by Grimaux (1882) on the one side, and to the polyaspartic acids by H. Schiff (1888, 1891, Recherches sur la synthèse des matières albuminoides et protéiques) on the other, and further the experiments of Schützenberger (1888, 1899) on the combination of various amino acids (leucines and leuceines) with urea by heating with phosphoric acid anhydride, of Lilienfeld's (1894) similar observations on the reaction of potassium bisulfate, formaldehyde and other condensing agents with a mixture of amino acid esters, and finally, of the report of Balbiano and Trasciatti (1900, 1901) of the conversion of glycocoll into a horn-like anhydride by heating with glycerol. (In the present quotation only the years, not the names of the journals are referred to by the authors).

After this introduction it would seem that no defined peptide bond had been produced prior to Fischers preparative work. However, he did not mention the experiments of Theodor Curtius (1857–1928) published from 1881 and 1882, who—unintentionally at first—had obtained and cleanly characterized benzoylglycylglycine (and higher homologs) by the interaction of benzoylchloride and the silver salt of hippuric acid (benzoylglycine). Fischer reported rather exactly on Curtius' contributions to the development of peptide chemistry in his great lecture of January 1906 pointing, however, to the difference between free peptides and *N*-benzoylpeptides which "für die Chemie der Proteine nur eine untergeordnete Bedeutung haben" ("have only secondary importance for the chemistry of proteins"). Nevertheless, we will begin our essay on the history of peptide syntheses by a discussion of the work of T. Curtius.

## 2.1 The Work of Theodor Curtius

Theodor Curtius (Plate 2) took his doctorate with Hermann Kolbe in Leipzig, who suggested re-examining the structure of hippuric acid to which he (Kolbe) had assigned the structure of a benzoic acid, substituted in the benzene nucleus by an amino-acetyl residue, $H_2NCH_2CO{-}C_6H_4CO_2H$. In order to prove or disprove the alternative structure proposed by Dessaignes, Curtius in 1881 tried to benzoylate glycine-silver with benzoylchloride and obtained, besides hippuric acid, products containing more than one glycine per benzoyl residue. One, the "$\beta$-Säure", could be characterized as benzoylglycylglycine, the structure of a second one, "$\gamma$-Säure", yielding a strong positive biuret reaction, indicating more than three peptide bonds in the molecule, could not at that time be resolved [3]. In the following year (1883), in the laboratory of Adolf Baeyer, in Munich, he obtained glycine ethyl ester from glycine, ethanol and HCl, observed the easy formation of dikopiperazine from the free ester, its polymerization to the, as it was later called, "biuretbase", and prepared hippuric acid ethyl ester which on heating with glycine to 180 °C among other products, yielded the above mentioned $\gamma$-Säure. Analog products were formed from aceturic acid (*N*-acetylglycine) ethylester. On preparing *N*-acetyl glycine by reaction of glycine-silver, with acetyl chloride, he

**Plate 2.** Theodor Curtius (1857–1928), Heidelberg

obtained, in analogy to the results with benzoyl chloride, *N*-acetyl-diglycine and higher homologs. Curtius, thus, had in his hands as early as 20 years before the first boom of synthetic peptide chemistry, *N*-benzoyl- and *N*-acetyl polypeptides, although not exactly characterized. Glycine ethylester was found to yield with nitrous acid a yellow oil, diazoacetic acid ester, a compound that was destined to play a great role in Curtius' future work. Amino acid esters and diketopiperazine were to play important parts in Emil Fischer's studies.

The intensive studies of diazo compounds brought Curtius, in 1887 to Erlangen, where he took charge of the department of inorganic chemistry, Curtius, in 1887, from his diazo compounds derived hydrazine, $H_2N\text{-}NH_2$, a compound, that was to be of great importance in later peptide chemistry. In 1884, publications now from Kiel, appeared in which syntheses of carboxylic acid hydrazides prepared from esters with hydrazine were reported, and syntheses of azides from hydrazides by their reaction with nitrous acid. In organic chemistry, the decomposition of acid azides followed by rearrangement of the "nitrene' yielding amines shortened by one C-atom is the well known Curtius-degradation. In peptide chemistry, where one tries to avoid this reaction, amino acid azides are often employed as "activated" amino acid derivatives for establishing a peptide bond with a second amino group of an amino acid or a peptide (ester). The azide method would have reached its present-day value at that early time, if it had been possible to remove the benzoyl residue from the hippuryl peptides or the acetyl group from aceturyl peptides without destruction of the peptide linkages formed by synthesis. Unfortunately the splitting off of the benzoyl group by electrolytic reduction on a mercury cathode (mediated by tetramethyl-ammonium) as described in 1965 by Leopold Horner [5] or an enzymatic deacylation were then not known. In that case, the history of peptide chemistry perhaps would have taken a different course. De facto, however, with the Curtius method a peptide chain could be elongated only by coupling at the carboxyl end, whereas in modern peptide synthesis peptides are built up mostly by coupling of activated building blocks to the free amino group of the chain to be elongated (see later). After all, Curtius and his associates, in Heidelberg (in 1902 and following years), successfully synthesized benzoylglycine peptides of defined length, e.g. benzoylpentaglycine ester according to Fig. 1.

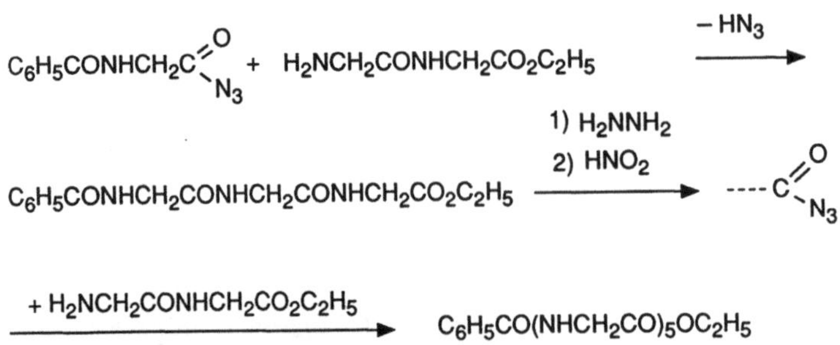


**Fig. 1.** Synthesis of benzoylpentaglycine by azide coupling

Under the influence of E. Fischer's extensive chemical studies on protein structure, Curtius extended his method to the synthesis of benzoylated peptides containing alanine and aspartic acid, but left the peptide field after 1905. The "γ-Säure" had been recognized as benzoylhexaglycine, the "Biuretbase" as tetraglycine ethylester. For a summary of his 23-years involvement see Ref. [6].

Most certainly, Curtius did not intend to enter protein chemistry when he started his work on hippuric acid and glycine. In the following years he was mainly fascinated by his discovery of diazo-fatty acid esters and the multitude of their reactions and so scarcely acquired interest in peptones and albuminoids. During the last decade of the past century, however, stimulated by E. Fischer's vigorous activity, Curtius resumed his studies, by which he contributed so much to modern peptide chemistry. It is somewhat ironical that Fischer's engagement, that had an immense echo at that time left less practical application for peptide chemistry than Curtius' invention of the azide coupling method. Emil Fischer's great merit is to have drawn the attention of the whole scientific world to the field of proteins, whose mystery could be revealed by application of chemical methods and to lend trust to chemists that they are able to synthesize complicated natural substances like peptides.

## 2.2 Emil Fischer's Work

To Emil Fischer (Plate 1), we owe the systematic attack on a field of natural substances that had previously been avoided by chemists. Before entering peptide synthesis he turned to the amino acids feeling that the enormous variety of proteins would never be clear and understandable if one could not handle with certainty the building blocks of these materials. In 1899 when Emil Fischer began his work 14 of the protein-forming amino acids had been characterized. Fischer added the two imino acids L-proline and L-hydroxyproline to this yet incomplete arsenal. Tryptophan, methionine and threonine were still undiscovered. They would complete the total number of 18, isolated from protein hydrolyzates; the amides asparagine and glutamine are additional amino acid residues, present in proteins but not in chemical hydrolyzates. All amino acids had been accessible as D,L-compounds by more or less satisfactory synthetic methods. This was the situation at the end of the 19th century.

Three topics in this field still awaited experimental work. The first was to work out methods to separate qualitatively the mixture of amino acids in a hydrolyzate, and, if possible at all, to determine their respective amounts. The second was to optimize the syntheses and to resolve racemic mixtures in order to obtain the natural enantiomers. The third aim was to modify the amino acids to make them useful for the intended coupling reactions.

For the separation and the putative detection of new amino acids Fischer introduced the "ester method". As Curtius already had described nearly 20 years before, amino acids were treated with HCl in absolute ethanol yielding the ester hydrochlorides from which Curtius had obtained the free esters by means of silver oxide. Fischer simplified the procedure by adding sodium hydroxide to the aqueous solution of the salt, and after saturation with potassium carbonate, extracting the free ester into diethylether. He found that the esters could be distilled under high vacuum and, according to their different volatilities, could roughly be separated by fractional distillation [7]. In this way the two imino acids were discovered in hydrolyzates of casein and gelatine, respectively.

The second task, resolution of synthetic D,L-amino acids has been solved by conversion of the neutral amino acids into real carboxylic acids by acylation of the amino group, either by the benzoyl or the formyl residue. The D,L-acyl amino acids then formed diastereomeric salts with optically active bases, mostly alkaloids, which differed in their solubility in various solvents, and so could be separated by recrystallization. This method is still in use, although enzymatic procedures, specific oxidation of the D-antipode in the presence of D-amino acid oxidase or enzymatic, stereospecific removal of *N*-acyl residues from D,L-*N*-acetyl-amino acids by acylase, are more convenient. Certainly, L-amino acids became accessible from nature by the ester method, but without synthetic material, extended experiments of peptide coupling would have been impossible.

Nowadays natural amino acids are commercially available at moderate prices and optically active amino acids are accessible by stereospecific synthesis.

The third approach, preparation of the derivatives of amino acids suitable for the intended coupling reactions was followed during the decade of peptide work. At first an amino protecting residue was to be found that after formation of peptide bonds would be removed under mild conditions without damaging the peptide formed. Since the benzoyl group of Curtius did not meet this requirement,

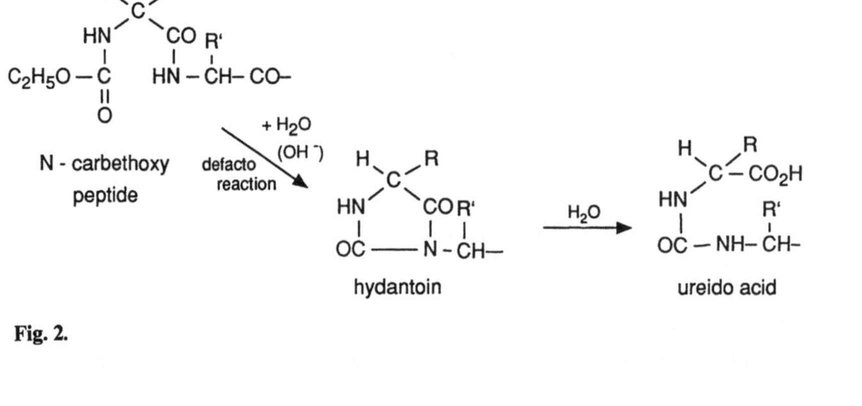


Fig. 2.

Fischer introduced a urethane, the ethoxy carbonyl group, $C_2H_5OCO$— that on removal by alkaline hydrolysis should yield an *N*-carbaminate, and on acidification would lose $CO_2$ releasing an $NH_2$-group.

Most unfortunately, these expectations were frustrated. A malicious alkali-catalyzed rearrangement converts *N*-carbethoxy peptides into derivatives of urea (via hydantoin) as Fritz Wessely (1897–1967) showed in 1927 [8].

Fischer, therefore, as Curtius with his benzoyl in the beginning of his studies, was able to perform coupling only at the carboxyl group, and for this purpose, he worked out the preparation of chlorides by reaction of the acids with thionyl chloride. Since, however, "the reaction, by no means, runs smoothly" Fischer switched to $PCl_5$ in the unusual solvent acetyl chloride. He showed that in this system, finely powdered free mono amino acids and even small peptides can also be transformed to their chloride hydrochlorides [9].

$$H_3N^+\text{-}CH(R)\text{-}CO_2^- + PCl_5 \longrightarrow H_3N^+\text{-}CH(R)\text{-}COCl\text{-}Cl^-$$

These extremely reactive substances could be coupled to amino acid or peptide esters in water-free solvents to yield peptides which again bore a free amino group, and so, theoretically would have been ideal components for the synthesis of longer peptides by elongation at the amino end. In practice, however, self-condensation of two or more of the activated components as a side reaction could not then and cannot be avoided so far.

The lack of an easily removable amino-protecting group led Fischer to the idea of introducing α-halogen fatty acid chlorides. These were coupled to the free amino group of the partner, the α-halogen fatty-acyl compound formed was reacted with ammonia to yield a new aminoacyl terminus [10].

$$Br\text{-}CH(R^1)\text{-}COCl + H_2N\text{-}CH(R^2)CO\text{—}$$

$$\xrightarrow{-HCl} BrCH(R^1)CONHCH(R^2)CO\text{—}$$

$$+ NH_3 \xrightarrow{-HBr} H_2NCH(R^1)CONHCH(R^2)CO\text{—}$$

Fischer improved the synthesis of α-halogen fatty acids (mostly bromo compounds) by starting from malonic acid derivatives. For syntheses of natural peptides optically active components were needed but the resolution of D,L-α-halogen fatty acids could not be performed in a simple manner by crystallization of alkaloid salts. Otto Warburg (1883–1970) to whom this problem was offered in 1905, obtained 10 percent optically pure laevorotatory, (−) α-bromopropionic acid (later recognized as L-enantiomer) after twenty recrystallizations of the cinchonine salt from water! On aminolysis with 25% ammonium hydroxide his product yielded optically pure D-alanine, the antipode of the natural L-compound. In this reaction a "Walden-inversion" at the asymmetric carbon atom had occurred. In a simpler way Warburg prepared the same L-α-bromopropionic acid from natural L-alanine by reaction with nitrosyl bromide, i.e. retaining the L-configuration. Since the Walden inversion occurs also on ammonolysis of α-halogen acyl peptides, in order to synthesize peptides containing L-amino acids it was necessary to prepare halogen acids from unnatural D-amino acids. These could be produced only by time-consuming resolution of synthetic racemic products via alkaloid salts (p. 28). With thionylchloride the α-halogen acid chlorides were formed which like, for instance, D-α-bromoisocaproyl chloride (from unnatural D-leucine) were needed in Fischer's peptide synthesis (Fig. 3). Parallel to his chloride method, Fischer also made use of the condensation of peptide esters, a reaction that for Curtius already had afforded the "biuret base" from glycine ethyl ester.

Emil Fischer and his numerous collaborators synthesized about 100 peptides within the first decade of the 20th century using the methods developed mainly in his laboratory, peptides containing from 2 to 18 amino acid residues. The variability was rather limited on account of the difficult accessibility of many

A $ClCH_2COCl + H_2NCH_2CONHCH_2CO_2H \longrightarrow ClCH_2CONHCH_2CONHCH_2CO_2H$

$\xrightarrow{+NH_3} H_2NCH_2CONHCH_2CONHCH_2CO_2H \xrightarrow{CH_3OH} \underset{\mathbf{1}}{H\,(Gly)_3OCH_3} \xrightarrow{heat}$

$H\,(Gly)_6OCH_3 \xrightarrow{sapon.} \text{Hexaglycine}$

B $D\text{-}BrCH(C_4H_9)CONHCH_2CONHCH_2CONHCH_2CO_2H \xrightarrow{\text{via acid chloride}} \xrightarrow[+\text{hexaglycine}]{}$

$D\text{-}\alpha\text{-bromopeptide} + NH_3 \longrightarrow L\text{-}H_2NCH(C_4H_9)CO(NHCH_2CO)_8NHCH_2CO_2H$ **2**

C $D\text{-}BrCH(C_4H_9)CO(NHCH_2CO)_2NHCH_2COCl + \mathbf{2} \longrightarrow D\text{-}\alpha\text{-bromopeptide}$

$\xrightarrow{+NH_3} L\text{-}H_2NCH(C_4H_9)CO(NHCH_2CO)_3NHCH(C_4H_9)CO(NHCH_2CO)_9OH$

Tetradecapeptide **3**

D - bromoisocaproyltriglycine chloride + **3** ⟶ D- α -bromopeptide

$\xrightarrow{+NH_3}$ L − H − Leu− $(Gly)_3$— L − Leu− $(Gly)_3$— L − Leu− $(Gly)_8$GlyOH

Octadecapeptide

amino acids and of the lack of experience with bifunctional amino acids. As an example by which Fischer's methods are illustrated very well, the synthesis of his longest peptide containing 18 amino acid residues [11] will be discussed below.

A Chloroacetyl chloride was reacted in aqueous alkaline solution with glycylglycine, obtained from diketopiparazine. After ammonolysis the tripeptide was esterified with methanolic HCl. The ester (1) on melting yielded the dimeric hexapetide ester that was saponified to hexaglycine.

B D-Bromoisocaproyl chloride (from D-leucine and NOBr) was coupled to triglycine, the bromoacyl derivative formed was converted to its acid chloride and then reacted with hexaglycine. The scarcely soluble product was treated with liquid ammonia to yield the decapeptide L-leucylnonaglycine (2).

C The decapeptide 2 was reacted with a 3–4 fold excess of the bromoisocaproyltriglycine chloride described above to yield the fourteen-membered $\alpha$-bromo acyl compound. After a 4–5 day ammonolysis in liquid ammonia the corresponding tetradecapeptide (3) was acylated by the D-bromoisocaproyltriglycine chloride once again with difficulty and the 18-membered product converted by treatment with liquid ammonia to the famous octadecapeptide with a molecular weight of 1213.

The decreasing reactivity with increasing chain length of peptides here again exemplified that nature ensures, "daß die Bäume nicht in den Himmel wachsen" ("that trees do not grow until they reach the sky"), as Curtius cited in his review on his own peptide work, and this also applies to Fischer's intensive and productive efforts. Fischer's large peptides may not have completely withstood today's criteria of purity, but they are milestones in the history of peptide synthesis. Most probably he himself felt uneasy about the homogeneity of his higher polypeptides, for after 1910 there were no longer any publications from his laboratory on synthetic polypeptides. Work in this field was continued by Emil Abderhalden (1877–1950), one of Fischer's most faithful, devoted, and diligent associates (they published 30 joint papers), who may be considered as the "biochemical hand" in Fischer's peptide laboratory.

As early as during his famous carbohydrate studies Fischer had gathered experience with enzymes (Fermente), with glycosidases from yeast and from bitter almonds (emulsin). In his peptide work he also made use of enzymes from gastric juice (pepsin) and pancreas (trypsin) in order to split proteins and peptones carefully, and to test the comparability of his synthetic polypeptides with the naturally occurring ones. In both these approaches Abderhalden was the principal collaborator. He continued the studies on substrates of proteolytic enzymes and extended these to create the "Abwehrfermente" (defense ferments) which were claimed in blood serum after parenteral administration of foreign proteins. In Fischer's laboratory enzymatic degradability among others, was a criterion for the proteinoid nature of the synthetic products. Since the larger members of them were precipitated by phosphotungstic acid or by ethanol or "salted out" from aqueous solution and all gave a positive biuret reaction, Fischer felt himself not very far from simple natural proteins. As mentioned before, the upper limit of the molecular mass of a protein then was thought to be around 4–5000, that means that in the mind of organic chemists a total synthesis would be, at least theoretically, possible, but as Fischer himself remarked with resignation—only by employing an army of diligent workers. Today we know that with Fischer's acylchloride method, and without mildly removable protecting groups, a synthesis of even such a small "protein" would not have been possible. Nevertheless, the great efforts were not in vain: in addition to a host of examples for the art of preparative chemistry, as Kurt Hoesch in his Fischer biography [12] says "...schaffen sie (the protein studies) auch unumstößliche Klarheit darüber, aus welchen Elementen und nach welchem Bindungsprinzip die Träger des Lebens letztlich zusammengefügt sind."[1]

## 2.3 Fischer Against Curtius

At a time when, unlike today, research topics were considered private property the tension created by the encounter of two successful peptide chemists, Fischer and Curtius, was higher than it appears from the literature. It was certainly not unintentional when Fischer in 1901 in the historical introduction of his first publication on a free dipeptide, glycylglycine, left without mention the synthesis of benzoylglycylglycine reported 20 years earlier by Curtius, although he could not omit the diketopiperazines described in the same paper. Curtius then, in his classical report on the preparation of benzoylpolyglycines by the azide method (1902), obviously by no mistake, gave no reference for Fischer's studies. In 1903 Fischer complained about this in a footnote and Curtius, in 1904, pointed to the omission by Fischer almost in the latter's words "regrettably my earlier experiments, although their objectives were clearly stated, have not been mentioned". In their later reviews the two rivals fulfilled their obligations to politeness but their personal relationship remained cool throughout their life. Thus in Fischer's autobiography a single reference is made to Curtius, as the author of an obituary on W. Koenigs. This is the more remarkable as Curtius began his carrier as the successor of Fischer and they knew and were on friendly terms with the same colleagues. Both started as assistants of Adolf Baeyer in Munich, Fischer (born in 1852) from 1879 to 1881, Curtius (born in 1857), after receiving his doctorate in Leipzig, from 1882 to 1886. They spent pleasant years in Munich, both in and out of the laboratory, with several young colleagues, for instance Wilhelm Koenigs, a friend of Fischer and fellow Rhinelander, Victor Meyer (whom, in 1998, Curtius succeeded as professor in Heidelberg) or Carl Duisberg, who became a powerful leader of the German chemical industry and kept friendly ties with both professors. He remained a life-long friend of Curtius and was near to Fischer particularly in the last decade of his life. It was Duisberg who in a memorial lecture at the German Chemical Society honored Fischer

[1] "...they (the protein studies) also create definitive clarity about the final building elements of the carriers of life and the principles by which these are bound together."

particularly for his contributions to the pharmaceutical industry. In 1881 Fischer was appointed "Extraordinarius" in Erlangen. He arrived there with his cousin and colleague Otto Fischer, under whose sponsorship Curtius became "Privat-dozent". (Curtius came to Erlangen in 1886, the year when Fischer moved to Würzburg.) Thus Fischer and Curtius had many mutual friends and acquaintances and yet clearly chose to ignore each other, even before Fischer developed an interest in peptides. The subsequent competition did not foster greater empathy between the two.

Fischer was a very private person. While he enjoyed company in his younger years, he cultivated real friendship with only very few people, with his cousin Otto, with Ludwig Knorr, his assistant in Erlangen for many years, or with W. Koenigs and later in life with C. Duisberg, both former fellow students. Curtius was similarly an introvert, his personality, as characterized by Heinrich Wieland in an obituary, "always clouded under the veil of a certain reservation". This might contribute to our understanding of the distance Fischer and Curtius kept from each other.

They grew up under similar circumstances. Both were born in the Rhineland, Emil in 1852 as the only son of the affluent and enterprising Laurenz Fischer in Euskirchen near Cologne, Theodor in 1857 as the son of the dye-manufacturer Julius Curtius in Duisburg. Emil Fischer was determined in his own words to "walk alone on the path of life... but the young woman did not give up" (in a letter to Adolf Baeyer); at the age of 36 he married Agnes Gerlach, their happy marriage was blessed with three sons. Two of them died during the First World War after their mother succumbed to meningitis as early as 1895. The bereft father was survived only by the youngest, Hermann Otto Laurenz (1888–1960) who after graduation under L. Knorr joined his father as post-doctoral associate in the years of 1912–1914 and later became Professor of Biochemistry in Toronto and then in Berkeley. Curtius remained unmarried. One can assume that his intense predilection for science in his younger years suppressed the sincere considerations and initiatives needed for intimate relationships.

Both scientists reached their final position at an early age. In 1892, still only forty, Fischer became successor of the famous August Wilhelm von Hofmann in the important chair of Professor of Chemistry at the University of Berlin. In spite of a flood of new responsibilities he continued in his scientific endeavors with unrelenting vigor until his death in 1919. Under the impression of having an incurable disease (presumably lung cancer as suggested by C. Duisberg in his eulogy) he took his own life. Theodor Curtius achieved success in 1898, at the age of 41, when he was appointed Professor of Chemistry at the University of Heidelberg, a position once occupied by Robert Bunsen and later by Victor Meyer. Here he further increased the intensity of his peptide studies and broadened his earlier work in the field of nitrogen-compounds. He died in early 1928.

The name of Th. Curtius is also associated with a chapter in the history of alpinism. In the Engadin, his second homeland, he undertook, in the company of mountain guides and baggage carriers, several difficult ascents, some of them first-climbs. Unlike E. Fischer he had no disciples who would have made their mark in peptide chemistry.

## 2.4 The School of Emil Fischer in the Peptide Field

Several coworkers participating in the peptide studies in Berlin kept the theme of their master alive after his death and advanced it further.

### 2.4.1 Emil Abderhalden

Emil Abderhalden (Plate 8), already mentioned on page 31, developed through energy, diligence and organizational talent into an important figure in physiological chemistry. Born in 1877, son of a teacher in Oberuzwil near St. Gallen, Switzerland, he studied medicine in Basel. As early as in his third semester he was given the task by the professor of physiology, Gustav von Bunge, of comparing the composition of the blood of various mammals. This led to his first scientific publication in 1897. After graduation in Basel in 1902 he moved to Berlin to join Emil Fischer. In the new environment he rapidly became well versed in chemistry, a science he recognized to be, next to medicine, a main pillar of physiological chemistry. Abderhalden deserves major credit for his sustained, never tiring efforts, pursued with conviction and energy, to unite the entire field of physiology, research and teaching, with the exact scientific methodology of chemistry and physics. The outcome of this effort appears in the numerous textbooks, handbooks, encyclopedia and editorships. The combination of chemistry with enzymology, an idea, realized by E. Fischer and further developed by Abderhalden, can be regarded as the foundation of the success of the field that subsequently became known as biochemistry.

In Berlin Abderhalden's name first appears together with that of Otto Diels as coauthors of a paper on the degradation of cholesterol, but after that he turned entirely to synthetic and enzymatic peptide chemistry, an area of which he was the recognized representative in Fischer's laboratory. Following his appointment in 1908 to professorship in physiological chemistry at the School of Veterinary Medicine in Berlin Abderhalden observed that polypeptides are cleaved by the blood serum of dogs if injected intraperitoneally (that is not into the blood stream) the day before, while control sera were "negative". The proteinases appearing in this case and later also in other examples, designated as "Abwehrfermente" (defense ferments or defense enzymes) which were subsequently detected in urine as well. They raised great hopes for theoretical biology and also for practical medicine, which however, in spite of many years of intensive work, mostly with the participation of his son Rudolf, were not fulfilled. Thus they did not succeed in the isolation and characterization of a defense proteinase. In later handbooks the term "Abwehrfermente" did not occur any more.

In addition to his activities as a teacher and researcher, Abderhalden was deeply involved in social problems as well. After the First World War he founded, for instance in Halle, a central office for the "Swiss Aid for German Children".

Through this organization food arrived in large quantities in the impoverished land and about 100,000 German children found care and were restored to good health in Switzerland.

In Halle, where after 1911 Abderhalden was Professor and Head of the Institute of Physiology, he continued the experiments in peptide synthesis by the bromoacyl chloride method with utmost consistency. In this effort, however, he surpassed the octadecapeptide of Fischer (p. 31) merely by a single L-leucine residue (that is a total of 19 amino acid residues) and this might have raised doubts in him about the general applicability of the approach. Enzymes that cleave peptides and proteins and the metabolism of amino acids in the framework of nutrition research remained his main interest.

Switzerland, where he was born, became also his last refuge. In the summer of 1945, as the war ended, the advancing American army evacuated technicians and scientists from the areas briefly occupied, when these, according to agreement, had to be abandoned. The US troops took the Abderhalden family to the West without giving them time to pack their belongings. They arrived in Darmstadt without means. As Swiss citizen he could return to the country where he was born and where he could have returned in 1920 under quite different circumstances had he followed an invitation to a professorship in Basel. As it happened, he had to be satisfied with the title of "Honorary Professor" at the University of Zurich and with an extremely modest salary. He never recovered the belongings and the property left behind. In 1950 he died in Zurich, after two strokes, bitter and disappointed.

### 2.4.2 Hermann Leuchs

Hermann Leuchs (Plate 29) wrote his doctoral thesis from 1900–1902, under the guidance of E. Fischer, on syntheses of serine, the four $\gamma$-hydroxyprolines and glucosamine. He was born in Nuremberg in 1879 as the second son of the 4 sons and 3 daughters of Friedrich Leuchs, a merchant and owner of a scent manufacturing company. Among his Frankish ancestors going back to the 14th century, one finds brewers and important family members, for instance a brother of Leuchs' great-grandfather Dr. Georg Leuchs, author of an article entitled "Germany at its deepest humiliation" for which in 1806 the bookseller Palm, on the order of Napoleon, was executed by a firing squad. Hermann L. attended the gymnasium in Nuremberg, then studied chemistry in Munich and from 1900 on in Berlin. After his graduation the PhD became in succession "Privatdozent", "Extraordinarius" and in 1926 Associate Director of the Institute of Chemistry of the University of Berlin.

In 1906 Leuchs observed, first on glycine, that *N*-carboethoxyamino acid chlorides at temperatures as low as 70 °C are transformed, with the loss of alkyl chloride, to the so called "Leuchs'schen Körper" (Leuchs' substances), *N*-carbamoic acid amino acid inner anhydrides. As he noted in the case of the glycine derivative these *N*-carboxyanhydrides (NCA-s) readily polymerize. For instance in the presence of a small amount of water they form poly-$\alpha$-amino acids consisting, as shown by later studies, of several hundred residues.

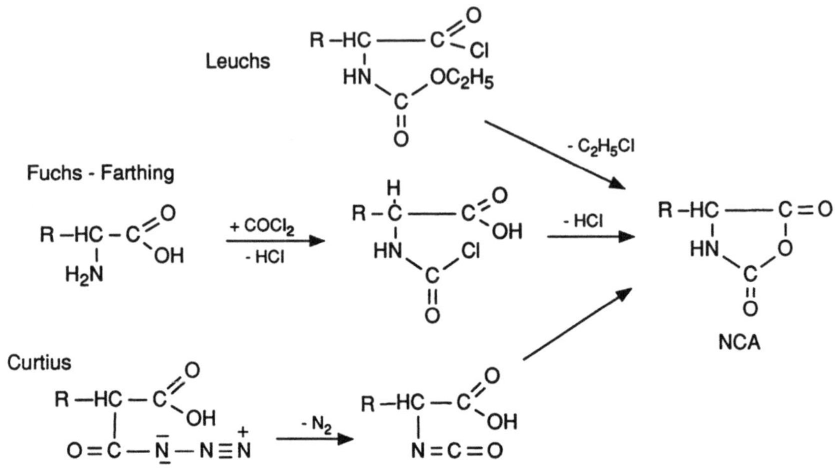


**Fig. 4.** Formation of inner *N*-carbonic carboxylic anhydrides

Leuchs' initial elan, fed by intelligence and power of imagination, which after successful (albeit by Emil Fischer inspired) peptide studies launched him into the difficult chemistry of strychnine alkaloids and also into problems of stereochemistry, diminished later, probably under the influence of some decisive inner experience unknown to us. After a while his life was consumed by work in the laboratory; he remained unimpressed by the rich cultural and scientific distractions offered by Berlin, that time in its heydays. After 1908 his principal field became strychnine chemistry and related alkaloids: 125 of his 178 publications deal with this topic.

He led a simple life, more and more isolated in the apartment which he had already occupied as a student and kept for many years. He looked upon the Hitler-movement first with suspicion and later with outright hatred. During the Nazi-era Leuchs witnessed the destruction of Nuremberg, the city where he was born, and of Berlin, where he lived. He sank into deep melancholy and in 1945, in the last days of the war he himself put an end to years of hopeless suffering.

The Leuchs' substances, NCA-s, oxazolidine-2,4-diones, can also be synthesized by the method of Fuchs (1910) from amino acids and phosgene, a method further developed by Farthing (1950). According to Curtius, NCA's can be generated from malonic acid half azides as well, through α-isocyanatocarbonic acids (Curtius degradation, Fig. 4).

### 2.4.3 NCA-s and Poly Amino Acids

The NCA-s are extremely reactive compounds (see Fig. 5). The inner anhydride grouping is cleaved by nucleophiles; the attack preferentially occurs at the CO group of the amino acid. With water an *N*-carboxy-amino acid is formed which spontaneously loses $CO_2$. The thus liberated amino group can then attack a second NCA molecule. The NCA-s react with HCl, for instance according to M.

Brenner and I. Photaki (1956), to afford the amino acid chloride hydrochlorides of Fischer (p. 29), with $H_2S$ they give aminothioic acids (Th. Wieland and K.E. Euler, 1958) and anilides with aniline as described by Curtius (1922) and by Wessely and Sigmund (1926) [17]. In the last mentioned reaction phenylalanine NCA yields mainly the dipeptide derivative phenylalanyl-phenylalanine anilide, the product of a secondary reaction, attack of the primarily formed phenylalanine anilide on Phe-NCA. If instead of aniline α-amino acids are brought into reaction then dipeptides are generated with a carboxyl attached to their amino group. These readily lose $CO_2$. The free amino group formed can attack another NCA molecule. Therefore the side reaction described in connection with aniline can not be avoided in the elegant method of peptide synthesis. An intentionally performed stepwise coupling of NCA molecules of the same amino acid leads to the polymerization observed already by Leuchs.

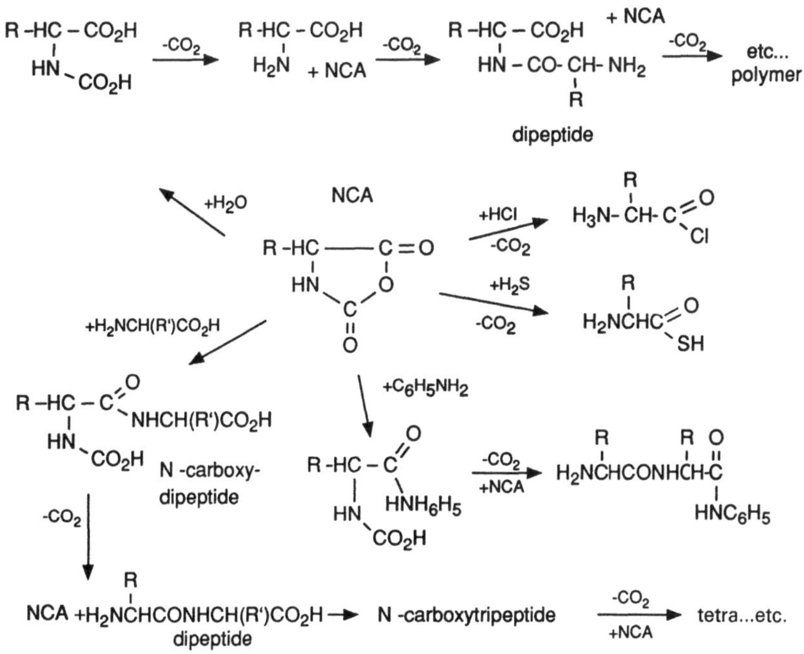


**Fig. 5.** Some reactions of NCA-s

Poly-α-amino acids [18a–c]

Interest in such macromolecules was awakened, not lastly on account of a 1947 paper of R.B. Woodward and C.H. Schramm, who, on storage of a solution containing NCA-s of DL-phenylalanine and DL-leucine in (moist) benzene, obtained films of a copolymer of a molecular weight of about 15000. This was the era when the synthetic polyamides (nylon, perlon) started on their triumphal march. These silk-like fibers posed a challenge for peptide chemists to prepare artificial macromolecules from α-amino acids, substances not unlike of natural silk, but without the unfavourable water-repellent properties of the hydrocarbon-chain containing lipophilic polyamides and yet with the mechanical strength of the latter. Of equal importance to these—never achieved—objectives was the concept of creating, with poly-α-amino acids, ideal models for the study of the physico-chemical properties of high molecular weight peptide chains in the solid state and in solution.

The most common method, by far, for the preparation of poly-α-amino acids, is the polymerization of NCAs. Before a discussion of this topic, a brief consideration of other procedures will be given which have been rarely used and which yield polypeptide chains shorter than obtained by the NCA method.

Free α-amino acids cannot, like amino acids with an amino group remote from the carboxyl (e.g. ε-amino caproic acid), be polymerized by heating at 150–200 °C to form polyamides. At that temperature, α-amino acids undergo degradative reactions. Therefore the method is not suitable for preparative experiments. After short heating periods, however, traces of peptides are formed from mixtures of amino acids, as Sidney Fox's group found in the fifties in their studies on the possible primordial formation of polypeptides. Early attempts to construct protein-like substances from glycine, for instance, by heating in a sealed tube with glycerol at 170 °C are mentioned on page 24 in this chapter.

Esters of α-amino acids and peptides generally undergo condensation more readily than the free amino acids and peptides. As already mentioned on p. 25, Th. Curtius, in 1904, found that glycine ethyl ester solidified even at room temperature to yield tetraglycine ethylester, the "biuretbase". Frankel and Katchalski (1939, 1942) extended these observations and studied the formation of various glycine peptides obtained by the polycondensation of methyl, ethyl and isobutyl esters of glycine for prolonged periods. Higher polyglycine esters were obtained, mainly from GlyOMe (average chain length up to 35 residues) as horn-like water insoluble substances [20]. Higher molecular weight polyglycine esters (and occasionally also mixed with other amino acids) are formed by polycondensation of tripeptide esters of which as an example the results of Pacsu and Wilson (1942) are monitored here. These authors described the formation of a polyglycine methyl ester with an average of 96 members on heating triglycine methyl ester at 102 °C for several hours [21]. In the context of similar studies the names of H. Brockmann, H. Musso, H.N. Rydon, P.W.G. Smith, G. Schramm, G. Thumm and L.A.E. Sluyterman, out of a long series, have to be mentioned. The classical azide coupling method of Curtius has also been used for the preparation of water-insoluble polyglycines, at first by M.Z. Magee and Klaŭs Hofmann in 1949 [22]. Oligomeric amino acids are formed from esters also by the action of proteolytic enzymes. See this topic in the next chapter.

Most effort has been made to investigate the polymerizarion of the *N*-carboxyanhydrides. The NCAs of all naturally occurring and of many artificial α-amino acids have been synthesized and subjected to various conditions of polymerization. Several reaction mechanisms have been formulated according to the reagents involved (Fig. 6).

In bulk polymerization at room temperature initiated by water, plausibly hydrolysis of the anhydride moiety and subsequent decarboxylation of the *N*-carboxyamino acid takes place, followed by reaction at the free amino group with the next NCA etc. (see Fig. 5). In polymerization by heating a primary formation of an α-isocyanato acid is conceivable according to Katchalski and Berger, that will polymerize by addition of a second isocyanato acid, decarboxylation under formation of the peptide bond and so on. Initiation of polymerization by primary or secondary amines will result in the formation of the amino acid amide which, as in the experiments of Wessely (p. 37) propagates the reaction. Highest polymerization degrees (about 1000 to 3000 monomers) are obtained with very strong bases ($NaOCH_3$, NaOH) in water free solvents, according to E.R. Blout

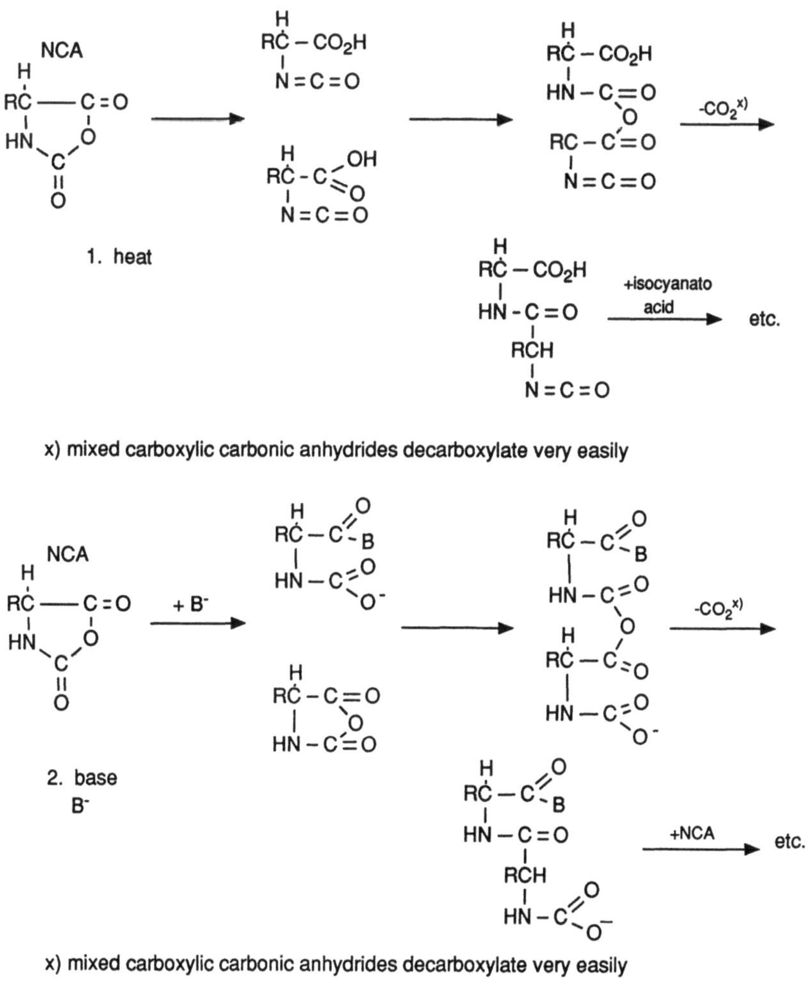


**Fig. 6.** Proposed mechanisms for polymerization of *N*-carboxy-amino acid anhydrides (NCA-s)

et al. (1954–1958). In this reaction the addition of the base to carbon atom 5 (or 2) of the NCA is assumed as the first step leading to a carbaminate that on further reaction opens the anhydride group of a second NCA etc. Polymerization can furthermore be initiated by salts, mainly LiBr (Ballard, Bamford, 1954–1956), and by tertiary amines as already Wessely's group had observed. Of the different mechanisms of initiation and polymerization which extensively have been discussed in E. Katchalski's and Sela's survey [18b], without the trivial $H_2O$ effect, only the heat polymerization (1) and that with strong bases (2) are presented in Fig. 6.

### 2.4.4 Properties of Poly-amino Acids

According to their different side chains the various poly-amino acids are of course different in their physical properties, first and foremost their solubility in water, concentrated salt solutions and organic solvents. The conformation of the polypeptide chain plays an equally important role. On polyglycine and polyalanine the so called α-forms (with high α-helix content) and the less water-soluble ß-forms (containing pleated ß-sheets) have been studied in depth (cf. p. 13). Particularly well investigated are the polyprolines with polymerization degrees of a few hundred. These, too, occur in two forms: polyproline I, with $[\alpha]_D^{25}$ up to +50°, is poorly soluble in water and forms, as shown by X-ray analysis, through *cis* peptide bonds a right-handed helix. It readily changes, with mutarotation, to form II, which is significantly more soluble in water, reveals $[\alpha]_D^{25}$-s up to −730° and generates via *trans* peptide bonds a left handed helix. As found by M. Rothe, who synthesized oligo-L-prolines with n = 2 to n = 15 [23a], later with n up to 40 [23b] helix formation begins already from n = 4, in the case of form I (*cis*, (+) rotation, right helix) in "bad" solvents (pyridine, alcohol), with form II (*trans*, (−) rotation, left helix) in "good" solvents (acetic acid, water) [23]. Polyprolines are also important because of their connection with collagen.

Of interest are also the polylysines. The preparation and properties of poly-L-lysine have mainly been studied by E. Katchalki's group (Plate 24). The ε-amino group of lysine was protected by the *Z*-group (to be described in the next chapter). ε-*Z*-lysine was converted to its NCA-derivative by the Fuchs-Farthing method ($COCl_2$). Polymerization of the NCA of ε-*Z*-L-lysine yields, depending on the initiators, products of up to several thousand residues. After deblocking with phosphonium iodide, polylysines are obtained which, as salts, are readily soluble in water. As polyelectrolytes (polycations) they form precipitates with polyanions as heparin, polyglutamic acids, nucleic acids, analogously to natural polyamines (histones, protamines). Most probably by similar ionic interaction, enzymes are inhibited by polylysine, e.g. pepsin at its pH optimum. The antibacterial and antiviral effect of polylysine may be explained, at least partially, by the same mode of interaction. The ε-amino-groups in the side chains can serve as anchoring sites for the fixation of small molecules, e.g. haptenes in raising of antibodies.

The polyglutamic acids are acidic polyelectrolytes. Since glutamic acid contains two carboxyl groups (α and γ), poly-α-, poly-γ- and mixed polyglutamic acids could be expected. Of these the γ-polymers are interesting. One, composed from glutamic acid exclusively, is known in nature. It is formed as capsule substance by certain bacteria, e.g. *Bacillus anthracis, B. licheniformis, B. subtilis.* Such a polymer with a molecular weight of about 50 000 was isolated by Ivanovics and Bruckner (Plate 14) in 1937, 1938 [25] from capsules and culture media of *B. anthracis.* It consists of D-glutamic acid residues, exclusively, which are linked only by γ-peptide bonds (Bruckner et al., 1952, 1953). Two years later a synthesis of this interesting polypeptide was carried out by the group of Bruckner (simultaneously with S.G. Waley). Poly-α-L-glutamic acid was synthesized (e.g. by Bruckner et al., 1952) from the NCA of γ-methyl ester or γ-benzyl ester of the amino acid with $n < 100$ and also with higher polymerization number, n up to 1000, by E.R. Blout et al. (1954) or C.H. Bamford (1956).

Oriented films of poly-α-L-glutamic acid and its esters were examined in 1951 via X-ray analysis by Pauling and Corey [26]. The reflexions related to a spacing of 1.5 Å, observed also by other investigators at about the same time, confirmed the existence of an α-helix predicted by Pauling et al. [27] for fibrillar proteins. (see Plate 34).

The reaction of the NCA of an amino acid with a second amino acid affords, as already mentioned, no homogeneous product because the dipeptide formed can further react with an additonal molecule of the NCA yielding a tripeptide. Only 20 years after the investigations of Wessely and his associates could I.L. Bailey demonstrate the suppression of the undesired reactions by performing the coupling between NCA-s and amino acid esters in the presence of triethylamine [28]. In this way the carboxyamino peptides are stabilized as salts and cleavage to $CO_2$ and peptide esters takes place only on gentle heating; hence the reaction can be repeated. Polymerization of NCA-s, however, is not completely eliminated in this approach. From 1963 on Ralph Hirschmann and Robert G. Denkewalter adapted the NCA method to a process carried out in aqueous solution and optimized the individual steps through the precise control of the pH and rapid execution of the coupling in the cold. Of the functional side chains of the amino acids only the ε-amino group of lysine and the SH-group of cysteine had to be blocked by suitable protecting groups. An ambitious goal, the synthesis of an enzyme, was attempted. The pancreatic enzyme ribonuclease A consists of 124 amino acid residues. The molecule is cleaved by subtilisin at a single peptide bond, near the carboxyl end of its chain: a mixture of the split-off "S-peptide" (20 residues) and the "S-protein" (104 residues) is produced. In combination the two products have full enzymic activity while separated they are inactive. Since the S-peptide has already been synthesized (Klaus Hofmann [29]) the total synthesis of the S-protein was sufficient for the production of an enzyme. Nineteen small peptides, containing in average 5 residues, were prepared and then linked together by an improved version (J. Honzl and J. Rudinger [30]) of the azide method to yield the S-protein. On addition of the S-peptide the resulting solution clearly exhibited the enzymic activity of ribonuclease A by cleaving ribonucleic acid. Occasionally also *N*-thiocarbamic acid anhydrides (TCA-s), which are less sensitive to hydrolysis then NCA-s, were applied. In their reaction instead of $CO_2$, carbonoxysulfide, COS, is liberated. The NCA method, in spite of its simplicity that seemingly permits the almost complete omission of blocking groups and activation of the carboxyl group, could not establish itself as a routine procedure for peptide synthesis. For a later total synthesis of ribonuclease A by H. Yajima see p. 239.

## References


1. E. Fischer, Untersuchungen über Aminosäuren, Polypeptide und Proteine, Ber. dtsch. chem. Ges. 39: 530–610 (1906)
2. E. Fischer, E. Fourneau, Über einige Derivate des Glykokolls, Ber. dtsch. chem. Ges. 34: 2868–2879 (1901)
3. Th. Curtius, Über die Einwirkung von Chlorbenzoyl auf Glycocollsilber (vorläufige Mitt.), J. prakt. Chemie [2] 24: 239–240 (1881); Über einige neue Hippursäureanalog constituierte, synthetisch dargestellte Amidosäuren, J. Prakt. Chemie [2] 26: 145–208 (1882)
4. Th. Curtius, Synthetische Versuche mit Hippurazid, Ber. dtsch. chem. Ges. 35: 3226–3233 (1902)
5. L. Horner, H. Neumann, Reduktive Spaltung von Säureamiden und Estern mit Tetramethylammonium. Benzoyl und Tosylrest als Schutzgruppe bei Peptidsynthesen, Chem. Ber. 98: 3462–3469 (1965)
6. Th. Curtius, Verkettung von Amidosäuren, J. prakt. Chemie 70: 57–108 (1904)
7. E. Fischer, Über die Hydrolyse des Caseins mit Salzsäure, Hoppe-Sayler's Ztschr. physiol. Chem. 33: 151–171 (1901)
8. F. Wessely, E. Komm, Zur Kenntnis der isomeren Glycylglycin-N-carbonsäuren, Hoppe-Seyler's Ztschr. physiol. Chem. 174: 306–318 (1928)
9. E. Fischer, Synthesen von Polypeptiden, IX. Chloride der Amino-säuren und ihrer Acylderivate, Ber. dtsch. chem. Ges. 38: 605–620 (1905)
10. E. Fischer, Synthese von Polypeptiden, IV. Derivate des Phenylalanins, Ber. dtsch. chem. Ges. 37: 3062–3071 (1904)
11. E. Fischer, Synthese von Polypeptiden, XVII Ber. dtsch. chem. Ges. 40: 1754–1767 (1907)
12. K. Hoesch, Emil Fischer, sein Leben und sein Werk, Ber. dtsch. chem. Ges. 54, Sonderheft (1921)
13. C. Duisberg, Emil Fischer und die Industrie, Ber. dtsch. chem. Ges. 52: 149–164 (1919)
14. K. Freudenberg, Theodor Curtius, Chem. Ber. 96: I-XXV (1963)
15. H. Hanson, Emil Abderhalden (1877–1950), Nova Acta Leopoldina, 36: 257–317 (1970)
16. F. Kröhnke, Hermann Leuchs, Chem. Ber. 85: LV-LXXXIX (1952)
17. F. Wessely, F. Sigmund, Untersuchungen über α-Amino-N-carbonsäureanhydride, Hoppe-Seyler's Ztschr. physiol. Chem. 159: 102–118 (1926) and preceding papers
18. a: E. Katchalski, Poly-α-amino acids in Advan. Protein Chem. 6: 123–185 (1951)
    b: E. Katchalski, M. Sela, Synthesis and chemical properties of poly-α-amino acids, Advan. Protein Chem. 13: 243–492 (1958)
    c: M. Sela, E. Katchalski, Biological properties of poly-α-amino acids, Advan. Protein chem. 14: 392–477 (1959)
19. C.H. Bamford, A. Elliot, W.E. Hanby, Synthetic Polypeptides, Academic Press, New York, 1956
20. M. Frankel, E. Katchalski, Poly-condensation of alanine ethylester, J. Org. Chem. 7: 117–125 (1942)
21. E. Pacsu, E.J. Wilson, Jr. Polycondensation of certain peptide esters, I Polyglycine esters, J. Amer. Chem. Soc. 64: 2268–2271 (1942)
22. M.Z. Magee, K. Hofmann, The application of the Curtius reaction to the polymerization of triglycine, J. Amer. Chem. Soc. 71: 1515–1516 (1949)
23. a: M. Rothe, R. Theysohn, K.-D. Steffen, M. Kostrzewa, M. Zamani, Helixbildung bei Prolinpeptiden, in Peptides 1969, 10th Europ. Pep. Symp. Abano Terme (E. Scoffone, ed.) North-Holland Publ. Co. Amsterdam, 1971, 179–188
    b: M. Rothe, H. Rott, I. Mazanek, Solid phase synthesis and conformation of monodisperse high molecular weight oligo-L-prolines, in Peptides 1976, 14th Europ. Pep. Symp. Wepion, Belgium (A. Loffet , ed.) Edit. de l'Université de Bruxelles, 1976, 309–318
24. J.P. Carver, E.R. Blout in Treatise on Collagen, Vol. I (G.N. Ramachandran, ed.) Academic Press, London New York, 1967
25. G. Ivanovics, V. Bruckner, Chemische und immunologische Studien über den Mechanismus der Milzbrandinfektion, I; Die chemische Struktur der Kapselsubstanz des Milzbrandbazillus und der serologisch identischen spezifischen Substanz des Bacillus mesentericus, Ztschr. Immunitätsforsch. 90: 304–482 (1937)

26. L. Pauling, R.B. Corey, The structure of synthetic polypeptides, Proc. Natl. Acad. Sci. USA 37: 241–250 (1951)
27. L. Pauling, R.B. Corey, H.R. Branson, The structure of proteins: two hydrogen bonded helical configurations of the polypeptide chain, Proc. Natl. Acad. Sci. USA 37: 205–211 (1951)
28. J.L. Bailey, A new peptide synthesis, Nature 164: 889 (1949)
29. K. Hofmann, M.J. Smithers, F.M. Finn, Studies on polypeptides, XXXV. Synthesis of S-peptide 1–20 and its ability to activate S-protein, J. Amer. Chem. Soc. 88: 4107 (1966)
30. J. Honzl, J. Rudinger, Nitrosylchloride and butyl nitrite as reagents in peptide synthesis by the azide method; suppression of amide formation, Coll. Czech. Chem. Comm. 26: 2333–2344 (1961)
31. Studies on the total synthesis of an enzyme, a) R.G. Denkewalter, D.F. Veber, F.W. Holly, R. Hirschmann, I Objective and strategy, J. Am. Chem. Soc. 91: 502–503 (1969)
b) R.G. Strachan, W.J. Paleveda, Jr., R.F. Nutt, R.A. Vitali, D.F. Veber, M.J. Dickinson, V. Garsky, J.E. Deak, E. Walton, S.R. Jenkins, F.W. Holly, R. Hirschmann, II. Synthesis of a protected tetratetracontapeptide corresponding to the 21–64 sequence of ribonuclease A, J. Am. Chem. Soc. 91: 503–504 (1969)
c) S.R. Jenkins, R.F. Nutt, R.S. Dewey, D.F. Veber, F.W. Holly, W.J. Paleveda, Jr, T Lanza, Jr, R.G. Strachan, E.F. Schoenewaldt, H. Barkemeyer, M.J. Dickinson, J. Sondey, R. Hirschmann, E. Walton, III. Synthesis of a protected hexacontapeptide corresponding to the 65–124 sequence of ribonuclease A, J. Am. Chem. Soc. 91: 504–506
d) D.F. Veber, S.L. Varga, J.D. Milkowski, H. Joshua, J.B. Conn, R. Hirschmann, R.G. Denkewalter, IV. Some factors affecting the conversion of protected S-protein to ribonuclease S', J. Am. Chem. Soc. 91: 506–507
e) R. Hirschmann, R.F. Nutt, D.F. Veber, R.A. Vitali, S.L. Varga, T.A. Jacob, F.W. Holly, R.G. Denkewalter, V. The preparation of enzymatically active material, J. Am. Chem. Soc. 91: 507–508

# 3 The Era After Emil Fischer. The Carbobenzoxy Group, Max Bergmann and His Scientific Circle

With the death of E. Fischer peptide research in Berlin came to a halt but the deep impression created worldwide by his accomplishments in chemistry remained an inspiration for several laboratories. The work of E. Abderhalden was discussed in the preceding chapter, the most important contribution to the further development of peptide chemistry came, however, from Fischer's principal coworker Max Bergmann (1886–1944) [1]. Son of a wholesaler merchant in Fürth, Bavaria, he started his studies in 1906 at the Technical University in Munich in botany but in 1907 he transferred to chemistry and to Berlin where, in Fischer's laboratory, he worked on his doctoral dissertation entitled "Hydrogen Persulfide" under the direction of Ignaz Bloch. After graduation in 1911 Fischer engaged him as assistant in his personal laboratory to work in the field of sugars, tannins, amino acids and peptides. In 1920, soon after Fischer's death, he was named Privatdozent and appointed Acting Director and head of the department of Organic Chemistry of the Kaiser Wilhelm Institute for Fiber Research in Berlin-Dahlem. In 1922 he moved to Dresden as Director of the newly founded Kaiser Wilhelm Institut for Leather research. There he broadened the area of his studies in the field of carbohydrates (including their polymers cellulose and chitin), tannins, amino acids and peptides.

In addition to studies of the reactions of amino acids, for instance with aldehydes and their transformations through the elimination of water, in 1926 M. Bergmann described as a new principle for the formation of the peptide bond, the opening of the azlactone ring by an amino acid in the presence of alkali. This afforded the peptide of a dehydro-amino acid that was converted by hydrogenation into a peptide, which was, however still blocked, as before, by an unremovable acyl (here acetyl) residue.

As mentioned on p. 4, M. Bergmann accepted the then prevailing diketopiperazine theory of protein structure and investigated for a few years the reactivity of this class of compounds. In these studies he enjoyed the collaboration of a young American, a National Research Council fellow, Vincent du Vigneaud and of Leonidas Zervas from Greece. With them he established the reactivity of two acyl groups linked to the same nitrogen atom (diacylamines) [2]. V. du Vigneaud, in his doctoral work at the University of Rochester (New York) led by J.R. Murlin, had just previously become acquainted with the polypeptide insulin. In Dresden, that time the center of peptide chemistry, he immersed himself deeper into this discipline and acquired the know-how which he would later, in 1953,

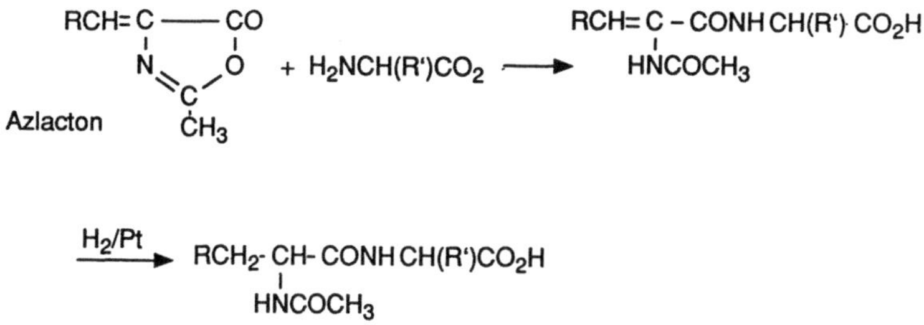


**Fig. 1.** Peptide synthesis from azlactone

convincingly employ in the first synthesis of a peptide hormone, oxytocin. This important event in the history of peptide chemistry and further details of the life of du Vigneaud are described in Chap. 7. Here, in Dresden, diketopiperazines, still suspected of being the building components of proteins, were still the targets of his research.

*N, N'*-Diacetyldiketopiperazine has two acyl residues on each nitrogen atom, an acetyl and an aminoacyl group. Both are linked to the nitrogen through "energy-rich" bonds; hence they are readily cleaved and can cause acylation. Thus, for instance glycine is smoothly acetylated by the *N*-acetyl residue to

**Plate 3.** Max Bergmann (1886–1944), (Photo Archives of the Rockefeller Univ., New York)

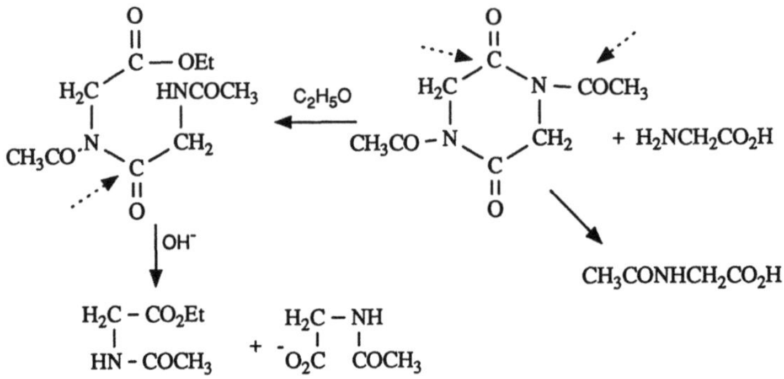

Fig. 2. Reactions of *N*, *N*-diacetyldiketopiperazine

*N*-acetylglycine; with alcohol and base alcoholysis of the other N—CO bond takes place with opening of the ring and formation of *N*-acetylglycyl-*N*-acetylglycine ethyl ester. In this compound the second diacylamino group is still intact, therefore its hydrolysis with alkali yields *N*-acetylglycine and *N*-acetylglycine (ethyl ester) [2].

If instead of the acetyl group, a residue with a functional group (OH, $NH_2$, SH) participates in diacylamine formation then interesting insertion and rearrangement reactions can be observed as described in a forthcoming section (p. 195).

In the course of peptide synthesis in Dresden, time and again the need for a group was noted that would protect the free amino end of the acylating amino acid and could be cleaved after the formation of the peptide without damage to the newly formed peptide bond. The ethoxycarbonyl group of E. Fischer, as shown on pp. 28–29, failed to fulfill these expectations. It was Bergmann, who in 1915, still with Fischer, found that *p*-toluene-sulfonylamino acids (tosyl-amino acids) can be non-hydrolytically detosylated with a mixture of hydroiodic acid and phosphonium iodide, but it remained for Rudolf Schönheimer (1898–1941), who later became known for the introduction of isotope-labeled substances for the study of metabolism in experimental animals ("The dynamic state of body constituents"), to use this method in 1926 for the synthesis of a few peptides [3]. The conditions required by the method, however, were not mild enough for general application. Meanwhile, in the nineteen twenties, several publications and patents of the E. Merck company appeared dealing with cleavage of the *O*-benzyl group by hydrogenation in the presence of platinum-metal catalysts. In 1928, Karl Freudenberg (1886–1983; from 1909 to 1914 coworker of E. Fischer) described the corresponding application of the benzylidene, $C_6H_5CH{=}$, and of the benzyl group in sugar and glycerol chemistry and in 1930 these methods were adopted by Bergmann's laboratory in the study of carbohydrates. In 1931, Leonidas Zervas used this approach in the synthesis of 1-benzoylglucose and it could well be that he and Bergmann came on this occasion to the idea of replacing ethyl in Fischer's ethoxycarbonyl group by benzyl and thus create the benzyloxycarbonyl, $C_6H_5CH_2$–OCO—, or as it was called at that time the carbobenzoxy (Cbz) group, which can be cleaved by catalytic hydrogenation under very mild conditions [4].

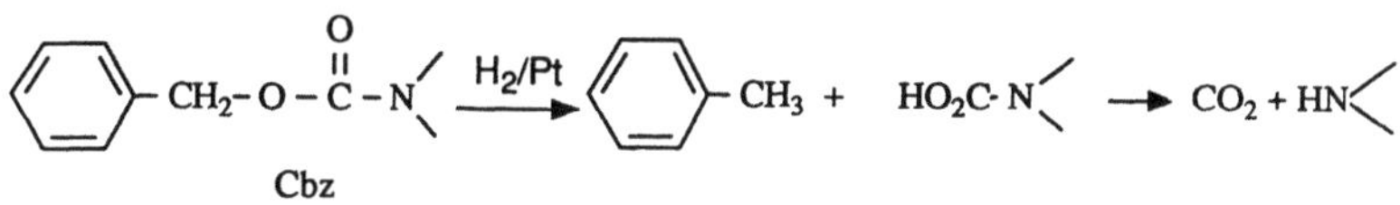

The Cbz (today Z) group brought about a decisive turn in the development of peptide chemistry and opened up a new era in synthesis. The new method was immediately put into action in Dresden for the preparation of peptides containing amino acids with reactive side chains. Until then, because of the unavailability of reversible blocking of side chain amino groups, such targets could not be attained. It became now possible to couple lysine in the form of $N^{\alpha}$, $N^{\varepsilon}$-dicarbobenzoxylysine methyl ester via the hydrazide and azide to esters of glutamic acid or histidine and to secure from the dipeptide derivatives through saponification and hydrogenation the desired free dipeptides [5].

If the discovery of the Z-protecting group failed to prompt a flood of peptide syntheses this was merely due to the small number of laboratories, all over the world, working on peptides in the thirties. Also, the political situation greatly deteriorated. Under the Nazi regime, Max Bergmann in 1934 with his wife Martha, the 19 year old Peter and 17 year old Ester, children from his first marriage, had to leave Germany. He was warmly received by the Rockefeller Institute for Medical Research in New York City and he was able, first as Associate Member, somewhat later, from 1937, as Member of the Chemical Laboratory, to continue peptide research in collaboration with Leonidas Zervas (Plate 55) who followed him from Dresden and with a few new coworkers of whom Joseph S. Fruton (Plate 18) and Heinz Fraenkel-Conrat were the first to join him. Europe, seething in the war, offered little to further the interest of

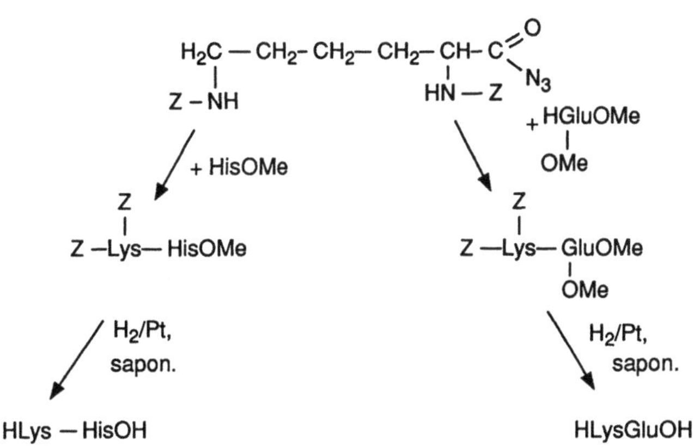

**Fig. 3.** First applications of the carbobenzoxy (Z) group

chemists in the peptide field which once again retreated into the background. After the outbreak of hostilities, in most European countries, fundamental research was reduced to a bare minimum.

## 3.1 Max Bergmann and the Rockefeller Institute

In New York, Bergmann continued his investigations of amino acids, peptides and proteins on a wider base. This was done through intensive work on analytical methods for the determination of amino acids and peptides and also through an enhanced examination of protein-cleaving enzymes with the necessary synthesis of peptides suitable for serving as model substrates.

The analytical studies were started without giving consideration to chromatography. This method was revived in the early thirties in the laboratory of Richard Kuhn at the Kaiser Wilhelm Institute for Medical Research in Heidelberg, mainly by Edgar Lederer, for the separation of carotinoid dyes but many years went by before it was learned how to apply it successfully for the separation of amino acids and peptides. Bergmann first addressed himself, with high intensity, to a search for precipitating reagents as specific as possible for individual amino acids. For groups of amino acids such compounds had been known for a long time, for instance phosphotungstic acid which in a strong acid solution had been used earlier to precipitate the basic amino acids lysine, histidine and arginine, or calcium hydroxide or barium hydroxide which form slightly soluble salts in alcoholic solution with aspartic acid and glutamic acid. Also well known was flavianic acid (2,4-dinitro-1-naphthol-7-sulfonic acid) introduced in 1924 by A. Kossel and R.E. Gross who employed it, with considerable success, to precipitate ariginine, or the well known reagent picric acid, as well as Reinecke acid (tetrarhodanato-diamminchromic acid), which, as found by J. Kapfhammer in the later twenties, forms slightly soluble salts with proline and histidine. Bergmann enlarged the list of suitable precipitants by additional chromium complexes, e.g. "rhodanilic acid" for proline and later, with his new colleagues William H. Stein and Stanford Moore, by several aromatic sulfonic acids. They worked out the "solubility product method" in which the (theoretical) constancy of the solubility product, the product of the molar concentrations of precipitant and amino acid in the supernatant, is used for the determination of the concentration of an amino acid in protein hydrolyzates. Briefly, two aliquots of the solution to be analyzed are saturated in parallel with a slightly soluble salt of the amino acid in question, the first one without, the second one with a definite small amount of the precipitant added. The comparison of the amounts of salts dissolved in each probe, after filtration, allows a calculation of the amount of amino acid present [6]. The amino acid composition of several proteins has been reported using this method. All methods of separation and analysis of amino acids, however, were superseded after perfection of chromatography had been achieved, mostly by the efforts of Stein and Moore in Bergmanns laboratory at the Rockefeller Institute. To Bergmann, regrettably, it was not given him to witness the breakthrough in protein chemistry that followed his analytical work in this area. He died in 1944 in New York, 59 years old, after long illness, but also after 10 years during which he, a modest, reserved man, had won many friends, particularly among his colleagues, and had learned to enjoy the cultural and geographical advantages of his new environment.

### 3.1.1 Chromatographic Separation of Amino Acid Mixtures [7]

Separation of a mixture can occur by washing it with a suitable solvent through a layer of an adsorbent on which the different components reversibly adhere with differing strengths. Such "chromatographic" separation through a column of calcium carbonate was demonstrated first by the Russian botanist M.S. Tswett (1903) with leaf pigments; the technique was resumed and further developed by Richard Kuhn and colleagues in the beginning of the thirties. For the fat-soluble class of compounds corresponding solvents such as benzene, petroleum ether etc. were used. Since amino acids and most peptides are hydrophilic substances, not soluble in these liquids, the original chromatographic systems could not be employed for their separation.

An additional difficulty was the invisibility, which, in contrast to the colored chlorophylls and carotenoids, prevented the visual observation of the process with amino acids. The first chemist to attempt chromatographic separation of amino acids and to study it in detail was Arne Tiselius in Uppsala (Sweden) from the early forties on. He has used principally active carbon as absorbent and worked mainly with water, sometimes adding acetic acid or phenol. Among his "frontal analysis", "displacement analysis" and "elution analysis" the latter can be regarded as a precursor of the technique finally and successfully in use at present. The differences in Freundlich-adsorption coefficients of the natural individual amino acids as the separating principle, however, are not sufficiently large to lead to clean fractions. Only the group of aromatic compounds (phenylalanine, tyrosine, tryptophan) which are distinctly more strongly adsorbed than the other amino acids, will be retained on rinsing the carbon column with water, and so can be separated from the rest.

Group separation by chromatography was worked out at the same time in other laboratories. Fritz Turba, in 1941, then in Prague with Ernst Waldschmidt–Leitz, systematically explored the use of the ion exchanging properties of naturally occurring earths tentatively tried before by H. Fuchs in Würzburg. Filtrol-Neutrol, a commercial product for the bleaching of e.g. oils, was found to bind from aqueous solution the basic amino acids arginine, lysine and histidine, whereas Floridines, from Florida, held back arginine and lysine but not histidine [8]. These minerals, in use also for softening of water (permutites) contain alkali metal ions which can be exchanged for the other cationic compounds. Anion-exchangers rarely occur in nature; in 1937 G.M. Schwab [9] showed that alumina, $Al_2O_3$, a widely used adsorbent, is a cation exchanger due to Na-ions bound to anionic $[(-O)_3{>}Al{-}OH]^-$ entities, and that the amphoteric matrix is converted to a Cl-containing anion exchanger by treatment with dilute hydrochloric acid, perhaps $\left[\begin{smallmatrix}-O\\-O\\-O\end{smallmatrix}\!\!>Al-Cl\right]^- H^+$. One of the present authors has found this "acid alumina column" very suitable for adsorbing, by exchange, the acidic amino acids aspartic and glutamic acid (and cysteic acid) from a mixture of amino acids [10]. At the same time, synthetic organic resins bearing either cationic ($-\overset{+}{N}R_3$) or anionic ($-CO_2^-$, $-SO_3^-$) groups, Amberlites, Dowex, Wofatite-s were being manufactured, with capacities surpassing those of the natural exchangers. With such resins, finally the breakthrough became possible to chromatographic systems in which the amino acids of a hydrolyzate of microgram amounts of proteins could be analyzed in less than an hour.

It was Stein and Moore who in the Rockefeller Institute continued the studies of conditions for the separation of all protein-amino acids and who finally arrived at a perfect solution by using a sulfonated polystyrene resin neutralized by Na-ions [11]. The amino acid mixture is placed on the column at pH 3 where the amino acids are positively charged and thus are bound to the column. The column is developed by gradually increasing the pH and ionic strength of the buffers with which the column is washed. These eluents gradually cause neutralization of the positive charges on the amino acids and so weaken the ionic linkages. The acidic amino acids are removed readily in the sequence aspartic acid, (a stronger acid than) glutamic acid, followed by the neutral amino acids, and finally by the basic amino acids. This method serves to separate the neutral

**Plate 4.** Stanford Moore, 1. (1913–1982) and William H. Stein (1911–1980). In the Rockefeller Institute, New York City

amino acids, too, because the resin material has not only ion exchanging but also adsorption properties. These are sufficiently selective as to effect the separation of the whole mixture, even isoleucine from leucine. The eluted amino acids are continuously quantitated by reaction with ninhydrin or, more recently also with fluorescamine and the color intensities recorded with an automatic instrument constructed together with D.H. Spackman. The method, then, was utilized for the amino acid analysis of several proteins leading to results which were incompatible with the periodicity rule mentioned on p. 4. Together with C.H.W. Hirs, Stein and Moore addressed themselves to the elucidation of the structure of a protein, the enzyme ribonuclease A, and with the help of their analyzer they solved the entire covalent structure of this globular protein, a single chain composed of 124 amino acid residues (for syntheses see pp. 41, 170). As highest among many recognitions they received was the 1972 Nobel Prize in Chemistry.

This successful evolution of amino acid analysis would not have been possible without the chromogenic reagent ninhydrin. This compound (tri-ketohydrindene hydrate I), prepared by Siegfried Ruhemann in Cambridge in 1910, generated a dark violet-blue coloration on the skin within several hours, and, as the discoverer described gave a dark blue coloration specifically with α-amino acids in a concentration of the reagent as low as one part in 15000 parts of water [12]. The color of "Ruhemann's purple" is due to a reaction with α-amino acids on short heating yielding the mesomeric anion II (Fig. 4). Ninhydrin became a widely used reagent in physiological chemical research. Abderhalden, as soon as one year after Ruhemann's publication extensively extended the reaction to "peptones" and used it later in his protein degradation studies. He also found that the

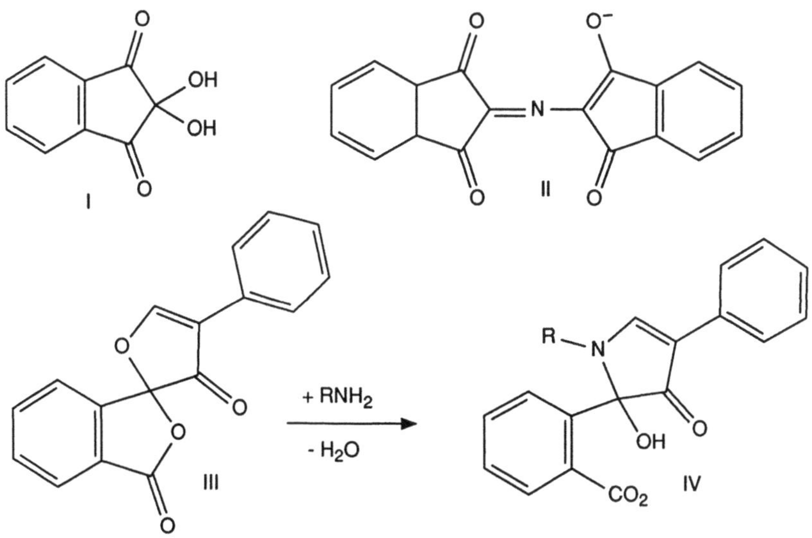


**Fig. 4.** Formulae of ninhydrine (I), Ruhemann's purple (II), fluorescamine (III) and fluorescent reaction product (IV)

blue-violet anion forms a red complex with mercury ions which can be extracted with toluene. Harding and MacLean probably were the first to use the ninhydrin-reaction, in 1916, for a colorimetric quantitation of α-amino acids [13]. In the Rockefeller Institute Stein and Moore standardized the color reaction, which orginally did not yield equal coloration and strength for every individual amino acid, for routine application in their chromatographic separation device.

An even more sensitive reagent for this purpose was introduced by S. Udenfriend and collaborators in 1972, M. Weigele's "fluorescamine" (III in Fig. 4) which at room temperature with primary amines quickly forms strongly fluorescent pyrrolinones (IV) fluorometrically measurable even in the picomole range [14].

On the way to the final concept Moore and Stein also made use of results which had been obtained in other laboratories. In 1941 A.J.P. Martin (Plate 31) and R.L.M. Synge (Plate 44) in England introduced *partition chromatography* [15]. In this method, in place of the normal solid adsorbent, an inert powder is used which can serve as a mechanical support for a liquid phase, mostly water. The column is run with a second liquid phase which is immiscible with the first. In such a chromatogram, partition between the two phases replaces adsorption or ion exchange as the physical properties on which the separations expend. At first the authors worked successfully with *N*-acetyl derivatives of amino acids and with precipitated silica gel (+ 50% water) as supporting powder and chloroform as the moving phase. Partition chromatography of free amino acids, the target aimed at, was reached by Synge who in 1944 found potato starch to be a suitable support for water and butanol as developing medium. Using their automatic fraction collector and the quantitative color reaction with ninhydrin, Moore and Stein in 1948 presented the starch method as a practical analytical procedure.

Whereas partition chromatography in this field was soon superseded by an ion exchange method (see above), it achieved great success with cellulose as a support, as *paper chromatography*. According to Consden, Gordon and Martin [16], a strip of filter paper bearing a dried drop of the amino acid mixture near the top, enclosed in a chamber is allowed to hang from a trough containing the solvent. The solvent syphons out of the trough by capillarity and flows slowly down the strip, transporting the individual amino acids at different rates ($R_F$ values) depending on the partition coefficient. On a rectangular sheet the one-dimensional chromatogram obtained in a first solvent can be developed in a second solvent in perpendicular direction generating a two-dimensional paper chromatogram. The positions of the individual amino acids are made visible by spraying with a dilute solution of ninhydrin and drying briefly in an oven (Fig. 5). The amino acid components of the antibiotic gramicidins have been clearly separated and identified for the first time by this technique. This extremely valuable method invaded all laboratories within a short time, and, among other important results, has led to the discovery of a multitude of new substances in natural sources, not only amino acids and peptides.

Filter paper, in a simple self-made device, as a support for electrolytes was recommended for the separation of amino acids as ionized substances due to their

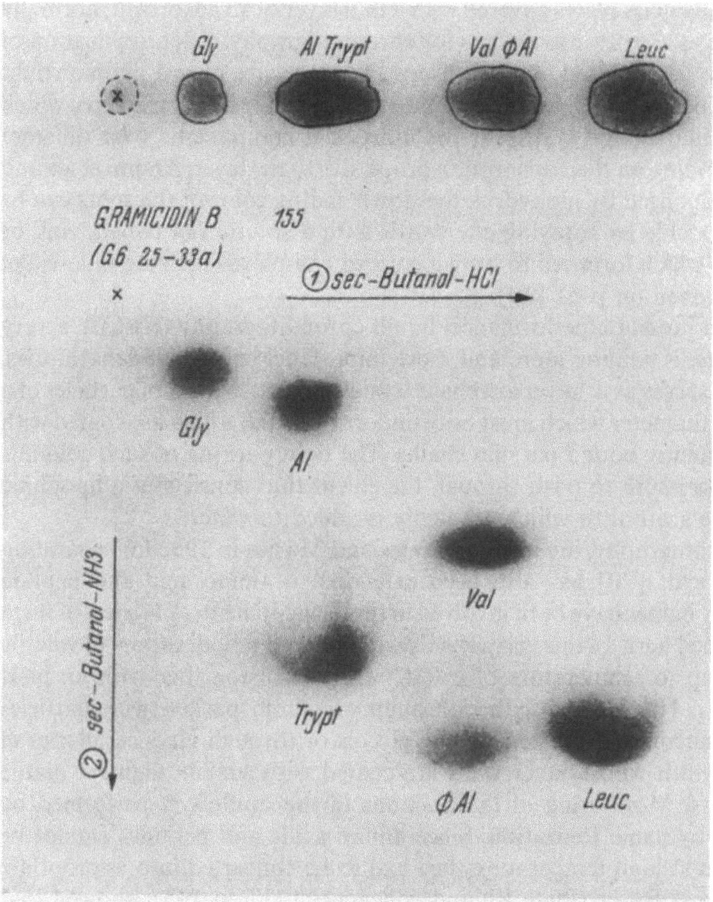


**Fig. 5.** Paper chromatogram of a hydrolyzate of gramicidin B, one dimensional on top, below two dimensional after visualization with ninhydrine. From L.C. Craig in Fortschritte der chemischen Forschung, 1, p. 315 (1949)

different migration velocity in an electric field by T. Wieland and E. Fischer [17]. Electrophoresis as a principle for the separation of differently charged amino acids has an early ancestor: In 1909 Ikeda and Suzuki took a patent for the preparative isolation of glutamic acid from protein hydrolyzates by an electrophoretic technique using a diaphragm device. Electrophoresis on a solid support other than paper was later described in starch layers and finally in synthetic gels, mainly polyacrylamide (PAGE). This technique is of invaluable utility for the analysis of mixtures of proteins (after reaction with sodium dodecylsulfate, SDS), and of nucleotides.

Also in the history of chromatography, "the better is the enemy of the good". During the past decades paper chromatography has been nearly totally superseded by *thin layer chromatography* (TLC) proposed by E. Stahl [18]. In this technique usually glass plates covered with a thin layer of an adsorbent, normally silica gel, serve as simple instruments for chromatography. After application of the mixtures to be analyzed as small drops the plates are placed in an airtight chamber with the lower edges immersed into appropriate solvent mixtures which ascend by capillarity and transport the individual components over different stretches, depending on their adsorption properties to the layer. Amino acids and peptides are visualized by ninhydrin; the slowly fading color of the spots can be rendered very stable by spraying afterwards with a solution of cobalt, zink or cadmium salts which form red to brown colored complexes with the blue-violet anion II mentioned on p. 51 [19].

Today high pressure (performance) liquid chromatography (HPLC), a very rapid procedure, is gaining more and more importance in peptide laboratories. Here, silica gel serves as a "reversed-phase" solid support, in form of particles of a few microns in diameter which must be of uniform size and which are coated with a layer of covalently bound paraffin chains. The tight packing of such columns requires high pressure to push through the eluent that consists of a lipophilic solvent such as acetonitril which gradually is added to water.

Gas chromatography, invented by James and Martin in 1952 for separating volatile fatty acids [20] has also been extended to amino acid and peptide research. Many names have been involved in this development, only a few of them can be mentioned here. In this very sensitive and rapid method, vapors of volatile components (up to temperature of 250 °C) are passed together with an inert carrier gas ($N_2$, $H_2$, He, Ar) either through a column packed with particles covered with silicones, polyesters or polyglycols or through glass capillaries of e.g. 0.25 mm width whose inner walls are coated with similar slightly volatile liquid polymers. Monitoring of the fractions in the outflow is performed by detectors, e.g. by flame ionization. Since amino acids and peptides cannot be vaporized even at high temperature they had to be converted into appropriate derivatives. Ernst Bayer (Plate 10) and colleagues [21], in 1957, showed that amino acid methyl esters (and Weygand's TFA-amino acid esters, see below) can be subjected to gas-chromatographic separation. The volatility was greatly enhanced by acylation of the polar amino groups, e.g. as acetyl derivatives and by trifluoroacetylation. Trifluoroacetic acid (TFA) as a reagent in peptide chemistry had been introduced by Fritz Weygand (1911–1969, Plate 48) in 1952. TFA-dipeptide methylesters were found by his group to be very well applicable for gas chromatographic separation [22]. Many other modifications at the amino group as well as at the carboxyl group have since been suggested for suitable derivatization of amino acids, TFA amino acid esters, however, have hold a leading position. This is true at least for the separation of enantiomeric amino acids and peptides.

The preparative separation of mixtures of D- and L-amino acids into the D- and L-enantiomers, already a problem in E. Fischer's laboratory, has been perfected nowadays but it still takes much time. In order to obtain analytical results as fast as possible, chromatographic methods were attempted early, and gas chromatography turned out to be the method of choice. In 1965, Emanuel

Gil–Av at the Weizmann Institute in Rehovot, Israel, showed that the selectivity of a gas chromatographic column can be extended to mobility-differences of enantiomers by using chiral stationary phases. For a stationary chiral phase, say L-, an L-compound will have an affinity (slightly) different from its D-antipode and so will migrate at a different rate. The trifluoroacetyl-L-dipeptide cyclohexyl esters initially used as stationary phase, although poorly volatile were lost slowly from the column at every run and therefore E. Bayer and W.A. König modified the stationary phase by the incorporation of chiral side chains (amino acid derivatives) into the high polymer polysiloxane matrix [23]. The second possibility for D- or L-discrimination consists of linking the compounds under investigation to a second one of defined chirality, say of L-configuration. A D,L-mixture then will yield an L,D- and an L,L-diastereomer. Since diastereomeric compounds behave differently in physical respects, also in chromatographic properties, their separation is possible by such methods. In amino acid analysis the compound whose optical composition is to be determined, is coupled to a second optically pure component, mostly an amino acid and the dipeptide is examined by chromatography. In the case of the presence of an enatiomer the two diastereomers formed, L,L- and D,L-, can easily be differentiated by the chromatographic method. Not only gas chromatography but also thin layer, liquid adsorption, partition or ion exchange chromatography methods have been adapted to the stereochemical problems discussed above.

Gel chromatography [24], a relatively new development, uses for separation the differences in the access of substances to pores or caves of a gel matrix. Smaller molecules which more readily enter such phases, will move more slowly than larger molecules. Among several gels only the well known Sephadex will be mentioned—products of the Swedish Pharmacia, Uppsala, derived from the initial work of Jerker Porath and Per Flodin. Sephadex is a dextran (1,6-polyglucosan) cross-linked by 1,3-glycerol ether bridges. The higher the cross-linking, the smaller is the average diameter of the mesh or pores and, simultaneously, the lower the accessibility for higher molecular weight substances. On a G-10 Sephadex column (strongly cross-linked), for instance, a clean chromatographic separation in water of the peptides H-Leu-Gly-Glu-PheOH, H-Gly-Glu-Phe-OH, H-Glu-PheOH and HPheOH (eluted in the sequence indicated) was possible. (Th. Wieland and H. Determann, 1967). Weaker cross-linked gels allow separation of proteins according to their molecular masses; this is the main area of their application.

As shown, a considerable part of the modern separation methods of amino acids and peptides originated in Max Bergmann's laboratory at the Rockefeller Institute. A different separation technique was introduced in the same institute at the same time by Lyman C. Craig, in the laboratory of J.A. Jacobs. Here, from 1933, he worked on ergot alkaloids developing a lasting interest in separation techniques. The well-known method of extraction of substances from an aqueous into a non-water-miscible phase like chloroform or vice versa was systematically elaborated by him to a multiple extraction and re-extraction procedure, countercurrent distribution [25]. By this method, performed in a specially constructed machine consisting of several hundred extraction units, isolation and purification of antibodies, nucleic acids as well as biologically active peptides became possible for amounts greater than by the normal chromatographic techniques. For a picture and biography of L.C. Craig see on Plate 16 and on p. 265.

### 3.1.2 Peptides and Enzymes

Synthetic work with peptides was resumed by M. Bergmann in a systematic way with Joseph S. Fruton (Plate 18), who joined him soon after his arrival at the Rockefeller Institute. The enormous advantages of the new, easily removable carbobenzoxy residue allowed—as before in Dresden—simple syntheses of peptides destined as model substrates for the characterization of protein-hydrolysing enzymes—following former interests of E. Fischer's school. At that time, in 1936 to 1937 the different points in a peptide chain were recognized where the main proteolytic enzymes specifically start cleaving, and definite evidence was found for genuine protein chains to contain no other linkages than CO—NH—bonds.

Continuing the tradition of Emil Fischer with enzymes in carbohydrate, protein and peptide research, Bergmann had already intensified the studies of, then still impure, proteinases in Dresden and in New York he embarked on the investigation of specificity of defined peptidases. This was the time shortly after pepsin had been crystallized as the first proteolytic enzyme by J.H. Northrop in 1930 and chymotrypsin and trypsin by M. Kunitz, 1932. Joseph S. Fruton utilized the potential of the new carbobenzoxy method to synthesize hitherto inaccessible peptides for testing as substrates for enzymatic hydrolysis. In the course of these studies Fruton published with M. Bergmann syntheses of the following simple substrates (arrow at fission point): For papain: acetylphenylalanylalanyglycine amide, Ac-Phe-Ala-Gly $\rightarrow NH_2$, for chymotrypsin: benzoyltyrosylglycine amide, Bz-Tyr $\rightarrow$ Gly-$NH_2$, for trypsin: benzoyllysine amide Bz-Lys $\rightarrow NH_2$ and for pepsin: benzoylaspartyltyrosine amide, Bz-Asp $\rightarrow$ Tyr-$NH_2$ [26].

In parallel the search was continued for peptidases yet unidentified, in which also Emil L. Smith, later a famous authority on protein research in the U.S.A. was successfully engaged. In Germany, similar work was set forth at the Dresden Institute by Max Bergmann's successor Wolfgang Grassmann (1898–1978), and by Ernst Waldschmidt-Leitz (both students of Richard Willstätter), who after his habilitation there (1924), in 1928 left Munich to hold a chair as professor in Biochemistry at the Technische Hochschule, later at the University of Prague until 1945. After his return to Munich he was there, as professor at the University, from 1953–1963.

The Kaiser-Wilhelm-Institut für Lederforschung in Dresden was completely destroyed by bombs in 1945. It was re-established by W. Grassmann, at first as a research laboratory in Regensburg, Bavaria, then as the Max-Planck-Institut für "Eiweiß- und Lederforschung" in Munich and has now been incorporated as a department of peptide chemistry, directed by Erich Wünsch, into the large Max-Planck-Institut für Biochemie in Martinsried, near Munich.

In the early days of enzyme research, stringent conclusions as to the exact catalytic sites of the respective protein- and peptide-splitting enzymes could not be drawn, but indications appeared on possible interactions of amino acid side chains of the substrate with those of the enzymes providing affinity binding. Closer information has been accumulated in subsequent studies from various laboratories [27]. In any case, in the early investigations, indications were obtained that for the action of papain on the substrate, a hydrophobic amino acid residue in the second position "before" the bond hydrolyzed greatly promotes catalytic activity; that for chymotrypsin, two hydrophobic side chains, here benzoyl and tyrosyl, at an analogous distance to that previously stated favor enzymatic efficiency; that trypsin attacks an amide (peptide) bond "behind" lysine; and that pepsin will split a peptide substrate "before" an aromatic side chain. Further experiments revealed that this "acid proteinase" in longer peptides preferentially splits a peptide bond between hydrophobic and/or aromatic amino acids (Phe, Tyr, Leu, Trp). A detailed overview on this chapter of peptide chemistry up to 1982 was given by J.S. Fruton [28]. In his review, however, in the context of specificity, primarily the opposite of peptide bond fission, proteinase-catalyzed formation of peptide bonds is discussed.

Enzymatic Peptide Synthesis

At the beginning of this century, Sawyalow gave the name "plastein" to the insoluble material which appeared upon the incubation of a soluble mixture of enzymatic digestion products of fibrin with rennin (a pepsin-like enzyme from calf-stomach). This reaction, later observed also with enzymes different from rennin was studied more intensively in the 1920s by Wasteneys and Borsook [29] who showed that the products of peptic hydrolysis of egg albumin at pH 1.6 gradually formed a precipitate when the concentrated solution was incubated with pepsin at pH 4.

This plastein was considered to be a protein largely because of its solubility properties. As mentioned in the Introduction (p. 3–4), at that time the nature of proteins was still far from being defined; the formation of a slightly soluble product by pepsin could have been a sort of coagulation without the real formation of peptide bonds. This may be the reason why the plastein reaction was not investigated further for more than twenty years, and completely elucidated only in the 1960s.

The problem of the enzymatic synthesis of peptide bonds assumed a new aspect in around 1937, when Heinz Fraenkel–Conrat [30], then in the laboratory of M. Bergmann, demonstrated the papain-catalyzed formation at pH 5 of sparingly soluble benzoylglycine anilide from benzoylglycine amide or from benzoylglycine (hippuric acid) and aniline as well as the condensation of benzoylleucine with leucine anilide yielding the nearly insoluble dipeptide benzoyldileucine anilide (Fig. 6).

By these and similar observations with proteolytic enzymes other then papain it appeared that proteinases established an equilibrium between carboxyl (or

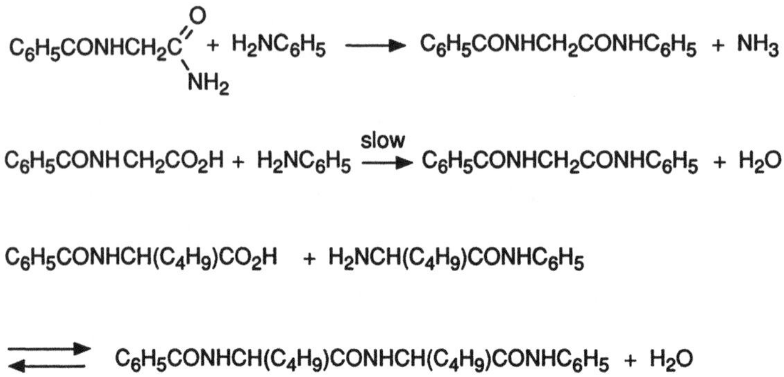

**Fig. 6.** Papain-catalyzed condensations

carboxamide) and amino groups of amino acids and peptides. This equilibrium in water normally lies over to the side of hydrolysis products; the reaction is exergonic by a free-energy change of about 1–2 kcal/mole. The equilibrium, however, is shifted to the right side by the (endergonic) precipitation of the sparingly soluble product(s).

There is a distinct dependence of the equilibrium constant of these systems on the ionization state (pH) of the components. Mechanistically, and thermodynamically, peptide synthesis occurs by negative free-energy (exergonic) change, if the carboxylate anion is protonated and the ammonium cation is deprotonated. Since, however, in aqueous solution of pH around the neutral point (5–8) the groups are ionized, prior to amide bond formation, energy is consumed by transfer of a proton from $—\overset{+}{N}H_3$ to $—CO_2^-$ (neutralization).

From this it is apparent that the gross enthalpy of peptide bond formation depends on the concentration of non-ionized participants, particularly of the carboxyl component, i.e. on the dissociation constants. Since the $K_{Diss}$ of the carboxyl group of a zwitterionic α-amino acid is relatively large ($pK_1$ 2.3), and this, generally also applies to the protonated amino group ($pK_2$ 9.7) as compared to a normal carboxylic acid (pK 4.7) and aliphatic amine (pK 10.8), respectively, in an equilibrium between two α-amino acids the concentration of coupling product is too small as to allow an enzymatic peptide formation. In contrast, in a system of two *peptides*, where the dissociation constants of carboxyl and amino groups are close to "normal" the equilibrium is more favorable. It can be shifted to the synthesis-side by increasing the concentration of one of the components (law of mass action) and by adding non-aqueous water-miscible solvents like glycerol or 1,4-butandiol by which the apparent pK of $RCO_2H$ is significantly shifted to higher pH values, as M. Laskowski and his associates demonstrated in detail [31].

Since proteinases have esterolytic properties as well, it is obvious that the reverse reaction, peptide bond formation from amino acid esters, is also feasible. This type of synthesis was first observed by M. Brenner et al. in 1950 with methionine isopropylester that on incubation with chymotrypsin yielded H-Met-

Met-OH and H-Met-Met-Met-OH [32]. In the following decades, aliphatic amino acid esters have been applied in a multitude of studies on enzymatic syntheses of peptides. In the living cell amino acids esterified to the 2′- or the 3′-hydroxyl group of the terminal adenosine in an aminoacyl transfer ribonucleic acid (tRNA) are the activated intermediates in protein synthesis by ribosomes.

The well known specificity of proteinases implies the use of specific amino acids (amides, esters) as acyl donors and—seldom—specific amino acid (derivatives) as acceptors in enzymatic peptide bond formation, since the same structural features of RCONHR′ that influence the rate of hydrolytic cleavage are also involved in the synthesis. Accordingly trypsin is well suited to the formation of a new -Arg-X or -Lys-X bond. As an example the transformation of the -Lys-AlaOH terminus of the B-chain of porcine insulin into -LysThrOBu$^t$ of human insulin may be mentioned. C-terminal Ala was removed by means of carboxypeptidase A, trypsin-catalyzed condensation of the des-alanine peptide with threonine tert. butylester gave 73% of the ester of human insulin [33] (see also p. 60).

The "plastein reaction", the origin of observations on enzymatic peptide-formation (p. 57) was re-investigated by Virtanen et al. in 1950/51 and systematically studied in the laboratory of one of the authors during the 1960s [34]. It was found that synthetic H-Tyr-Ile-Gly-Glu-Phe-OH at pH 5 in the presence of pepsin was converted to an insoluble oligomer with an average amino acid number of 13, higher values (up to 35 amino acids) were obtained with other pentapeptides of related structure, e.g. H-Tyr-Leu-Gly-Glu-Leu-OH. The minimal chain length of monomeric plastein-forming peptides was four amino acids, the amino-terminal residue could be L-Phe but not D-Phe, and its carboxy-terminal residue could be replaced by L-Phe, but not by D-Phe, nor L-Ile, L-Val or L-Ala. The preference for hydrophobic L-residues on both sides of the newly formed peptide bond corresponds to the specificity of pepsin in the catalysis of hydrolytic peptide bond fission. It should be added that the peptide bond syntheses with pepsin consist of condensation reactions i.e. elimination of water between the reaction partners, transfer of the whole peptidyl residue from a hydroxyl- to an amino group rather than of transpeptidation i.e. transfer of a peptide component from a peptide bond to the acceptor (Fig. 7).

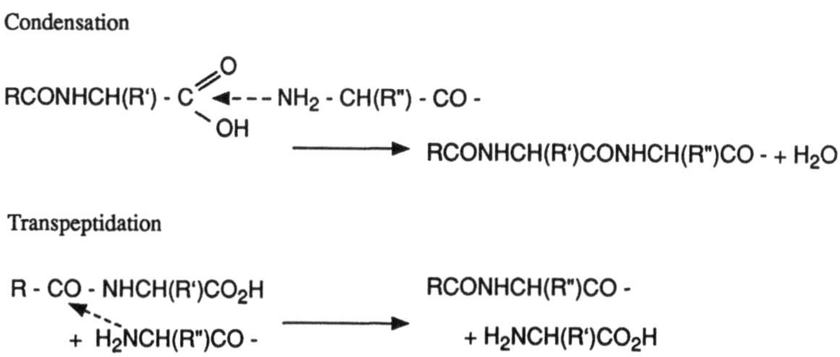


Fig. 7. Peptide bond formation by condensation or transpeptidation

Transpeptidation reactions with proteolytic enzymes have been frequently observed, too; the first example of Fraenkel–Conrat mentioned at the beginning of this section belongs to this type of reactions. A remarkable instance for a transpeptidation reaction is the transformation of pork insulin into human insulin by carboxypeptidase A [35]. This enzyme splits a carboxyl-terminal amino acid from the end of a polypeptide chain and also catalyzes the transpeptidation at the same linkage. Thus, in the B-chain of porcine insulin the terminal alanin is exchanged by threonine on incubation with the enzyme and a large excess of this amino acid.

Among many other peptide splitting enzymes such as (bacterial) subtilisin and thermolysin, (vegetable) papain, ficin and bromelain, (mammalian) cathepsin and others, the yeast enzyme carboxypeptidase Y finally deserves special mention. The enzyme is an exopeptidase, like carboxypeptidase A i.e. it catalyzes, rather unspecifically, the hydrolytic fission of the carboxy-terminal α-amino acids from a peptide chain. J.T. Johansen and his associates at the Carlsberg laboratory in Copenhagen showed about 10 years ago that CPD-Y is an effective catalyst of peptide bond synthesis [36].

The enzyme works optimally at pH > 9 with N-protected amino acid esters as acyl donors, and amino acids or amino acid amides as nucleophiles (acceptors). At that pH value it has high esterase activity; an amino acyl-enzyme complex will be formed rapidly which transfers the acyl residue to the acceptor molecule (transpeptidation), and since at alkaline pH the rate of peptide cleaving is minimal, the product formed will be released in excellent yield. As an example the last coupling step in a synthesis of the opioide peptide Met-enkephalin that started with benzoylarginine ethylester may be shown (Fig. 8).

The N-terminal benzoyl arginine eventually was removed by trypsin that would not attack another peptide bond of the pentapeptide. Thus *N*-benzoyl-L-arginine is a mildly, enzymatically, removable N-protecting group that can also be used in chemical peptide syntheses. The carbobenzoxy group (Z), with which the present chapter was initiated, is a protecting tool likewise utilized in enzymatic peptide synthesis; Z-amino acids (esters) are suitable acylating components for the enzymatic formation of Z-peptides, and the same is true for other easily removable protecting groups to be considered in the following paragraphs.

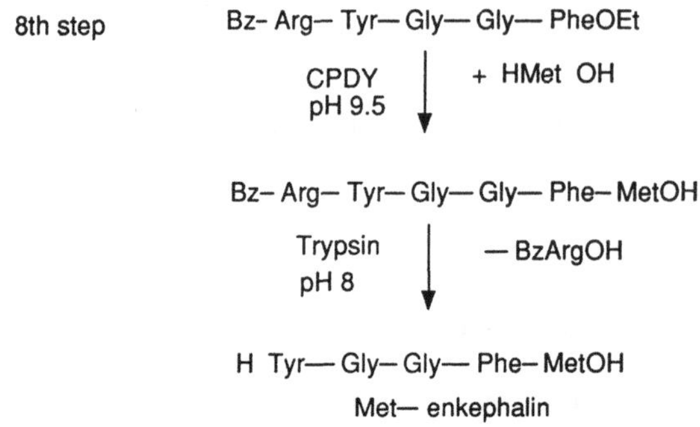


**Fig. 8.** Last steps in a synthesis catalyzed by carboxypeptidase Y

It can be regarded as an advantage of enzymatic peptide synthesis that protection of functional groups of side chains (OH of tyrosine, serine, threonine, —$NH_2$ of lysine, guanido group of arginine, imidazolyl group of histidine) is not necessary, even unwellcome, because side chain protection would render the peptide intermediate insoluble in water, the medium of enzymatic reactions.

Most of the enzymatic peptide forming reactions are strictly stereo-specific for L-amino acids so that racemization that often accompanies chemical coupling of optically active amino acids does not occur. The formation of only the L-amino acid derivative out of a D,L-mixture, in an enzymatic formation e.g. of an anilide from a D,L-Z-amino acid ester and aniline, makes proteolytic enzymes useful reagents for the resolution of racemic mixtures. This is supplementary to the enzymatic stereospecific deacylation of D,L-*N*-acylamino acid mixtures where exclusively the L-derivative will be deacylated.

The imino acids proline and hydroxyproline are practically excluded so far from our consideration either as acyl or as acceptor components since the enzymes commonly used in the studies discussed have no specific affinity for imino acids.

Since the beginning of systematic experimentation toward the exploitation of enzymatic peptide bond formation for preparative purposes much has been achieved. After more than a quarter of a century it is now possible to attempt a comparison with chemical methods and this will not necessarily be in favor of the enzymatic methods. Perhaps the expectations were raised too high in connection with the already briefly mentioned advantages: racemization-free coupling in water (the natural medium of proteins) without the necessity of blocking side chain functions and therefore good prospects for the synthesis of real proteins. Enzymatic synthesis, however, just as the chemical approach, is a method for the coupling of two components, which as in the chemical method, must be suitably derivatized. The reaction does not proceed entirely in the desired direction and therefore it is necessary to work up the mixture and to purify the product. Also, water can be a rather unfavorable solvent in which starting materials are often poorly soluble while enzymes might be rapidly denatured on addition of water-miscible organic solvents. The application of two phases, as proposed by H.D. Jakubke and his associates [37], aqueous buffers for the enzyme and (water soluble) reactants and an organic phase that extracts the product from the equilibrium mixture, deserves serious consideration. The same authors published recently a comprehensive status-report on enzymatic peptide synthesis [38]. In principle the chemical-preparative synthesis of small peptides with advanced coupling methods (Chapter 4) appears to be simpler while the enzymatic approach has certain promise for success in the coupling of larger segments.

During these discussions we have leaped forward many years in the history of peptide chemistry. We discussed the situation after the invention of the Z-group by Bergmann and Zervas in 1932 and the continuation of peptide research first in Dresden then in New York. There the analysis of amino acids and peptides advanced rapidly and this led to the present high-performance chromatographic method of Stein and Moore. At the same time, with the work of Fruton, a systematic investigation was started on the specificity of proteolytic enzymes with the help of substrates which were synthesized, in part, by using the Z-group. The search for additional suitable protecting groups then attracted an ever-increasing number of peptide chemists.

## 3.2 New Easily Removable Specific Protecting Groups

In the enumeration and description of new chemical methods in the following section some preparative details will also be discussed but they are by no means directions for the laboratory. Only the most important methods of preparation, properties, advantages and disadvantages of the reagents will be described. The broad area of preparative peptide chemistry has been covered in numerous books, a recent on dealing with laboratory practice [39]. The most comprehensive collection of data on synthetic methods in peptide chemistry up to 1973 was provided by Erich Wünsch (Plate 51) and his coworkers, published in two volumes of Houben-Weyl's handbook "Methoden der organischen Chemie" [40]. A series, "The peptides" started in 1979 by E. Gross and J. Meienhofer and after the tragic death of E. Gross in 1981 continued with S. Udenfriend [41], attempts to keep peptide chemists up to date. The larger field that includes topics beyond analysis and synthesis as well is made continuously accessible through published Proceedings of various meetings beginning with the first European Peptide Symposium in Prague, in 1958, the American Peptide Symposia since 1968 and the Japanese Peptide Symposia from 1963 [42]. The five progress reports that followed the reawakening in the development of peptide chemistry in 1950 through the fifteen years that followed should be of historical value [43].

Chemical synthesis of peptides from their constituent amino acids requires two types of protecting groups: temporary ones which are needed at the amino group of a building block (amino acid or peptide) the carboxyl group of which will serve for coupling it to the amino component, and at the $\alpha$-carboxyl group of the amine component; and semi-permanent ones for side-chain functional groups, which endure the removal of the temporary groups and are detached only after the peptide bonds have been established. Together with the availability of chemically different residues, the diversity of deprotection methods plays an important role in the selection and in the combination of protecting groups. Functions in side chains which can or, sometimes, must be reversibly blocked are, besides the $\varepsilon$-amino group of lysine, the OH-groups of serine, threonine and tyrosine, the SH-group of cysteine the carboxyl groups ($\beta$-) of aspartic acid and ($\gamma$-) of glutamic acid, and if indicated also the carboxamide groups ($CONH_2$) of asparagine and glutamine. The guanido moiety of arginine as well as the imidazol ring of histidine sometimes requires reversible protection and this is true in many cases also for the indole ring of tryptophan and the thioether group of methionine. Under the guidance of these aspects, modern peptide chemistry advanced considerably after the invention of the carbobenzoxy group [44].

The Z-group plays an important role even today. The reagent usually applied for its introduction, benzyl chlorocarbonate, $C_6H_5CH_2—O—CO—Cl$ (also benzyl chloroformate, carbobenzoxy chloride) is commercially available. Beside the chloride, further reagents were proposed in which the benzyloxycarbonyl group is linked not to a chlorine atom but rather to a different leaving group (X), for instance a suitably substituted phenol or others ($C_6H_5CH_2$—O—CO—X). None of these perhaps more stable reagents could prevail over the chloride.

Acylation of the amino or imino nitrogen is performed, as in many similar acylations, by thorough agitation of Z-chloride with the solution of the amine component in aqueous alkali. The removability of the Z-group under mild conditions by hydrogenation in the presence of a palladium or platinum catalyst yielding toluene and a *N*-carbonic acid (*N*-carboxylic acid; carbamoic acid) which then spontaneously disintegrates to $CO_2$ and the amino component (see on p. 47) is its most important advantage over earlier known protecting groups, an advantage that after the era of Th. Curtius and E. Fischer became the starting point of a new epoch in peptide chemistry.

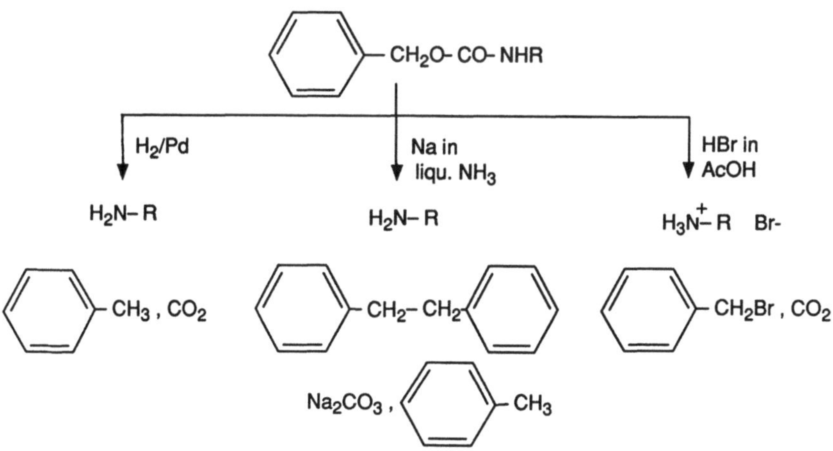


**Fig. 9.** Removal of the Z-groups

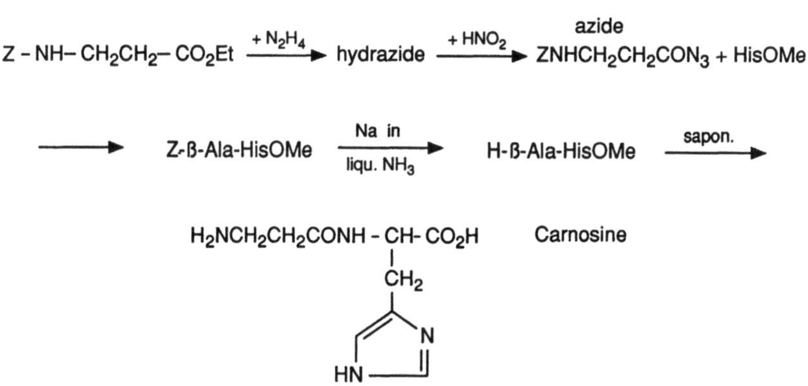


**Fig. 10.** Synthesis of Carnosine [46]

Sulfur, for instance in cysteine, can poison the catalyst, a disadvantage that prompted attempts to find alternative methods of reduction. The solution of sodium in liquid ammonia offered itself for this purpose.

Cleavage of the *N*-tosyl group with this reagent was developed by du Vignaud already in 1930 [45]. Before the invention of the carbobenzoxy protection the tosyl group mentioned on page 46, played a certain role in peptide chemistry but then gradually disappeared from the scene. The appearance of diphenylethane among the cleavage products of Z-peptides points to a mechanism involving free radicals. The method was successfully used by du Vigneaud in a synthesis of carnosine [46] depicted in Fig. 10.

### 3.2.1 Acidolysis

A further reagent for the relatively selective removal of the Z-group is a concentrated (33%) solution of hydrobromic acid in glacial acetic acid. This mixture, introduced by Ben Ishai and A. Berger in 1952, removes the benzyloxycarbonyl group at room temperature within an hour through the $H^+$ catalyzed attack of the $Br^-$ anion on its benzyl moiety which is converted thereby to benzyl bromide with the liberation of an unstable carbamoic acid (Fig. 9).

Subsequently in 1959, Friedrich Weygand broadened the possibilities for the acidolytic cleavage of the Z-group by recommending heating with anhydrous trifluoroacetic acid. In the process instead of benzyl bromide, benzyl trifluoroacetate forms.

The cleavage reagent known the longest, a mixture of hydroidodic acid and phosphonium iodide, discovered in E. Fischer's laboratory for the removal of the *p*-toluenesulfonyl residue (with the formation of thiocresol) and then systematically applied in peptide synthesis by R. Schönheimer in 1926, proved itself effective

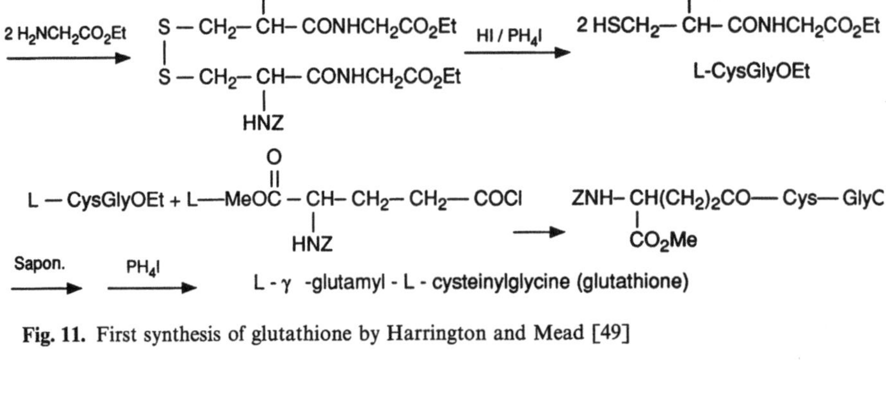


**Fig. 11.** First synthesis of glutathione by Harrington and Mead [49]

in the removal of the *Z*-group as well. The epoch making synthesis of glutathione [47] by C.R. Harington and T.H. Mead in 1935, in which the sulfur present in the molecule excluded catalytic hydrogenation, became possible with the hydrogen iodide reagent. Dicarbobenzoxy-L-cystine was converted with $PCl_3$ to the di-acid chloride and the latter coupled with glycine ethyl ester. Cleavage of the Z-groups and reduction of the disulfide bond with phosphonium iodide afforded L-cysteinylglycine ethyl ester that was acylated with Z-L-glutamic acid $\alpha$-methyl ester $\gamma$-chloride to the tripeptide derivative from which on a second treatment with $PH_4I$ glutathione was obtained. Removal of the Z-group with this reagent probably occurs by acidolysis due to the hydrogen iodide present in equilibrium rather than by reduction.

The first synthesis of glutathione demonstrates another important advantage of the Z-protection, to wit, that racemization at the $\alpha$-carbon atom of the acylating amino acid, here cysteine, is greatly diminished or prevented in its presence. As Bergmann and Zervas stated in their famous paper [4], carbobenzoxy amino acids retain their optical activity when treated with acetanhydride or chlorinating agents, reactions which lead to extensive racemization in *N*-acetyl or *N*-benzoyl amino acids. Later, peptide chemists learned that it is a general property of urethane-type blocking groups to prevent racemization at the $\alpha$-carbon atom during peptide coupling, for reasons to be discussed elsewhere.

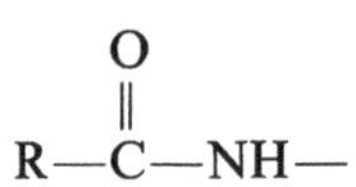

acyl-

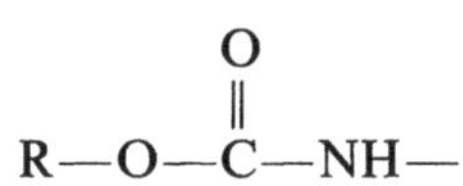

urethane-type

The acidolytic cleavability of the Z-group, in the following years gave rise to the elaboration of a multitude of more or less easily removable analog urethan-type protecting groups in many laboratories all over the world. It was found that the reactivity can be varied within wide limits by suitable structural modifications. The introduction into the aromatic ring of electron-withdrawing substituents like $NO_2$ reduces the rate of acidolytic fission by a factor of about ten, two chlorine atoms in 2, 6 position reduce it by as much as a thousand, whereas electron-donating substituents like $OCH_3$ increase it more than 100 fold; both effects are due to the stability of the carbonium ion, supposed as intermediate. Such differences in reactivity are useful for the selection of temporary vs semi-permanent protection—not only of $\alpha$-nitrogen, but also of side chain functions.

The arsenal of acid labile protecting groups was very effectively enlarged by the invention of the *tert*-butyloxycarbonyl group (Boc) by L.A. Carpino, and by McKay and Albertson in 1957 [48]. The initially difficult introduction of the Boc residue into amino acids was facilitated by the use of Carpino's *tert*-butyloxycarbonylazide which was successfully applied by Robert Schwyzer in peptide synthesis in 1959. Several other Boc-derivatives have been suggested since then, e.g. the promising Boc-fluoride by E. Schnabel and colleagues (1968).

Today the anhydride $(Boc)_2O$, *tert*-but-OCO-O-COO-*tert*-but, is most frequently employed.

The Boc-group surpasses the Z-group in its rate of acidolytic fission by a factor of $10^3$. This is, however, not the reason for its enormously wide application in peptide synthesis but the reason is rather its resistance against catalytic hydrogenation, reduction by Na in liquid ammonia and strong alkali. It is, therefore, the ideal partner of the Z- or of modified Z-groups which can be removed under conditions where Boc-protection remains intact.

The *tert*-butyl group by its readiness to form a carbonium ion is also prone to acidolysis from ester or ether bonds, although less than in Boc, particularly in ethers, $ROC(CH_3)_3$, by an order of magnitude. The benzyl residue, as in Z, is also acid-labile, but the conditions of its fission from ether- or ester linkage are more rigorous than for *tert*-butyl. Since benzyl-esters and benzyl-ethers are easily split by catalytic hydrogenation and by $Na/NH_3$, absolute selectivity compared to *tert*-butyl-protected functions can be obtained. The *tert*-butyl rest offers a further advantage to peptide synthesis by its resistance towards alkaline saponification in *tert*-butyl esters. Aspartic acid and glutamic acid side chains can thus be protected during a whole synthesis of the peptide chain whereas carboxyl groups taking part in condensation reactions are temporarily protected as methyl or ethyl esters and deprotected by alkali.

Obviously, the *tert*-butyloxycarbonyl residue after its successful start has been modifed in the following years in order to increase its acid-lability. The directive idea was to increase the stability of the respective intermediate carbonium ion, e.g. by adding the resonance-stabilizing benzene ring. In fact Sieber and Iselin [49] in 1968 showed that the phenyl-propyl-2-oxycarbonyl group, a "di-methylated" benzyloxy-carbonyl (Z) group (therefore also Dmz or Poc) is removed from an amino acid 700 times faster than the Boc group, and that this effect is heightened another four times on adding a second phenyl group in biphenylylpropyl-2-oxycarbonyl (Bpoc). Fifty percent fission from amino-N by 80-percent acetic acid at 22–25 °C occurs with

Boc
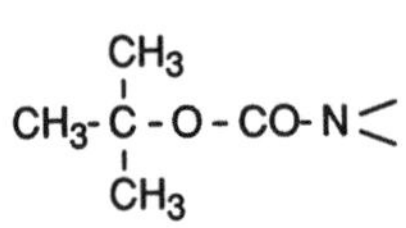

in 58 days

Dmz (or Poc)
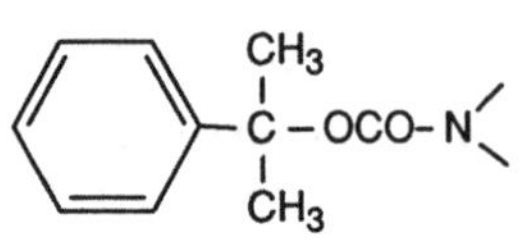

in 2 hours

Bpoc
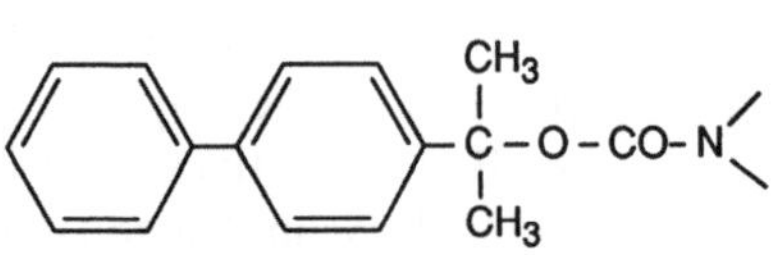

in o.5 hours

Ddz 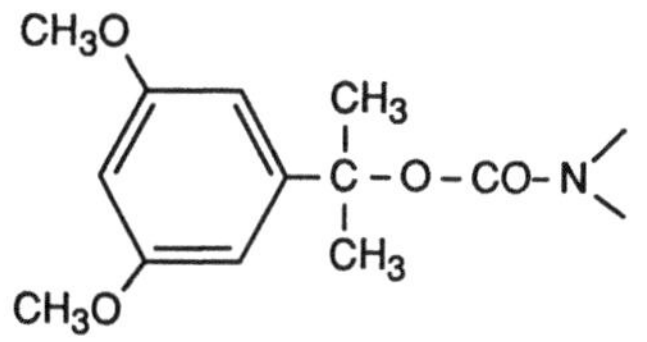 about 1 hour

As a curiosity the 4,4′-dimethoxybenzhydryloxycarbonyl-residue, $(pCH_3OC_6H_4)_2$ CHOCO-, will be mentioned, which under the above conditions is removed with a half-reaction time t/2 of 1 minute, too sensitive for practical use.

For introduction of these residues, among other derivatives, the corresponding azides are suitable reagents. The equally highly sensitive α,α-dimethyl-3,5-dimethoxybenzyloxycarbonyl group (Ddz) synthesized by Birr et al. [50] in 1972 e.g. is split off at room temperature by 5-percent trifluoroacetic acid within 8 minutes, immediately by 10 percent aqueous trifluoroacetic acid. The acid-labile hypersensitive protecting groups have extended the arsenal of selective tools over an area of ten orders of magnitude; the 2,6-dichlorobenzyloxycarbonyl rest of Erickson and Merrifield (1968) can practically only be split off by the most drastic reagent, anhydrous liquid hydrogen fluoride, which was introduced into peptide chemistry in 1967 by Sakakibara (Plate 37) et al. [51]. This acid is also able to cleave the *N*-tosyl group formerly only removable by Na in liquid $NH_3$ (p. 64) or by $HI/PH_4I$ (p. 46).

In the last few years, interest of biochemists has been increasing in *glycopeptides*. Carbohydrate side chains, attached to hydroxyl groups of serine and threonine or to the side chain amide of asparagine are responsible, for instance, for the recognition of proteins by various receptors. The synthesis of peptides of that type makes great demands on selective methods of protection. Among other special reactions a particularly mild cleavage of the allyloxycarbonyl group will be mentioned. This residue, normally cleavable by strong acids, is removed by treatment with the palladium complex $[(C_6H_5)_3P]_4Pd$, which catalyzes under neutral conditions an allyl transfer to nucleophides (Nu) like morpholine or dimedone added in excess (H. Kunz) [52].

$$H_2C{=}CH{-}CH_2{-}O{-}\underset{\underset{O}{\|}}{C}{-}N\!\!< \xrightarrow[Pd(O)]{+Nu} H_2C{=}CH{-}CH_2{-}Nu + CO_2 + HN\!\!<$$

Briefly, it must be mentioned that besides the urethane-type compounds there exist other acid-labile protecting residues, one of them known for even longer than the Z-group. This is the triphenylmethyl rest, introduced into amino acids as early as in 1925 by Burckhardt Helferich (1987–1981).

Helferich did his doctorate with Emil Fischer and worked at his institute from 1910–1920 as an assistent and collaborator on syntheses of glycosides using the trityl protecting group and on the specifity of carbohydrate-splitting enzymes. As a lecturer (1920), for a short time he was associate director at the chemical department of the Kaiser-Wilhelm-Institut für Faserforschung (fiber research) in Berlin, together with Max Bergmann. He left only one year later for Frankfurt. Helferich's interest in amino acids certainly originated from this time, his main research, however, was in the field of carbohydrate chemistry including the associated enzymology. After his time in Frankfurt he became professor at the University of Greifswald, and in 1930 of Leipzig. After the war he was appointed director of the Chemistry Institute of the University of Bonn.

The trityl group (Trt) was later employed rather frequently in peptide synthesis by Schwyzer and Sieber in the 1950s but has been replaced by the Boc group, not least because of the instability of *N*-trityl compounds in the presence of only traces of acids. For its use, as protecting residue of SH-groups see later. Nowadays a revival of the use of the trityl residue seems possible as an analytical tool: di-*p*-methoxytritylchloride and tri-*p*-methoxy-tritylchloride, colored compounds, are utilized to react quantitatively covalently, but fully reversibly, with unreacted amino groups in solid phase peptide synthesis where 100% control of reactions at amino groups is cardinal (p. 103). After reaction with yet free amino groups in the resin-bound peptide and thorough washing, solvolysis with very dilute trichloroacetic acid generates the trityl dye which is measured quantitatively in the solution in a flow spectrometer [53]. Beyond trityl, among highly acid sensitive groups finally the *ortho*-nitro-phenylsulfenyl group (*o*)$NO_2C_6H_4S$ of Gördeler and Holst (1959) will be mentioned which is cleaved rapidly by very dilute inorganic acids, but so far has not found wide-spread application.

### 3.2.2 Photolysis

The Ddz-group mentioned on p. 67 was intended not only as a group readily removable by acidolysis but additionally by *photolysis*. In the Heidelberg laboratory in the 1960s the idea of an automatic peptide synthesizer was pursued in which the growing peptide after each step of prolongation by an N-protected amino acid, should be deblocked continuously in a flow system in order to expose its deblocked amino group in the next coupling step to a new resin-bound activated amino acid. As a suitable mode of deblocking, without addition of foreign reagents, only photolysis came into question and photolytically cleavable compounds were required [54]. J.W. Chamberlin was the first, in 1966, to suggest the 3,5-dimethoxybenzyloxycarbonyl residue as a group removable by uv irradiation. Cleavage of this Z-$(OCH_3)_2$-group by irradiation with a high pressure mercury lamp occurred in solution running through a quartz tube which was wound around the cylindrical lamp, but a higher decomposition rate was desirable and so the above-mentioned even more sensitive Ddz-residue was introduced. A similarly photo-sensitive group, 2-nitro-4,5-dimethoxy-Z was used by Patchornik (Plate 25) and colleagues in 1970; the general concept, however, could not compete with the other successful methods of peptide synthesis in use.

### 3.2.3 Silico-Organic Compounds

Among the residues which are split off extremely easily the trimethylsilyl group is worth mentioning. The employment of *silico-organic compounds* in peptide chemistry was studied extensively by Leonhard Birkofer and his colleagues in the 1960s [55]. The trimethylsilyl group $(CH_3)_3Si$, easily introduced into amino-, carboxyl-, hydroxyl-, and thiol-groups by reaction with trimethylchlorosilane or hexamethylsilazane, $(CH_3)_3SiNHSi(CH_3)_3$, is split off from nitrogen and oxygen by solvolysis even at room temperature, whereas *S*-silyl compounds are stable under the same conditions. This is, however, not the practical use of silyl groups: they are also displaced by electrophilic parts of reaction partners such as activated carboxyl compounds, e.g. in acylation reactions. The *N*-silyl group reacts with acyl chlorides instantaneously to give the carbonamide bond; since silylchloride is eliminated, no base is needed, as would be in acylation of a free amino or imino group. Likewise, silyl esters react with acyl chlorides and via formation of silyl chloride yield mixed anhydrides, even in the absence of a base, and—important in peptide chemistry—the alkyl silyl residue will facilitate solubility in organic solvents.

Another elegant reaction is the use of the 2-trimethylsilylethyl residue for protection of the carboxyl group by Sieber and colleagues in 1977 (for Ref. see e.g. in [44]). In some cases, saponification of methylesters requires alkaline conditions too strong to leave intact a sensitive peptide. Trimethylsilylethyl esters, however, in a type of $\beta$-elimination, are decomposed at neutral pH by fluoride ions to yield trimethylfluorosilane, ethylene, and the carboxylic acid.

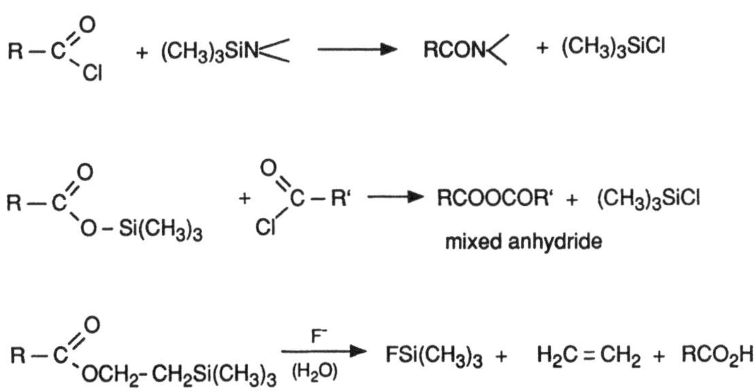


Ethylene and the carboxylic acid

### 3.2.4 Alkaline Reagents

A very important deblocking procedure ever since the beginning of peptide chemistry is the use of *alkaline* reagents. Saponification of alkyl esters, i.e. deblocking of carboxyl groups without destroying established peptide bonds is a usual reaction; it was the first deblocking reaction invented by Emil Fischer also for N-protecting groups, the *N*-ethoxycarbonyl group, but, unfortunately, alkali, here, led to an unforseen reaction (p. 28). After that a long interval passed until finally the hydrazinolytic removal of the *phthalyl group* was brought back to new life in two laboratories, by King and Kidd in England and by J.C. Sheehan and Frank in the U.S.A. In this reaction, as described as early as 1895

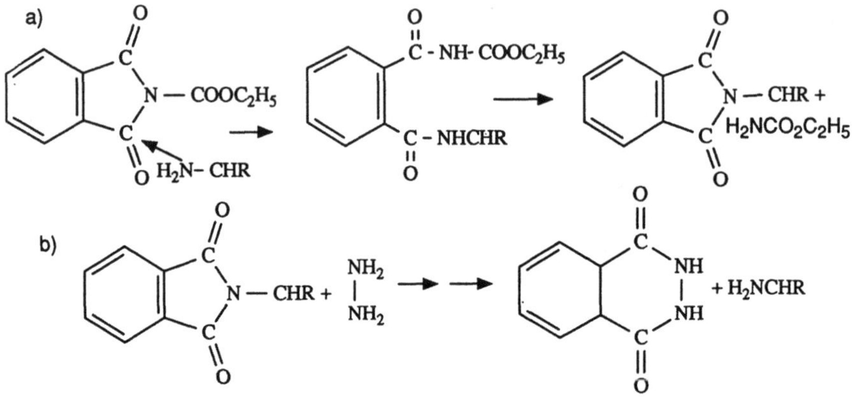

**Fig. 12.** Introduction (a) and removal (b) of the phthalyl group at amino-nitrogen

by Radenhausen, the phthalyl group is converted to the stable heterocycle tetrahydrophthalazine-1.4-dione. The phthalyl residue is best introduced according to Nefkens (1960) by the reaction of the amino compound with *N*-ethoxycarbonylphthalimide (Fig. 12).

In 1952, F. Weygand started his studies with trifluoroacetic acid (TFA) in peptide chemistry. TFA-amino acids and TFA-peptides showed themselves to lose the TFA residue by treatment with dilute aqueous hydroxides or with piperidine. The volatility of TFA esters allowing the separation of amino acids and peptides by gas chromatography was mentioned on page 54.

More importance as reversible protecting groups was gained by molecules prone to alkali-induced $\beta$-elimination. Since the first report of Kader and Stirling who described 2-(4-toluenesulfonyl-methyl) ethoxycarbonyl as an amine protecting group in 1964 [56], a number of appropriately $\beta$-substituted ethyl oxycarbonyl groups have been proposed for carboxyl as well as for amino protection. Since the groups are completely stable under acidic conditions, optimal selectivity against the acidolytically removable protecting residues is given.

The principle of the "safety-catch" can be illustrated very well at the sulfonyl-containing group mentioned above (Fig. 13). In the thioether-analog the readiness for $\beta$-elimination is much less than in the sulfone state. Therefore the unoxidized residue will tolerate alkaline conditions during a peptide synthesis without danger and in order to be removed, must only be oxidized.

Sulfonyl-containing groups have also been proposed for carboxyl protection, i.e. as esters, readily cleavable by alkali. This principle can be usefully applied in solid phase peptide synthesis where the growing peptide chain is anchored as an ester to the support (p. 108) and finally is split off by means as mild as possible.

$\beta$-Elimination is not restricted to sulfonyl-containing groups. Out of many compounds presented e.g. in Ref. [44] only the fluorenylmethoxycarbonyl group of Carpino and Han [57] will be mentioned. This moiety, Fmoc, owes its easy $\beta$-elimination (by piperidine, morpholine) to the tendency to form a conjugated system by loss of a proton and the N-residue (Fig. 14).

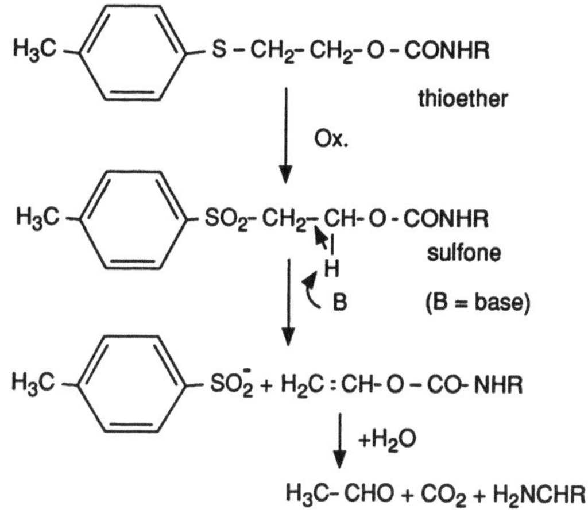


**Fig. 13.** Safety catch principle

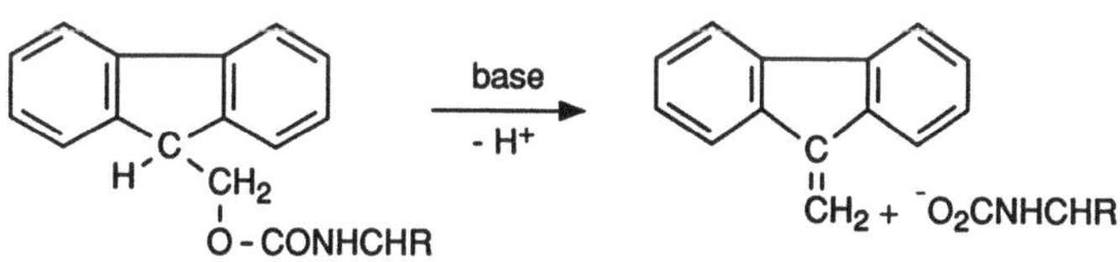


**Fig. 14.** Removal of the Fmoc group

The Fmoc group has proven very valuable during the past decade of peptide chemistry, particularly in the solid phase technique [58].

### 3.2.5 Protection of the SH-Group

Protection of side chain functions has been briefly dealt with several times in this chapter, for details the reader is referred to the comprehensive literature [39–41]. The SH-group, however, because of its peculiar features, will be discussed briefly. Its protection in peptide synthesis is considered to be essential on account of its high nucleophilicity and its ease of oxidation to disulfides. The benzyl group was the first one to be used by du Vigneaud as early as the beginning of 1930s [45, 46]. The fission of the thioether linkage by Na in liquid $NH_3$ was also described by the same author. It resists very strong acids except liquid hydrogen fluoride and hence was considered to be an ideal protecting group for a long time. The rough methods of deprotection, however, turned out more and more noxious with increasing complexity of the peptides to be synthesized. Therefore protecting groups with more favorable properties were sought which cannot be enumerated in this short history. The problem of cysteine-*S*-protection comprises 110 pages in the great review by Wünsch et al. [40]. Instead of the benzyl residue, in 1956 Zervas and Theodoropoulos in Athens and, independently, the laboratory of L. Velluz in Paris used triphenylmethyl chloride to prepare *S*-trityl-cysteine and peptides. The tritylthioether linkage is split only by rather strong acids

(HBr/acetic acid) or by heavy metal ions ($AgNO_3$, Hg(II) salts). A certain popularity was enjoyed by the acetamido-methyl residue, $CH_3CONHCH_2$—, recommended by D.F. Veber from R. Hirschmann's laboratory in 1968. The group although an acetal type is rather stable towards acids; it is removed by mercury (II) acetate in aqueous acetic acid, $Hg^{2+}$ is precipitated by hydrogen sulfide. The number of S-protecting groups is increased by the well known carbobenzoxy (Z) residue. The Z-group is very tightly bound to the S-atom: cleavage does not even occur with HBr in acetic acid, a reagent used for Z-N fission, but succeeds with a base, e.g. $NaOCH_3$ in methanol. A further mode of protection, specific for thiol groups is the formation of mixed disulfides with *tert*-butylthiol according to E. Wünsch and R. Spangenberg [42, Italy]. Since *tert*-butylthiol, for steric reasons, does not form a disulfide on oxidation by air, but can split disulfide bonds by thiolysis, a concomitant oxidation of cysteine and *t*-Bu-SH will generate S-*tert*-butylcysteine.

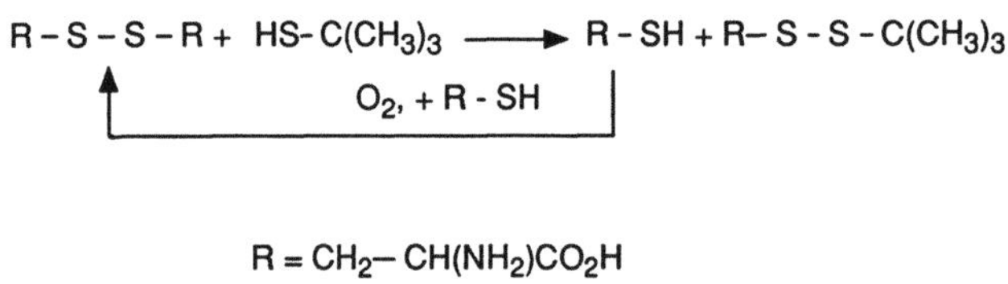


**Fig. 15.** Formation of S-tert-butyl cysteine

Restoring of the cysteine-SH is achieved by reduction, e.g. with 1,4-dithioerythrol (Cleland's reagent) or by sulfitolysis. Disulfides on reaction with nucleophiles (Nu) like rhodanide, cyanide or sulfite are split with formation of an SH- and an -SNu moiety. The thiosulfate compounds formed by $SO_3^{2-}$, so called "Bunte salts" were important in the insulin studies in Aachen (Helmut Zahn and colleagues). They react, like thiorhodanides, -S-SCN, with free thiols to form disulfides.

The cysteine side chain fulfills various functions in many biological active peptides. Whereas one of them in a polypeptide can exist as free thiol, others are destined to be parts of defined disulfide bridges, either intramolecularly (see vasopression on p. 49) or as crosslinks of two peptide chains as in immunoglobulins or in insulin (see on p. 57). The synthesizing chemist, therefore, must have a choice of selectively removable protecting groups, as shown above, and have at his disposal strategies for employing the whole arsenal in combination with the *N*- and *O*-protecting groups principally described in this

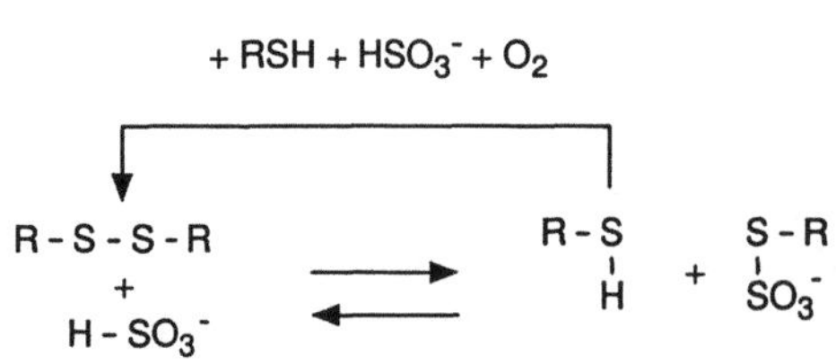


**Fig. 16.** Reversible hemi-sulfitolysis of a disulfide; in presence of oxygen total sulfitolysis

section. The development, the evolution, up to the modern era in which finally the three disulfide crosslinks in synthetic insulin were established by pinpointed reactions by P. Sieber et al., 1974, cannot be evaluated here. The history is marked by names like Leonidas Zervas, Iphigenia Photaki, Helmut Zahn, Panayotis G. Katsoyannis, Wang Yu (Shanghai), H.N. Rydon, R.G. Hiskey and the Swiss group of P. Sieber, B. Kamber, B. Riniker et al. For a comprehensive review see Ref. [59].

## References


1. Biography with a complete collection of over 350 publications: B. Helferich, Max Bergmann (1886–1944). Chem. Ber. 102: 1-XXI (1969)
2. M. Bergmann, V. du Vigneaud, L. Zervas, Acylwanderung mit Spaltungs-vorgängen bei acylierten Dioxopiperazinen, Ber. dtsch. chem. Ges. 62: 1990–1913 (1929)
3. R. Schönheimer, Ein Beitrag zur Bereitung von Peptiden, Hoppe-Seyler's. Z. physiol. Chem. 154: 203–224 (1926).
4. M. Bergmann, L. Zervas, Über ein allgemeines Verfahren der Peptidsynthese, Ber. dtsch. chem. Ges. 65: 1192–1201 (1932)
5. M. Bergmann, L. Zervas, J.P. Greenstein, Synthese von Peptiden des D-Lysins. D-Lysyl-D-glutaminsäure und D-Lysyl-1-histidin. Ber. dtsch. chem. Ges. 65: 1692–1696 (1932)
6. M. Bergmann, W.H. Stein, A new principle for the determination of amino acids and application to collagen and gelatin. J. Biol. Chem. 28: 217–232 (1939)
7. A. competent review on this topic is included in the article of A.J.P. Martin, R.L.M. Synge, Analytical chemistry of the proteins, Advan. Protein Chem. 2: 1–81 (1945)
8. F. Turba, Chromatographie der basischen Aminosäuren an Bleicherden, L. Über das Adsorptionsverhalten von Eiweißspaltprodukten. Ber. dtsch. chem. Ges. 74: 1829–1838 (1941)
9. G.M. Schwab, G. Dattler, Anorganische Chromatographie, II. Säuretrennung, Angew. Chem. 50: 691–692 (1937)
10. Th. Wieland, Quantitative Trennung von Aminosäuren durch Austauschadsorption an Aluminiumoxyd, Hoppe-Seyler's Z. physiol. Chem. 273: 24–30 (1942)
11. D.H. Spackmann, W.H. Stein, S. Moore, Automatic recording apparatus for the use in the chromatography of amino acids, Analyt. Chem. 30: 1190–1206 (1958)
12. S. Ruhemann, Cyclic di- and tri-ketones, J. Chem. Soc. 99: 1438–1449 (1910); -, Triketohydrindene hydrate ibid. 99: 2025–2031 (1910)
13. V.J. Harding, R.M. MacLean, A colorimetric method for the estimation of α-amino acid nitrogen, II. Application to the hydrolysis of proteins by pancreatic enzymes. J. Biol. Chem. 24: 503–514 (1916)
14. S. Stein, P. Böhler, J. Stone, W. Dairmen, S. Udenfriend, Aminoacid analysis at the picomole level. Arch. Biochem. Biophys. 155: 203–213 (1973)
15. A.J.P. Martin, R.L.M. Synge, A new form of chromatogram employing two liquid phases. 1. Theory of chromatography, 2. Application to the microdetermination of the higher monoamino-acids in proteins. Biochem. J. 35: 1358–1368 (1941)
16. R. Consden, A.H. Gordon, A.J.P. Martin, Qualitative analysis of proteins: a partition-chromatographic method using paper. Biochem. J. 38: 224–232 (1944)
17. Th. Wieland, E. Fischer, Über Elektrophorese auf Filterpapier. Trennung von Aminosäuren und ihren Kupferkomplexen. Naturwissenschaften 35: 29–30 (1948)
For a short history of precursors of this technique see Th. Wieland, K. Dose, Electrochromatography (Zone Electrophoresis, Pherography) in Methods in Chemical Analysis, W.G. Berl ed., Vol. III Academic Press, New York, 1956, pp. 29–70
18. E. Stahl, Dünnschichtchromatographie, Springer Verlag Berlin New York, Heidelberg, 1962
19. E. Kawerau, Th. Wieland, Conservation of amino acid chromatograms, Nature 168: 77–79 (1951)
20. A.T. James, A.J.P. Martin, Gas-liquid partition chromatography: separation and microestimation of volatile fatty acids from formic acid to dodecanoic acid. Biochem. J. 50: 679–690 (1952)
21. E. Bayer, K.H. Reuther, F. Boon, Analyse von Aminosäure-Gemischen mittels Gasverteilungschromatographie, Angew. Chem. 69: 640 (1957)

22. F. Weygand, B. Kolb, A. Prox, M.A. Tilak, L. Tomida, N-Trifluoracetylaminosäuren, XIX. Gaschromatographische Trennung von N-TFA-dipeptidmethylestern, Hoppe-Seyler's Z. physiol. Chem. 322: 38–51 (1960)
23. see W.A. König, The practise of enentiomer separation by capillary gas chromatography, Hüthig-Verlag Heidelberg Basel New York, 1987
24. G.K. Ackers, Analytical gel chromatography of proteins, Advan. Protein Chem. 24: 343–446 (1970)
25. L.C. Graig, Contercurrent distribution and some of its applications, parts I, II, III. Fortschr. Chem. Forsch. 1: 292–324 (1949)
26. M. Bergmann, J.S. Fruton, The specificity of proteinases, Advan. Enzymol. 1: 63–98 (1941)
27. For a recent essay see H. Neurath, Proteolytic enzymes, past and present, Federation Proc. 44: 2907–2913 (1985)
28. J.S. Fruton, Proteinase catalyzed synthesis of peptide bonds, Advan. Enzymol. 53: 239–306 (1982)
29. H. Wasteneys, H. Borsook, The enzymatic synthesis of protein, Physiol. Rev. 10: 110–145 (1930)
30. M. Bergmann, H. Fraenkel-Conrat, The enzymatic synthesis of peptide bonds, J. Biol. Chem. 124: 1–6 (1938)
31. G.A. Homandberg, J.A. Mattis, M. Laskowski, Jr., Synthesis of peptide bonds by proteinases. Addition of cosolvents shifts peptide bond equilibria towards synthesis. Biochemistry 17: 5220–5227 (1978)
32. M. Brenner, R.H. Müller, R.W. Pfister, Eine neue enzymatische Peptidsynthese, Helv. Chim. Acta 33: 568–591 (1950)
33. K. Morihara, T. Oka, H. Tsuzuki, Semisynthesis of human insulin by trypsin-catalyzed replacement of Ala-B 30 by Thr in porcine insulin, Nature 280: 411–413 (1979)
34. H. Determann, K. Bonhard, R. Köhler, Th. Wieland, Plasteinreaktion, VI. Einfluβ der Kettenlänge und der Endgruppen der Monomeren auf die Kondensierbarkeit, Helv. Chim. Acta 46: 2489–2509 (1963) and preceding publications
35. M. Bodanszky, J. Fried, Process for preparing human insulin, United States Patent 3276961, 1966
36. F. Widmer, K. Breddan, J.T. Johansen, Carboxypeptidase Y as a catalyst for peptide synthesis in aqueous phase with minimal protection. Peptides 1968, Proc. 16th Europ. Peptide Symp. (Scriptor, Copenhagen, 1981) pp. 46–55
37. P. Kuhl, A. Könnecke, G. Döring, H. Däumer, H.D. Jakubke, Enzyme catalyzed peptide synthesis in biphasic aqueous-organic systems. Tetrahedron Lett. 21: 893–896 (1980)
38. H.D. Jakubke, P. Kuhl, A. Könnecke, Grundprinzipien der proteasekatalysierten Knüpfung der Peptidbindung. Angew. Chem. 97: 79–140 (1983)
39. M. Bodanszky, Y.S. Klausner, M.A. Ondetti, Peptide Synthesis 2nd Ed., John Wiley & Sons New York London Sidney Toronto, 1976. M. Bodanszky, Principles of Peptide Synthesis, Springer Verlag Berlin Heidelberg New York Tokyo, 1984. M. Bodanszky, A. Bodanszky, The Practice of Peptide Synthesis, Springer-Verlag, 1984
40. E. Wünsch et al., Synthese von Peptiden in Houben-Weyl Methoden der organischen Chemie Vol. 15 1/2 (E. Müller Ed.) Georg Thieme Verlag Stuttgart, 1974
41. The Peptides Vol. 1 (1979) to Vol. 8 (1987) ed. by E. Gross (until Vol. 5), S. Udenfriend, J. Meienhofer, Acad. Press, Inc.
42. Peptide Symposia

    The European Peptide Symposia:

    Prague (Chechoslovakia) 1958. Coll. Czech, Chem. Comm., Special Issue 24: 1–160 (1959)

    Munich (Germany) 1959. Angew. Chem. 71: 741–743 (1959)

    Basle (Switzerland) 1960. Chimia 14: 366–418 (1960)

    Moscow (Soviet Union) 1961. Zh. Mendeleyevskovo Obshch. 7: 353–486; Coll. Czech. Chem. Comm. 27: 2229–2262 (1962)

    Oxford (England) 1962. Young, G.T. (ed.) "Peptides", Pergamon Press, Oxford (1963).

    Athens (Greece) 1963. Zervas, L. (ed.) "Peptides", Pergamon Press, Oxford (1966)

    Budapest (Hungary) 1964. Bruckner, V. u. Medzihradsky, K. (eds.) Acta Chim. Acad. Sci. Hung. 44: 1–239 (1965)

    Noordwijk (Netherlands) 1966. Beyerman, H.C., van de Linde, A.W., Massen van den Brink, W. (eds.) "Peptides", North-Holland, Amsterdam (1967)

    Orsay (France) 1968. Bricas, E. (ed.) "Peptides 1968", North-Holland, Amsterdam (1968)

    Padua (Italy) 1969. Scoffone, E. (ed.) "Peptides 1969", North-Holland, Amsterdam (1971)

    Vienna (Austria) 1971. Nesvadba H. (ed.) "Peptides 1971", North-Holland, Amsterdam (1972)

    Reinhardsbrunn (German Democratic Republic) 1972. Hanson, H. u. Jakubke, H.-D. (eds.) "Peptides 1972", North-Holland, Amsterdam (1973)

Kiryat Anavim (Israel) 1974. Wolman, Y. (ed.) "Peptides 1974", Kater Press, Jerusalem (1975)
Wepion (Belgium) 1976. Loffet, A. (ed.) "Peptides 1976", Editions de l'Université de Bruxelles (1976)
Gdansk (Poland) 1978. Siemion, I.Z. u. Kupryszewski, G. (eds.) "Peptides 1978", Wydawnictawa Universytetu Wroclawskiego (1979)
Helsingør (Denmark) 1980. K. Brunfeldt (ed.) "Peptides 1980", Scriptor Copenhagen (1981)
Prague (Czechoslovakia) 1982. K. Blaha, P. Malon (eds.) "Peptides 1982", de Gruyter Berlin New York (1983)
Djurönäset (Sweden) 1984. U. Ragnarsson (ed.) "Peptides 1984" Almquist und Wiksell International, Stockholm (1985)
Porto Carras, Chalkidiki (Greece) 1986. D. Theodoropoulos (ed.), "Peptides 1986", de Gruyter, Berlin New York (1986)
Tübingen (F.R. Germany) 1988. E. Bayer, G. Jung (eds.) "Peptides 1988", de Gruyter, Berlin New York (1989)
The American Peptide Symposia:
Weinstein, B. u. Lande, S. (eds.) "Peptides: Chemistry and Biochemistry", Proc. 1st Amer. Peptide Symp., Yale University, 1968, Marcel Dekker, New York (1970)
Lande, S. (ed.) "Progress in Peptide Research", Vol. II, Proc. 2nd Amer. Peptide Symp., Cleveland, 1970, Gordon and Breach, New York, London Paris (1972)
Meienhofer, J. (ed.) "Chemistry and Biology of Peptides", Proc. 3rd Amer. Peptide Symp., Boston, 1972, Ann. Arbor Science Publ. Inc., Michigan (1972)
Walter, R. u. Meienhofer, J. "Peptides: Chemistry, Structure and Biology" Proc. 4th Amer. Peptide Symp., New York, 1974, Ann Arbor Science Publ. Inc., Michigan (1975)
Goodman, M. u. Meienhofer, J. (eds.) "Peptides", Proc. 5th Amer. Peptide Symp., San Diego, 1977, Wiley, New York (1977)
Gross, E. u. Meienhofer, J. (eds.) "Peptides: Structure and Biological Function", Proc. 6th Amer. Peptide Symp., Georgetown, 1979, Pierce Chemical Comp., Rockford, III (1979)
Rich, D., Gross E. (eds.). "Peptides", Syntheses-Structure-Function, Proc. 7th Amer. Pept. Symp. Madison, 1981. Pierce Chemical Comp., Rockford, III. (1981)
Hruby, V., Rich, D. (eds.) "Peptides, Structure and Function" Proc. 8th Amer. Peptide Symp. Tucson 1983, Pierce Chemical Comp. Rockford, III. (1983)
Deber, C., Hruby, V., Kopple, K. (eds.) "Peptides: Structure and Function", Proc. 9th Amer. Peptide Symp., Toronto, 1985, Pierce Chemical Comp. Rockford, Ill. (1985)
Marshall, G., Deber, C. (eds.) "Peptides, Chemistry and Biology", Proc. 10th Amer. Peptide Symp. St. Louis, 1987, ESCOM Publ., Leiden (1988)
The Japanese Peptide Symposia:
Japanese Peptide Symposia from 1962–1975 were published only in Japanese language.
Nakajima, T. (ed.) "Peptide Chemistry 1976", Proc. 14th Symposium on Peptide Chem., Hiroshima, 1976, Protein Research Foundation, Osaka (1977)
Shiba, T. (ed.) "Peptide Chemistry 1977", Proc. 15th Symposium on Peptide Chem., Osaka, 1967, Protein Research Foundation, Osaka (1978)
Izumiya, N. (ed.) "Peptide Chemistry 1978", Proc. 16th Symposium on Peptide Chem., Fukuoka, 1978, Protein Research Foundation, Osaka (1979)
Yonehara, H. (ed.) "Peptide Chemistry 1979; Proc. 17th Symp. on Peptide Chemistry, Osaka, Japan (1980)
Okawa, K. (ed.) "Peptide Chemistry 1980", Proc. 18th Symp. Peptide Chem., Nishinomiya 1980, Protein Res. Found. (1981)
Shioiri, T. (ed.) Peptide Chemistry 1981, Proc. 19th Symp. Peptide Chem. Nagoya 1981; Protein Res. Found. (1982)
Sakakibara, S. (ed.) "Peptide Chemistry 1982", Proc. 20th Symp. Peptide Chem., Toyomaka 1982, Protein Res. Found (1983)
Munekata, E. (ed.) "Peptide Chemistry 1983", Proc. 21th Symp. Peptide Chem., Tsukuba, 1983; Protein Res. Found. (1984)
Izumiya, N. (ed.) "Peptide Chemistry 1984", Proc. 22nd Symp. Peptide Chem., Fukuoka, 1984. Protein Res. Found. (1985)
Kiso, Y. (ed.) "Peptide Chemistry 1985", Proc. 23nd. Symp. Peptide Chem. Kyoto, 1985. Protein Res. Found. (1986)
Shiba, T., Sakakibara, S. (eds.), "Peptide Chemistry 1986", Proc. Japan Symp. Peptide Chem., Kobe, 1986, Protein Res. Found. (1987)
Ueki, M. (ed.) "Peptide Chemistry 1988", Proc. 24th Symp. Peptide Chem., Tokyo, 1988. Protein Res. Found. (1989)

43. Th. Wieland, Peptidsynthesen, Angew. Chem. 63: 7–14 (1951); II. Angew. Chem. 66: 507–512 (1954); III. with B. Heinke, Angew. Chem. 69: 362–371 (1957); IV. Angew. Chem. 71: 417–425 (1959); V. with H. Determann, Angew. Chem. 75: 539–551 (1963)
44. Review: J.L. Fauchère, R. Schwyzer, Differential protection and selective deprotection in peptide synthesis, in The Peptides, ed. E. Gross, J. Meienhofer Vol. 3, 203–252, Acad. Press, Inc. 1981
45. V. du Vigneaud, L.F. Audrieth, H.S. Loring, The reduction of cystine in liquid ammonia by metallic sodium, J. Amer. Chem. Soc. 52: 4500–4504 (1930)
46. R.H. Sifferd, V. du Vigneaud, Synthesis of carnosine, with some observations on the splitting of the benzyl group from carbobenzoxy derivatives and from benzylthioethers. J. Biol. Chem. 108: 753–761 (1935)
47. C.R. Harington, T.H. Mead, Synthesis of glutathione, Biochem. J. 29: 1602–1611 (1935)
48. L.A. Carpino, Oxidative reactions of hydrazine, II. Isophthalimides. New protective groups on nitrogen. J. Amer. Chem. Soc. 79: 98–101 (1957); F.C. McKay, N.F. Albertson, New amine-masking groups for peptide synthesis, J. Amer. Chem. Soc. 79: 4686–4690 (1957)
49. P. Sieber, B. Iselin, Selektive acidolytische Spaltung von Aralkyloxycarbonyl-Aminoschutzgruppen. Helv. Chim. Acta 51: 614–621 (1968)
50. Ch. Birr, W. Lochinger, G. Stahnke, P. Lang, Der α,α-Dimethyl-3,5-dimethoxybenzyloxycarbonyl (Ddz)Rest, eine photo- und saurelabile Stickstoff-Schutzgruppe für die Peptidchemie, Liebigs Ann. Chem. 763: 162–172 (1972)
51. S. Sakakibara, Y. Shimonishi, Y. Kishida, M. Okada, Removal of protective groups by anhydrous hydrogen fluoride, Peptides 1966, Proc. 8th Eur. Peptide Symp. (North-Holland Publ. Company-Amsterdam 1967) pp. 44-49
52. H. Kunz, Synthese von Glycopeptiden, Partialstrukturen biologischer Erkennungskomponenten, Angew. Chem. 99: 297–311 (1987)
53. M. Horn, C. Novak, A monitoring and control chemistry for solid-phase peptide synthesis, Amer. Biotech. Lab. 5: 12–16 (1987)
54. Th. Wieland, Ch. Birr, Some experiments with resin-activated amino acid derivatives. Peptides 1966, Proc. 8th Eur. Peptide Symp. (North-Holland Publ. Company Amsterdam 1967) pp. 103–106. See here references to the parallel attempts of M. Fridkin, A. Patchornik, E. Katchalski.
55. L. Birkofer, F. Müller, Peptide syntheses without racemization using silyl esters, Peptides 1968, Proc. 9th Eur. Peptide Symp. (North Holland Publ. Company Amsterdam 1968) pp. 151–155.
56. A.T. Kader, C.J. Stirling, Elimination-addition III. New procedures for the protection of amino-groups, J. Chem. Soc. 1964: 258–266
57. L. Carpino, G. Han, The 9-fluorenylmethoxycarbonyl amino-protecting group, J. Org. Chem. 37: 3404–3409 (1981)
58. G. Fields, R.L. Noble, Solid phase peptide synthesis utilizing 9-fluorenylmethoxycarbonyl amino acids, Int. J. Peptide Protein Res. 35: 161–214 (1990)
59. R.G. Hiskey, V. Rao, W.G. Rhodes in Protective grouping in organic chemistry (ed. J.W.F. McOmie), Plenum Press, London, 1973) 235–296

# 4 A Second Breakthrough: New Methods for the Formation of the Peptide Bond

To some extent it is surprising that in the first half of this century so little attention was paid to methods of coupling. A reasonable explanation for this neglect and delay could be the effectiveness of the acid azides of Curtius and the acid chlorides of E. Fischer in the formation of the peptide bond. Most of the desired bonds could be secured without fail and for a long period the principal obstacle in the development of peptide synthesis remained the lack of suitable, readily removable blocking groups. When this barrier was finally removed in the 1930s (cf. Chap. 3) the incentive needed for intensive research toward improved coupling methods was still missing. The elucidation of the structure of excitingly interesting peptides, such as insulin, oxytocin or angiotensin, all within a few years in the early 1950s, provided the necessary stimulus for studies in the methodology of synthesis. Yet, even before these tangible objectives became apparent a certain inspiration was offered by the explosive growth of knowledge in biochemistry at about the same time. The reactive intermediates recognized in biological processes involving acylation revealed a degree of sophistication not seen in the methods generally used in the organic laboratory. Hence biomimetic procedures became both attractive and challenging.

Derivatives of phosphoric acid were repeatedly found to be reactive intermediates in acylation. For instance acetyl phosphate, produced in the reaction between adenosine triphosphate and acetate, yields, when attacked by suitable nucleophiles, their acetyl derivative:

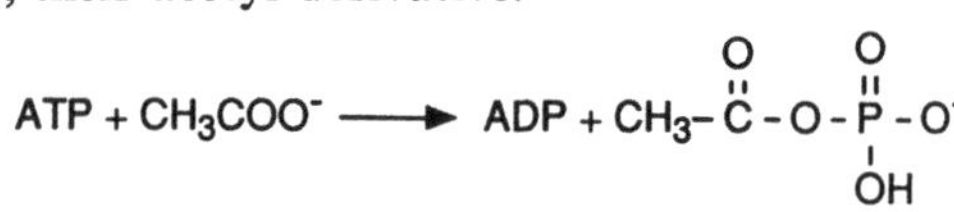

It was understandingly tempting to simulate this kind of acylation and H. Chantrenne, in 1947, proposed [1] the use of mixed anhydrides such as in peptide

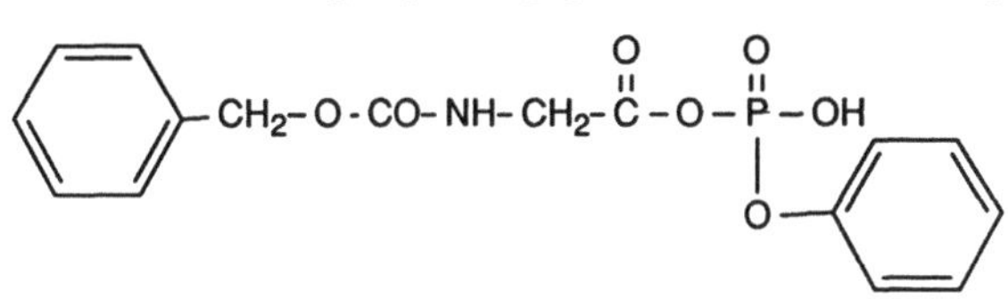

synthesis, while J.C. Sheehan (Plate 41) and V.S. Frank, in 1950, suggested [2] the application of reactive intermediates of the type:

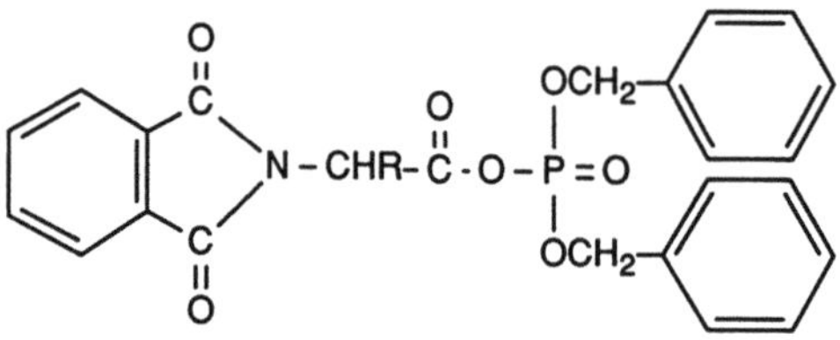

These mixed anhydrides were adopted for the actual synthesis of peptides in other laboratories, but, as we intend to show it later, the idea to activate the caboxyl group in the form of phosphorous derivatives reemerged time and again and more recently yielded practical results.

In 1951, Feodor Lynen (1911–1979) and his coworker E. Reichert demonstarted that *S*-acetyl coenzyme A is a more generally implicated form of "active acetate" than acetyl phosphate that was recognized in this role by Fritz Lipmann in 1940. The thiol ester character of *S*-acetyl CoA called the attention of Th. Wieland to energy-rich *S*-acyl compounds as promising intermediates for the formation of the peptide bond. In 1951, the same year when the isolation of *S*-acetyl CoA was published [3], Wieland and his coworkers described [4] the preparation of thiophenyl esters of benzyloxycarbonyl-amino acids and benzyloxycarbonyl-peptides and their application in the synthesis of blocked peptides:

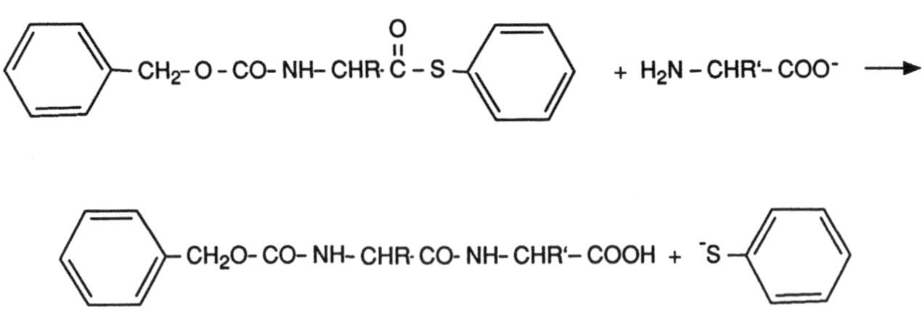

Thiophenyl esters are reactive enough to allow, when the reactants are present in sufficiently high concentration, the coupling of amino acids and peptides at practical rates without enzymic catalysis. While not generally applied in the synthesis of complex peptides, they stimulated the study of additional biomimetic acylating agents [5] such as *S*-aminoacyl-CoA, aminothioic acids ($H_3\overset{+}{N}$—CHR—$COS^-$), *S*-aminoacyl-cysteines and *S*-aminoacyl cysteamines ($H_2N$—CHR—CO—S—$CH_2CH_2NH_2$). Last but not least thiophenyl esters of blocked amino acids became the starting point of an important development in the methodology of peptide synthesis: they were the prototypes of active esters that will be discussed in some detail in this chapter.

## 4.1 Anhydrides

It is not quite obvious why the simple, classical method of acylation with acid anhydrides, for instance acetylation with acetic anhydride, was adopted rather late in peptide synthesis. We believe that the economy of the process in which only half of the valuable blocked amino acid is utilized in peptide bond formation while the other half is regenerated appeared unsatisfactory in the eyes of the

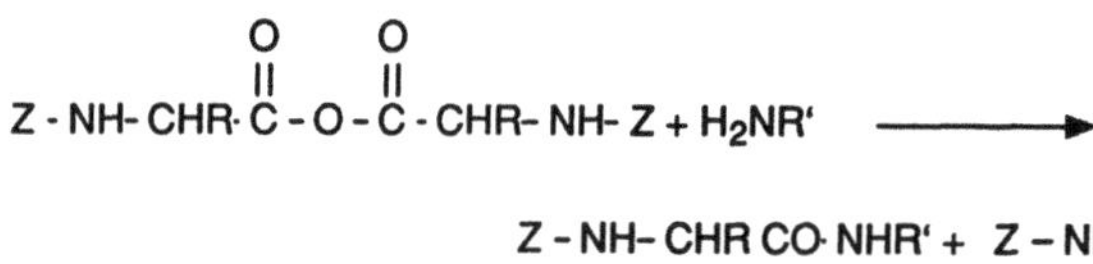

researchers of the 1950s. Even when the formation of symmetrical anhydrides was noted [2,6], this was not followed by recommendation for their practical application in synthesis. The situation changed in more recent times when protected amino acids became relatively inexpensive and also commercially available and simultaneously important biologically active peptides, needed in rather small amounts, became the targets of synthesis endeavors. Prior to this period a wide and vigorous search for suitable mixed (or unsymmetrical) anhydrides kept many laboratories engaged.

In mixed anhydrides an electron-withdrawing grouping (X) "activates" the carboxyl group of the acylating amino acid:

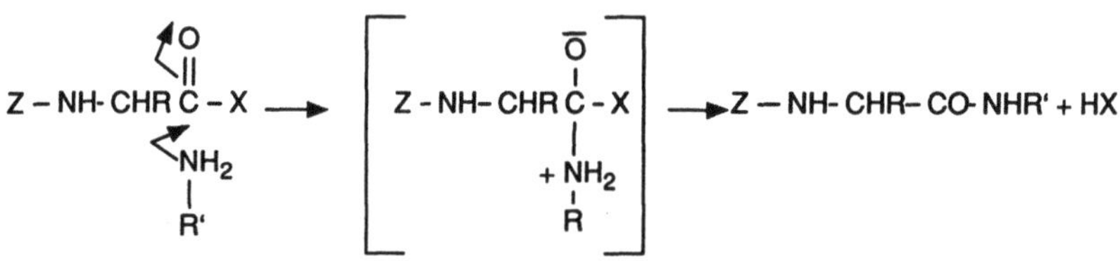

The activating moiety can be greatly varied, hence it appeared reasonable to seek a mixed anhydride method that provides high acylation rates and is not conducive to side reactions.

In a fundamental study in 1950, Wieland and his associates [6] refer to a suggestion made in 1882 by K. Kraut and F. Hartmann, namely that in the formation of the first peptide derivative, benzoylglycylglycine in the hands of Curtius (cf. p. 25) a mixed anhydride intermediate should be assumed. The same two authors proposed as early as 1865 a mixed anhydride intermediate in the preparation of aceturic acid (*N*-actylglycine) from silver glycinate and acetyl chloride:

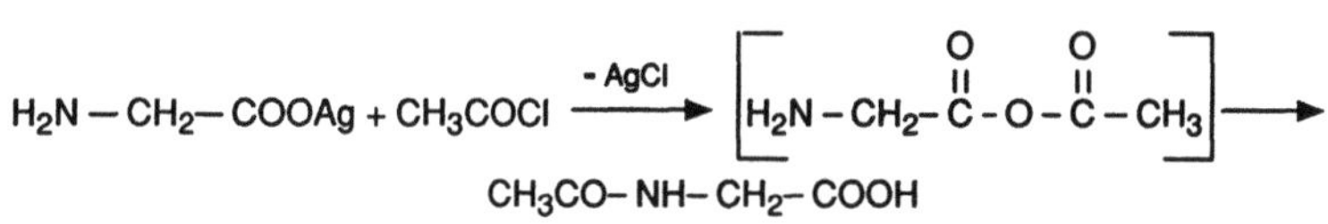

Wieland and his coworkers reinvestigated in depth the potential of mixed anhydrides. They prepared anhydrides from blocked amino acids (and from blocked peptides) with acetyl chloride and benzoyl chloride and applied them for

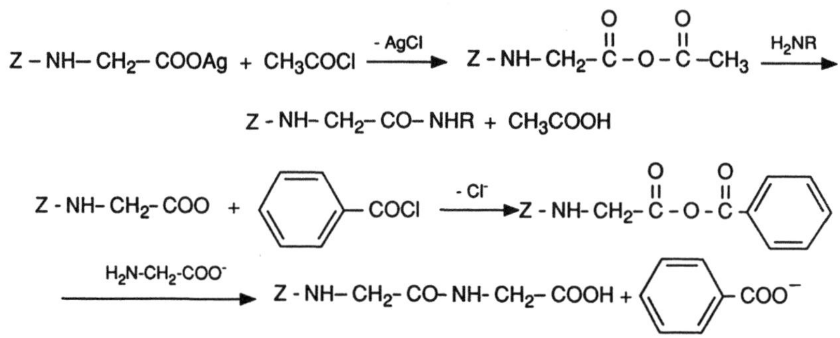


the acylation of amino acids and peptides.

An important aspect of the mixed anhydride method becomes apparent in its actual use in peptide synthesis. On attack by the nucleophilic component in the reaction mixture acylation occurs, but in addition to the desired peptide derivative also a second, undesired product must be expected:

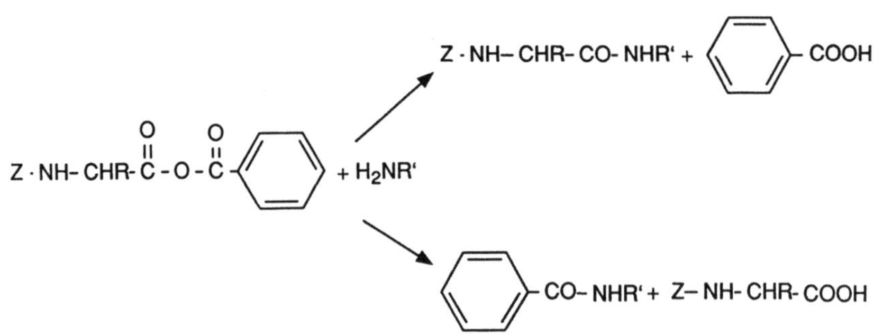


The ratio of the desired to the undesired (second) acylation product is a function of the relative sensitivities of the two carbonyl groups to nucleophilic attack. Clearly it is desirable that the carbonyl carbon in the activating moiety should be much less electrophilic than the carbon atom in the carbonyl group of the (blocked) amino acid that has to be incorporated into the target peptide. Neither the benzoyl nor the acetyl group are ideal in this respect, therefore the Wieland team continued the investigations which then resulted, in the following year, in a novel type of mixed anhydrides, anhydrides with alkyl carbonic acids, which turned out to be practical and in some respects still unsurpassed acylating agents [8]. The new anhydrides were readily obtained from the reaction of the blocked amino acid (or peptide) by the addition of a tertiary amine, for instance *N*-ethylpiperidine, followed by an alkyl chlorocarbonate (chloroformate). The mixed anhydride is then used "in situ", that is without isolation, for the acylation of the amine component:

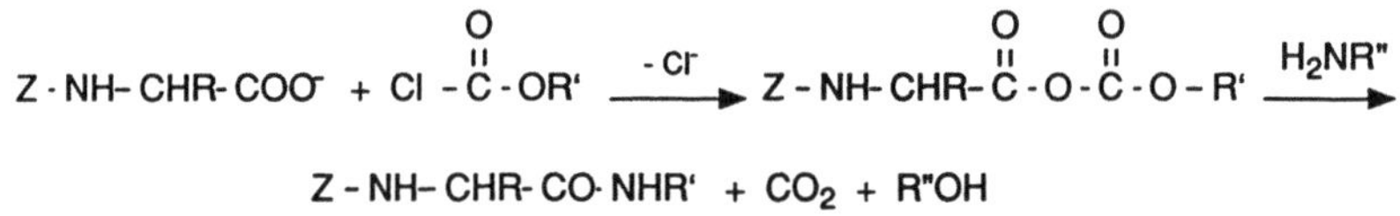

In order to keep side reactions at a minimum alkyl-carbonic acid mixed anhydrides are prepared and used for acylation in the cold. Nevertheless, both reactions require a very short time. The byproducts of coupling, $CO_2$ and an alcohol (R′OH), do not interfere with the isolation of the desired peptide. These features readily explain the popularity of the approach, but the principal advantage of alkylcarbonic acid mixed anhydrides lies in the fact that they give rise to only small amounts of the undesired second acylation product, a urethane:

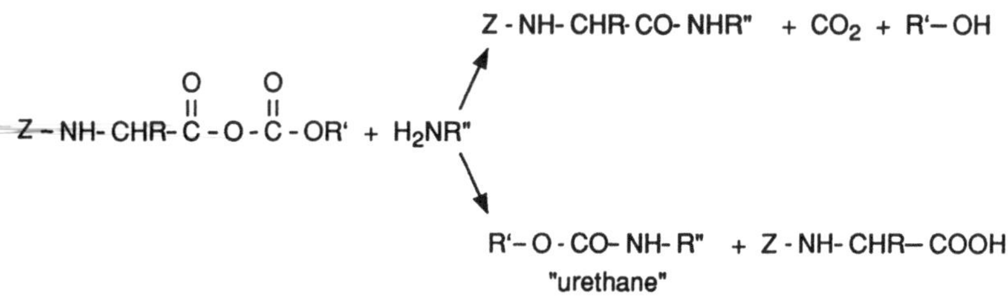

However, in the activation of hindered amino acids, such as valine or isoleucine, the reactivity of the carbonyl group is diminished by the electronic and steric effects of the nearby branched aliphatic side chain and urethane formation is significantly higher. This side reaction can be counteracted through the careful selection of the alkyl group in the activating reagent. Ethyl chlorocarbonate introduced, also in 1951, by R.A. Boissonnas [9] is less than ideal in this respect. Better results can be obtained by the application of isobutyl chlorocarbonate proposed by J.R. Vaughan in the same year and used ever since with considerable success [10]. Urethane formation in acylation with isobutylcarbonic acid mixed anhydrides is usually less than 1%, but it still amounts to several percent in the activation of (blocked) valine and isoleucine. Further improvement might be attainable by the application of isopropyl chlorocarbonate [8] or *sec.* butyl chlorocarbonate as activating reagent. Many interesting and also important aspects of the alkyl carbonic acid mixed anhydride method, such as the choice of reagents (including the tertiary amine), solvents, reaction time, the possible side reactions, among them racemization, were carefully reviewed by Meienhofer [11].

In the development of mixed anhydrides the year 1951 was indeed an "annus mirabilis". In England G.W. Kenner (Plate 27) invented the interesting and efficient method of peptide bond formation via anhydrides of acylamino acids with sulfuric acid [12]. The reactive intermediates prepared from the blocked amino acid and the anhydride of sulfuric acid applied in the form of a complex

with dimethylformamide

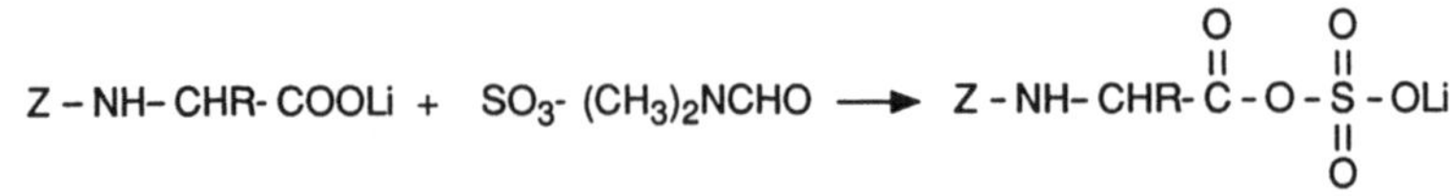

could be used without isolation, for instance, for the acylation of amino acids dissolved in water at pH 9. In spite of the excellent yields obtained with this method it could not compete with the alkyl carbonate mixed anhydride procedure, perhaps because of the additional operation needed for the preparation of an anhydrous salt of the carboxyl component. At the same time, in the USA, G.W. Anderson and his associates introduced [13] derivatives of 3-valent phosphorus, anhydrides of various diesters of phosphorous acid. Their first reagent, diethyl chlorophosphite was followed by several other diester mono-

chlorides of phosphorous acid and finally by the anhydride, tetraethyl pyrophosphite, $(C_2H_5O)_2P{-}O{-}P(OC_2H_5)_2$, that found important application in the first synthesis of oxytocin (cf. p. 140).

An interesting feature of the method is the alternative execution of the reaction in which the reagent is added to the amine component rather than to the carboxyl component. It would appear that a case of N-activation is at hand, but further experiments revealed that the actual acylating intermediates are still mixed anhydrides:

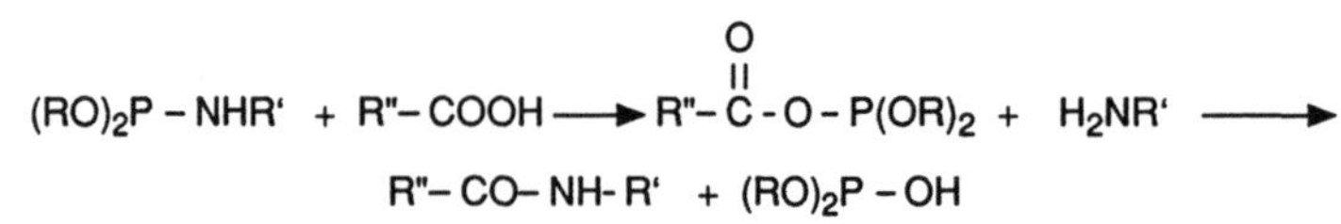

The relative reactivities of the two acyl groups in mixed anhydrides of two carboxylic acids,

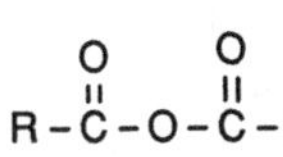

was established in the studies of Emery and Gold [14] who found that attack of the nucleophile takes place mainly at the less hindered carbonyl group or on the CO group with lower electron-density at its carbon atom. These principles were considered by Vaughan and Osato [15] in their design of new mixed anhydrides suitable for peptide synthesis. From a series of fatty acid chlorides they selected isovaleryl chloride for activation. The resulting isovaleric acid acylamino acid mixed anhydrides

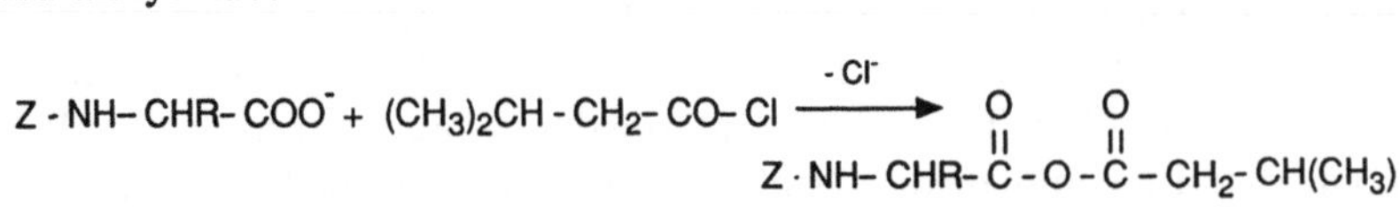

proved themselves as valuable intermediates in the oxytocin synthesis just mentioned, but for no obvious reason, they were not widely adopted by the practitioners of peptide synthesis. This situation remained unchanged even after the approach was extended [16] to include anhydrides of acylamino acids with trimethylacetic (pivalic acid):

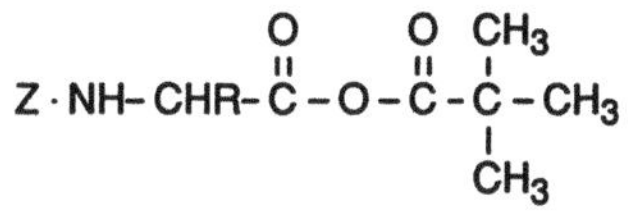

In the decade that followed the discovery (or perhaps rediscovery) of the mixed anhydride method, many alternatives, a great variety of mixed anhydrides, were proposed for peptide synthesis, too numerous to be discussed in the framework of this book. Fortunately, the development of the mixed anhydride idea during this period was thoroughly reviewed by Albertson [17].

Before concluding this section dealing with anhydrides of acylamino acids we have to return to symmetrical anhydrides. Their already mentioned formation through disproportionation of mixed anhydrides

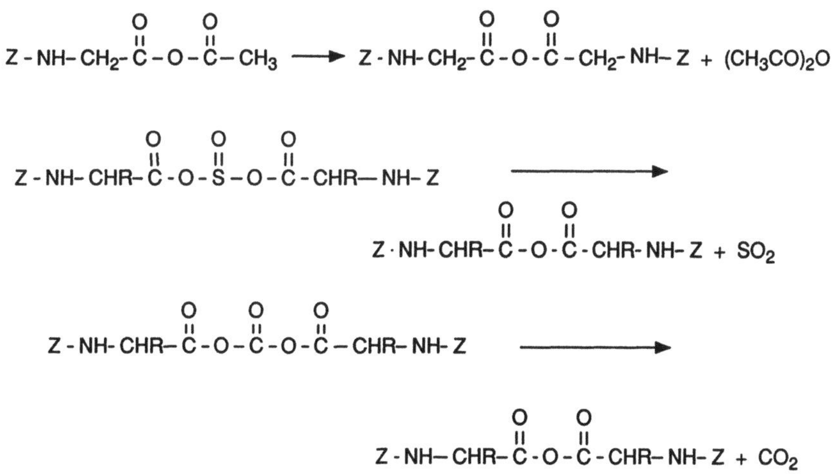

while observed [6, 8] was not immediately exploited for the formation of the peptide bond. A convenient method was found later for the preparation of symmetrical anhydrides in the reaction of carbodiimides with protected amino acids:

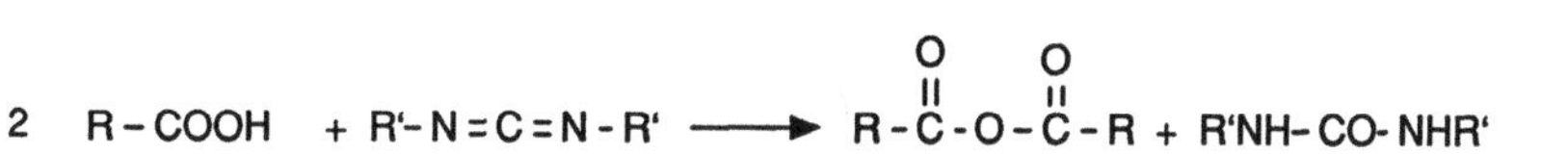

The obvious advantage of symmetrical anhydrides over many, although certainly not all mixed anhydrides is the absence of a second acylation product. They therefore became the acylating agents of choice in numerous laboratories.

However, a discussion of symmetrical anhydrides in somewhat more detail seems to be appropriate rather in the section dealing with coupling reagents, among which carbodiimides play a principal role.

## 4.2 Active Esters

Acylamino acid thiophenyl esters prepared by Th. Wieland and his coworkers via mixed anhydrides were the first realization of the active ester concept: enhancement of the electrophilic character of the carbon atom in the ester carbonyl by an electronwithdrawing "alcohol", in this case thiophenol [4].

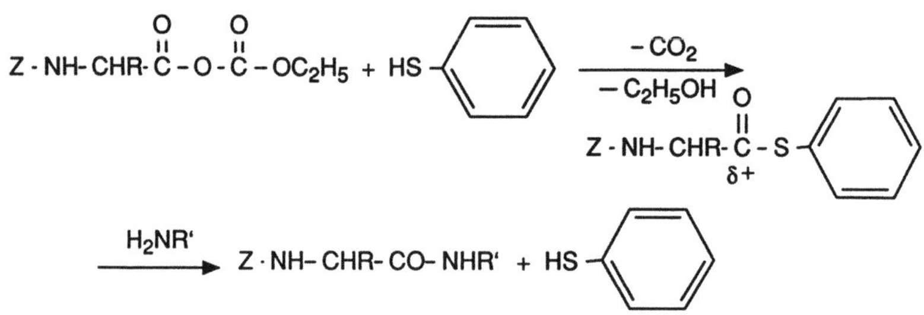

Without such activation simple esters, for instance methyl esters, react with nucleophiles at a low rate. The often applied ammonolysis of methyl esters

$$R-\overset{\overset{\displaystyle O}{\|}}{C}-OCH_3 + NH_3 \longrightarrow R-CONH_2 + CH_3OH$$

requires concentrated solutions of ammonia (e.g. in methanol) and several hours or even days for completion. The polycondensation of amino acid methyl esters proceeds at a low rate [18] unless catalyzed by a strong base [19] (see p. 38).

A few years after the first publication on acylamino acid thiophenyl esters [4] the peptide research team of the CIBA laboratories in Basel, led by Robert Schwyzer, described a systematic study of the aminolysis of methyl esters substituted with various electron-withdrawing groups [20]. From a series of esters examined with respect to their reaction rates in aminolysis cyanomethyl esters were selected as best suited for peptide synthesis. Cyanomethyl esters were readily prepared through the reaction of acylamino acid salts with chloroacetonitrile and they showed satisfactory rates in the acylation of amino acid esters

$$Z-NH-CHR-COO^- + ClCH_2CN \xrightarrow{-Cl^-} Z-NH-CHR-\overset{\overset{\displaystyle O}{\|}}{C}-OCH_2CN \xrightarrow{H_2NR'}$$

$$Z-NH-CHR-CO-NHR' + HOCH_2CN$$

particularly when both the activated ester and the amine component were present in sufficiently high concentration. This condition can easily be met in the synthesis of small peptides, but not when a blocked amino acid is used for the acylation of a longer chain or when sizeable segments have to be condensed. In such demanding syntheses, activation higher than found in cyanomethyl esters seems to be necessary.

Noting the fairly pronounced reactivity of phenyl esters in aminolysis Bodanszky concluded [21] that the enhanced aminolysis rates observed with thiophenyl esters [4] are due only in part to their thiol ester character and perhaps to a major extent to the fact that they are aryl esters. This conclusion was based on the fundamental studies of Gordon, Miller and Day [22] who found extremely high rates in the ammonolysis of phenyl and vinyl esters. To further increase the electronic effect operative in phenyl esters, their nitro-derivatives were prepared and examined. For practical application *p*-nitrophenyl esters were

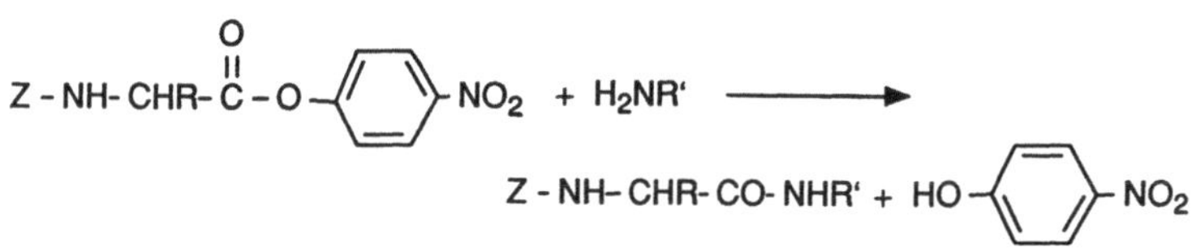


selected although subsequently the 2,4-dinitrophenyl esters, first eliminated because of their extreme reactivity, were found to be preferable for the activation of *N*-acyl-serine, -threonine and -nitroarginine, while *ortho*-nitrophenyl esters turned out to be superior to the *para* substituted derivatives, especially under crowded conditions such as present in the acylation of amine components attached to an insoluble polymeric support (cf. Chap. 5). The usefulness of nitrophenyl esters was demonstrated in stepwise syntheses: oxytocin (p. 145), lysine-vasopressin (p. 149) and secretin (p. 167) were prepared through the exclusive application of this method.

In 1955, almost simultaneously with the first publication on nitrophenyl esters, Farrington, Kenner and Turner proposed [23] the use of *p*-nitrothiophenyl esters in peptide synthesis but in spite of the high reactivity of these intermediates their recommendation found few followers.

The substituted aryl esters prepared and examined as potential acylating agents are too numerous to be described here. They are discussed in a comprehensive review article [24]. Yet, the studies of J. Pless and R.A. Biossonnas [25] should be mentioned even in this brief presentation. Based on an extensive comparison of substituted phenyl esters, they chose the 2,4,5-trichlorophenyl esters of *N*-acylamino acids as practical acylating agents, a choice fairly well accepted by several laboratories. The powerful pentafluorophenyl esters introduced by J. Kovacs, L. Kisfaludi and M.Q. Ceprini in 1967 [26] has gained importance in recent years through their application in solid phase peptide synthesis.

A novel type of acylating reagents, *O*-acyl derivatives of *N*-hydroxyphthalimide, were introduced in 1963 by Nefkens and Tesser [27]. The designation "active ester" might be practical but in reality these highly reactive compounds are mixed anhydrides of acylamino acids and hydroxamic acids. A convenient feature of the method is that the by-product of the reaction, *N*-hydroxyphthalimide

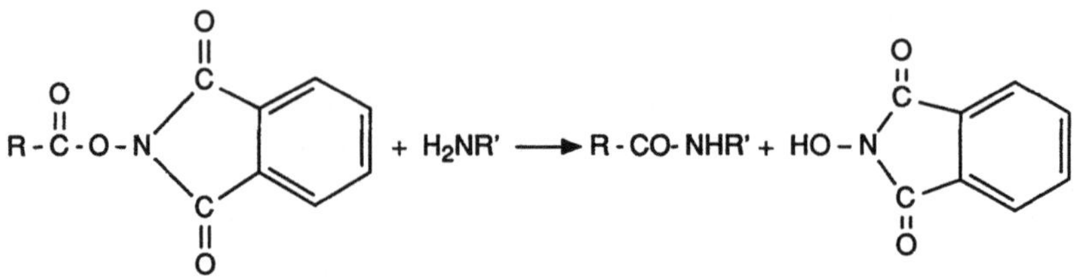

is soluble in aqueous bicarbonate and thus readily removed from the reaction mixture. In the subsequently proposed esters of *N*-hydroxysuccinimide this feature is further enhanced, because *N*-hydroxysuccinimide is soluble even in water [28]. Interestingly, while the hydroxyphthalimide "ester" method received little attention esters of hydroxysuccinimide were adopted by many investigators, became commercially available and are used in numerous laboratories. Many more hydroxamic acids were recommended for the activation of blocked amino acids (cf. Ref. 24) but it should be pointed out here that hydroxylamine derivatives without an *N*-acyl group also have activating effect [29]. In *O*-acyl *N*-dialkyl-hydroxylamines, such as *O*-acylaminoacyl-piperidines, the incoming nucleophile participates in the activation of the in itself perhaps not too reactive intermediate:

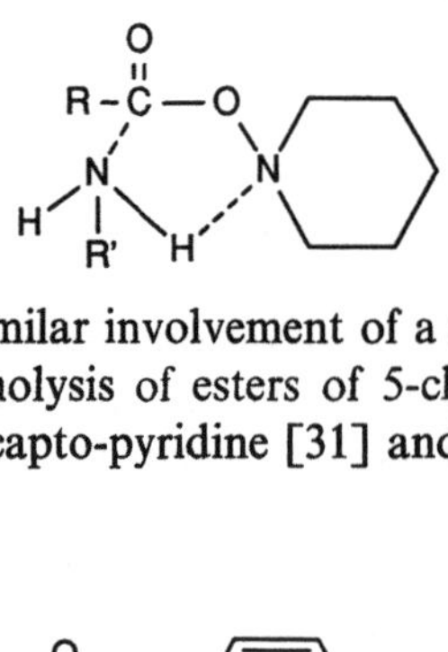

A similar involvement of a nearby N atom must be assumed also in the rapid aminolysis of esters of 5-chloro-8-hydroxyquinoline [30], 2-hydroxy- and 2-mercapto-pyridine [31] and of 1-hydroxybenzotriazole [32]:

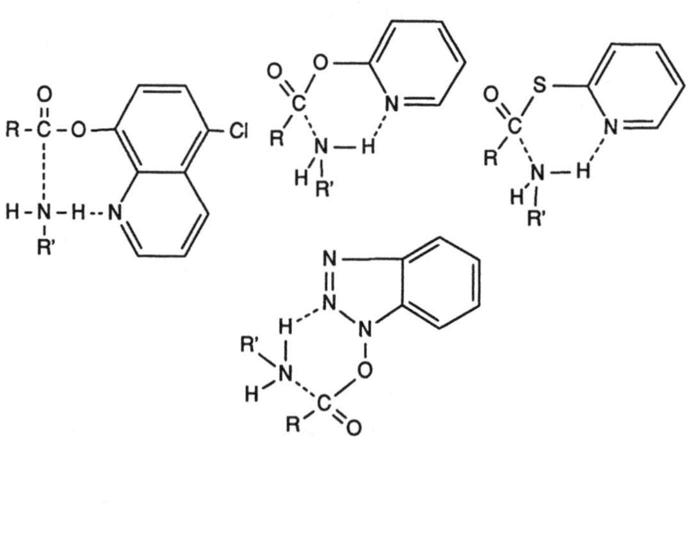

Preparation of active esters, initially achieved by acylation of thiophenol and other phenols with mixed anhydrides [4], was carried out later with the help of carbodiimides as condensing agents [33]. We shall return to this approach in the next section, dealing with coupling reagents. Coupling with active esters can be performed under mild conditions, mostly without side reactions. The progress of acylation can be monitored by the determination (e.g. by UV absorption) of the by-product, for instance a substituted phenol, released in the reaction. Yet, in spite of such advantages, active esters did not become the most preferred acylating agents in peptide synthesis, perhaps because of some shortcomings inherent in the method. One of these, the need to remove the by-product formed from the leaving group, prompted the development of insoluble active esters [34] in which the eliminated phenol etc. remains attached to the polymeric support. A more serious difficulty, the often only moderate reactivity of some esters, can be overcome through the use of esters with more than one electron-withdrawing substituent, for instance 2,4-dinitrophenyl or pentafluorophenyl esters. An alternative solution for this problem is catalysis of the aminolysis reaction. In addition to the modest catalytic effect of weak acids, such as acetic acid, and bases, a pronounced acceleration of acylation was observed [35] in the presence of imidazole. *N*-Acyl imidazoles are effective acylating agents.

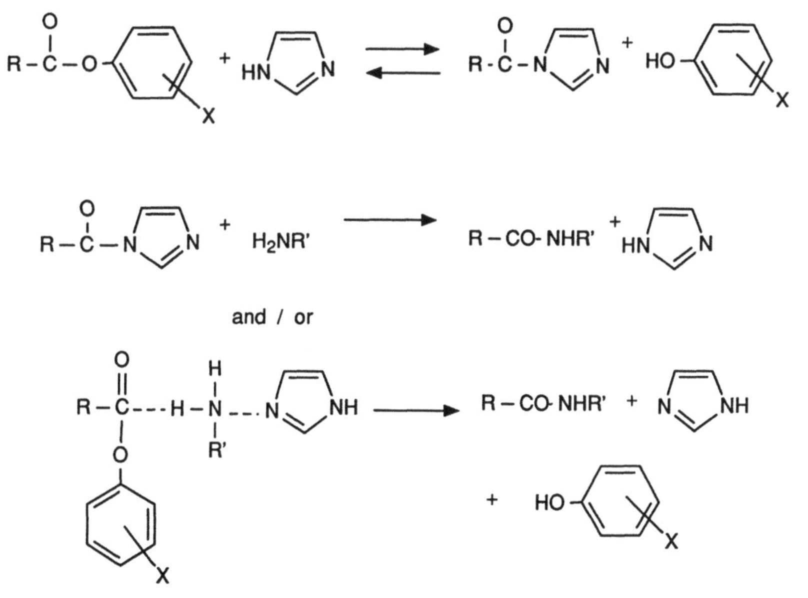


and also of pyrazole, 1,2,4 triazole [36] and tetrazole. In 1973 W. König (Plate 20) and R. Geiger (Plate 19) reported that 1-hydroxybenzotriazole, first proposed as an additive that can reduce racemization in coupling with DCC [37], is also an efficient catalyst, at least in dimethylformamide, of the aminolysis of aryl esters; 4-hydroxy-quinazoline-3-oxide is even more powerful in this respect [38]. In order to explain these catalytic effects first ternary complexes were assumed between the aryl ester, the amine component and the catalyst, but the additional increase of the reaction rate in the presence of base suggests an alternative mechanism, base induced transesterification:

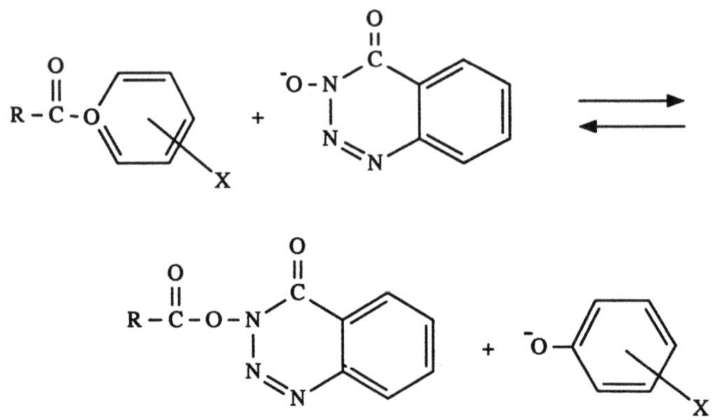

## 4.3 Coupling Reagents

The year 1955 was a turning point in the history of peptide chemistry. A new approach to coupling emerged, two methods in which, instead of activation of the carboxyl component followed by acylation of the amine, a reagent is added to the mixture of the carboxyl and amine components in order to achieve their coupling. One of these "coupling reagents" ethoxyacetylene was described by J.F. Arens in the Netherlands [39]. The new method perhaps with an acyl enole intermediate appeared to be quite attractive because the by-product of the coupling reaction, ethyl acetate, is rather innocuous and does not interfere with the isolation of the product. Unfortunately the reaction proceeds too slowly for practical purposes.

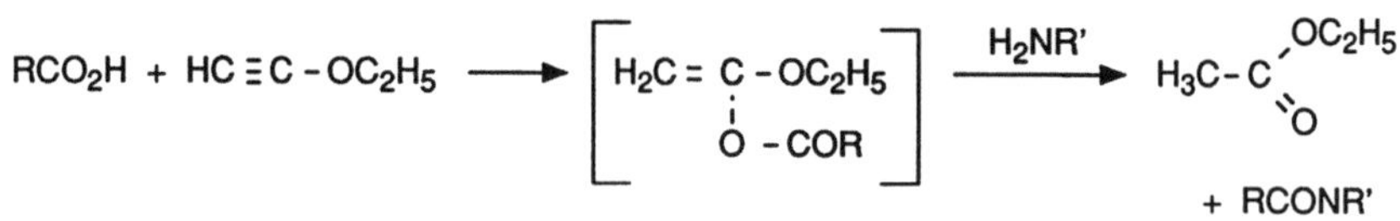

The second reagent, dicyclohexylcarbodiimide (DCC, DCCI), introduced by J.C. Sheehan (Plate 41) and G.P. Hess for the formation of the peptide bond [40] brought about a revolution in synthesis. Carbodiimides (R—N=C=N—R) had been known for a long time [41] and their reactivity had been exploited before for the preparation of esters and anhydrides. According to H.G. Khorana [42] the condensation reaction starts with the addition of the carboxyl component to one of the double bonds in the carbodiimide to yield the reactive

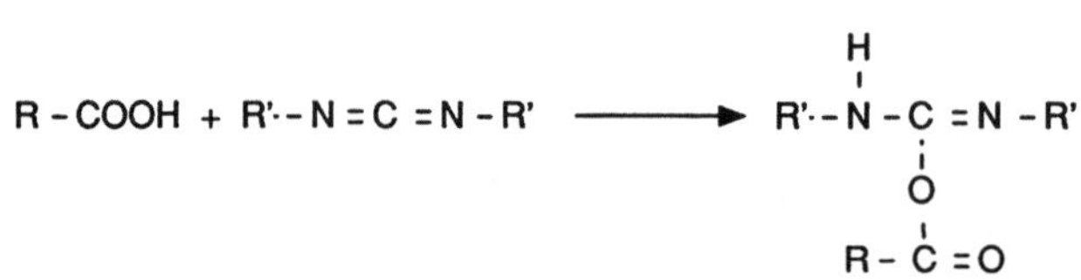

O-acylisourea derivative. From here on two alternative pathways lead to the desired amide. In the first of these the amine component itself attacks the reactive

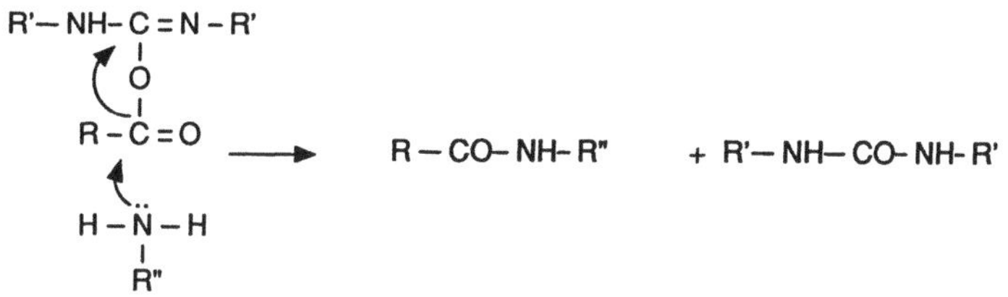

intermediate while in the second pathway a yet unreacted molecule of the carboxyl component is the attacking nucleophile and a symmetrical anhydride is produced which, in turn, acts as the acylating agent:

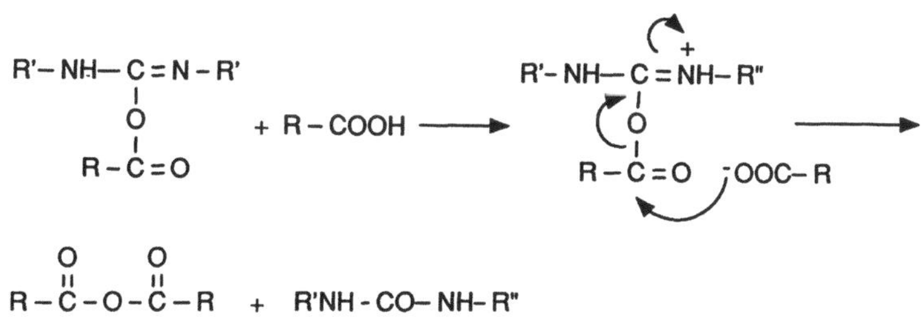

Although the cyclic *O*-acyl-isourea

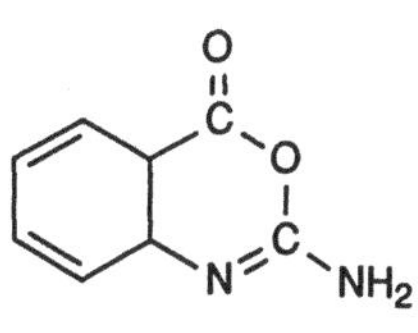

could be secured in 1963 by Doleschall and Lempert, the reactive intermediates formed from acylamino acids with carbodiimides so far eluded all attempts at their isolation. Nevertheless it is clear that no activation of the amine component occurs in coupling reactions mediated by carbodiimides. Therefore the method is novel only with respect to the technique of execution but not in principle: it still involves activation of the carboxyl component; only this takes place in the presence of the nucleophile. For some not quite obvious reason the coupling-reagent-idea remains fascinating for many peptide chemists and a long series of compounds appeared through the years, designed, prepared and proposed for this purpose, yet none of them proved itself superior to DCC. Of course the properties of the ideal coupling reagent are rather narrowly defined. It should not (irreversebly) react with the amine component, should not produce a reactive intermediate with a nucleophilic center that can compete with the amino component in its attack on the electrophilic center and last but not least it should not generate a center basic enough to cause racemization via proton abstraction from the chiral carbon atom of the activated amino acid (cf. p. 96). These requirements are difficult to meet and even carbodiimides are less then perfect in these respects. Thus through the attack of a nucleophilic center within its molecule the reactive *O*-acylisourea intermediate gradually undergoes rearrangement to an *N*-acylurea derivative (or ureide):

This mostly minor side reaction takes up major properties if the amino component is hindered or if it is a weak nucleophile. Also, the *O*-acyl-isourea derivatives are probably fairly strong bases and this might be the cause of racemization, perhaps through enolization, often observed when DCC was used for the coupling of peptide segments.

Since the soon extremely popular DCC method was proposed by John C. Sheehan, Professor of Chemistry at the Massachusetts Institute of Technology (MIT) in Cambridge, Massachusetts, it was not too surprising that Robert B. Woodward (Plate 50), Professor at Harvard University, a rival of MIT in the same town, after earlier flirts with peptide chemistry (p. 37) tried to develop another, perhaps even more spectacular method of coupling. In 1961 Woodward and his coworkers announced, in two short papers [43], the preparation and use of a new coupling reagent, an isoxazolium salt:

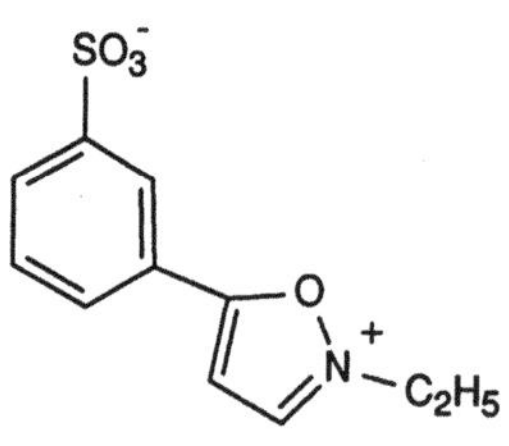


The generation of acylating agents from "Woodward's Reagent K" proceeds, as shown in Claisen's laboratory in Kiel early in the century [44], via base-induced opening of the isoxazole ring to a ketene-imine (resembling carbodiimides) and continues through the addition of the carboxylic acid to one of the double bonds: A mere inspection of the structure of the reactive intermediate reveals, however, an internal nucleophilic center, the nitrogen atom, that makes it prone to suffer a rearrangement similar to the one known in the case of the *O*-acyl-isourea intermediate in coupling with DCC[1]. Unfortunately the tendency of the new

[1] The formation of an *N*-acetylamide as the end product of the reaction of an isoxazolium salt with sodium acetate was described as early as 1910 (cf. Ref. 44). Also, an analogous intramolecular *O*—*N* acyl migration was noted in the reactive enol ester intermediate obtained with keteneimines, proposed as coupling reagents by C.L. Stevens and E.M. Munk in 1958

reagent to cause racemization is also pronounced. Hence, the perhaps exaggerated expectations, due to the association of the reagent with a famous name, were not fulfilled and the method was soon abandoned.

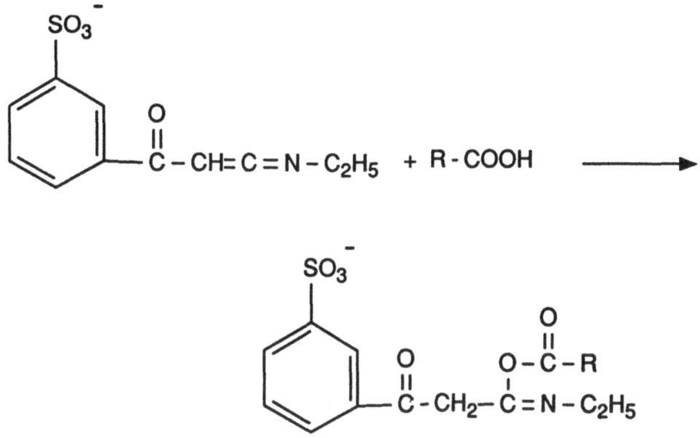


A novel and efficient coupling reagent, carbonyldiimidazole (CDI) was discovered [45] by H.A. Staab in Heidelberg, in 1957. The reactive intermediates, acylimidazoles were known before as potent acylating agents [5, 35]. In spite of

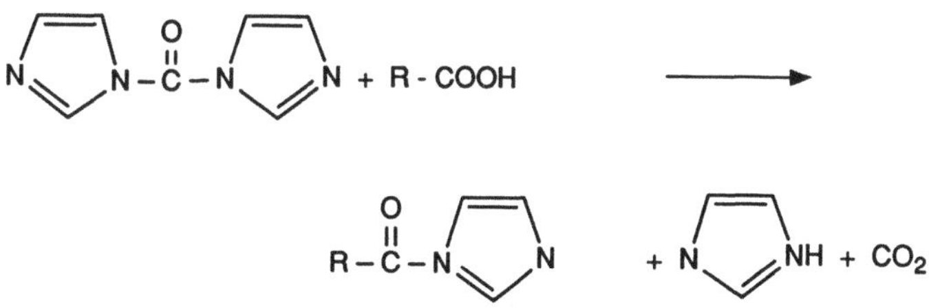


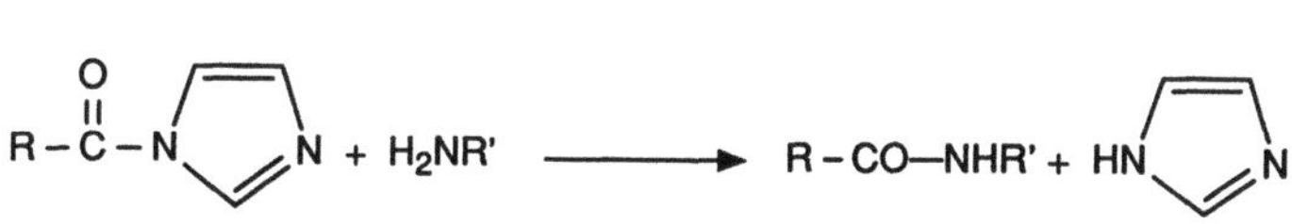


the good results achieved with CDI, it could not compete with DCC because CDI is expensive and sensitive to moisture. The coupling reagent l-ethoxycarbonyl-2-ethoxy-1,2-dihydroquinoline (EEDQ), introduced in 1968 by the Canadian investigators Belleau and Malek [46], is less demanding. It is a crystalline solid, obtained from ethyl chlorocarbonate, quinoline, triethylamine and ethanol. EEDQ converts the carboxyl component to the ethylcarbonic acid mixed anhydride, which is the actual acylating agent. In coupling with the help of EEDQ no tertiary amine needs to be added to the reaction mixture, hence the risk of racemization and of other base induced side reactions, is reduced. It should be noted, however, that EEDQ is not entirely inert toward amines: a definite, albeit slow, formation of ethyl urethanes can be observed.

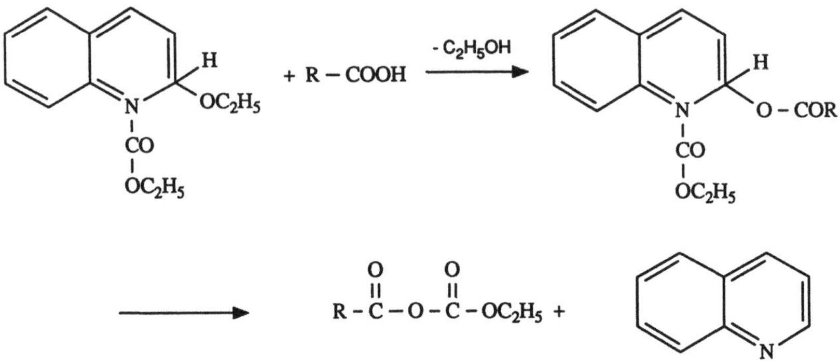


The biomimetic phosphoric acid mixed anhydride concept stimulated further search for activating reagents that produce derivatives of acylamino acids (and peptides) in more simple ways than the ones mentioned earlier in the Chapter (on p. 78)[1]. An interesting attempt in this direction is the so called Bates' reagent [47] which on reaction with carboxylic acids yields acyloxyphosphonium salts,

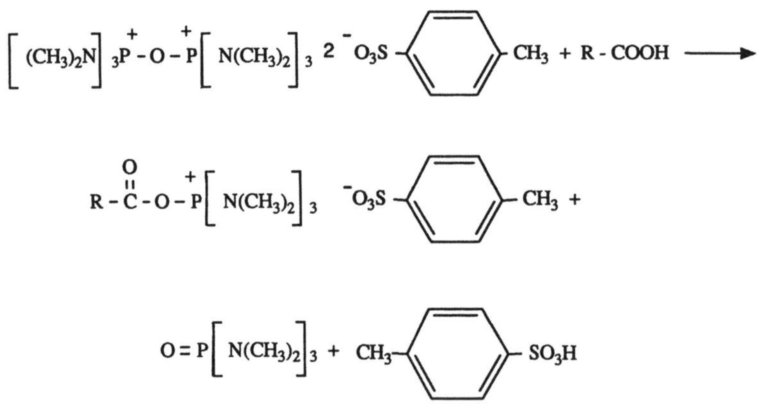


essentially a new type of mixed anhydrides. The more recently recommended and fairly well received BOP reagent (benzotriazolyl *N*-oxytridimethylamino-phosphonium hexafluorophosphate) of Castro and his associates [48] generates the highly reactive 1-hydroxybenzotriazole ester (cf. Ref. 32) of the carboxyl component.

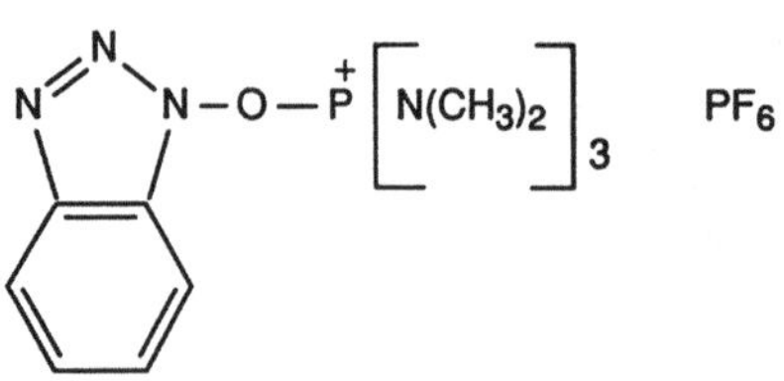


[1] Several more phosphorus derivatives, briefly tried and then abandoned,—for instance, $PCl_3$, $POCl_3$, tetraethyl pyrophosphate and the related diethyl chloroarsenite—could be regarded as coupling reagents

The coupling reagents discussed in the preceding paragraphs produce acylating agents which are not substantially different from and sometimes identical to the reactive intermediates obtained with conventional activating methods. The advantage presumably gained by adding a reagent to the mixture of the two components to be coupled rather than to the carboxyl component alone is at least questionable. To clarify this point it could be enlightening to consider once more certain aspects of coupling with carbodiimides, particularly dicyclohexylcarbodiimide, the most popular of all coupling reagents.

The excellent reactivity of the *O*-acyl-isourea intermediate in coupling with DDC is a mixed blessing. It must be credited for the speed and efficiency of acylation but also blamed for the various side reactions accompanying the process. The formation of N-acyl-ureas and loss of chiral purity in the activated residue have already been mentioned yet several more undesired reaction, for instance dehydration of the carboxamide group in the side chain of asparagine residues to a cyano group, were noted time and again. In 1970 König and Geiger proposed [32] the addition of 1-hydroxybenzotriazole (HOBt) to the reaction mixture and brought about thereby a dramatic impovement in the process. Almost all its shortcomings were if not eliminated at least greatly reduced in extent. The best explanation that can be offered for the remedial effect of HOBt is the intermediate formation of HOBt esters, which in turn acylate the amine component:

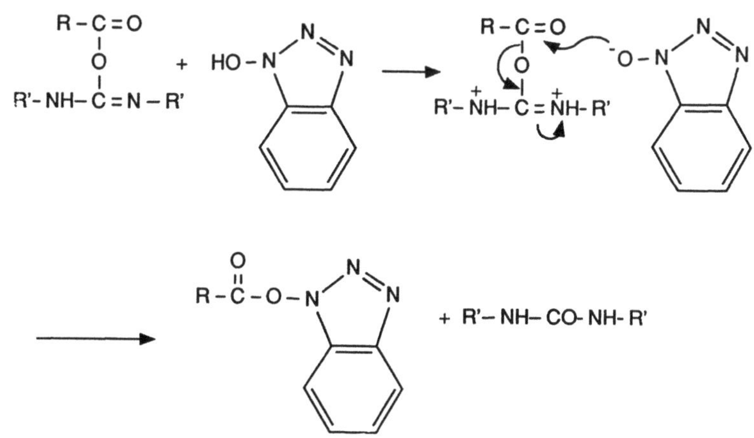


The beneficial influence of HOBt continues throughout the coupling reaction because it is regenerated in the coupling step. Being a weak acid, of about the strength of acetic acid, HOBt protonates the basic center in the *O*-acyl-isourea intermediate and maintains thereby chiral purity. The most important feature of the "additive" is, however, the reduction of the life-time of the "overactivated" (a term coined by Max Brenner) *O*-acylisourea by converting it to the less reactive but still very potent HOBt ester. Therefore, instead of the merely operational expression "additive" the more descriptive term "auxiliary nucleophile" might be also more appropriate. An even more pronounced racemization preventing effect was found if 3-hydroxy-3,4-dihydro-4-benzotriazinone was added to the reaction mixture [49]. It seems to be important to stress here that instead of adding an auxiliary nucleophile to the mixture of the two components to be coupled by DCC it is equally possible, or perhaps even preferable, to prepare the active esters which form in the reaction in isolated, pure form and to use them for acylation in a separate step. Crystalline acylamino acid esters of 1-hydroxybenzotriazole, 3-hydroxy-3,4-dihydro-4-benzotriazinone and of 3-hydroxy-3,4-dihydro-4-quinazolinone have been prepared and used in practical syntheses.

Looking back on a quarter of century elapsed since the introduction of coupling reagents it appears that they may have been a passing vogue in peptide synthesis. Most of the reagents proposed for this purpose harbor imperfections, such as a slow but measurable reaction with the amine component. Even the best reagents, DCC, EEDQ, CDI, are no exceptions in this respect, therefore it is not obvious why they should be used for activation in the presence of an amine. In fact, it is quite practical to execute the activation step prior to coupling merely by changing the order of addition of the reactants. Thus while carbodiimides, and DCC particularly, remain the most popular tools for the formation of the peptide bond, they are used in the "activating reagent mode" rather than in the "coupling reagent mode". In solid phase peptide synthesis (cf. Chap. 5) it became common practice to use a mixture of two equivalents of the blocked amino acid and one equivalent of the carbodiimide for acylation. This is a sound approach because the basic and also overactivated *O*-acyl isourea intermediate is neutralized as it forms and the quasi-intramolecular attack by the carboxylate leads to rapid formation of the symmetrical anhydride, the actual acylating agent, as shown on page 89, hence both O→N migration (cf. p. 90) and racemization via base induced enolization (cf. p. 96) are drastically reduced. The role of DCC in its original application, the condensation of peptide segments, underwent a similar change. With the advent of additives coupling is now based on acylation with active esters. Also, the utilization of DCC for the preparation of active esters which are then isolated in pure form and used as such, increased to almost exclusivity. Clearing: carbodiimides are not "coupling reagents" any more.

## 4.4 All That Can Go Wrong, Will: Side Reactions During Coupling

Murphy's law certainly prevails in peptide synthesis. Some important side reactions, such as the formation of urethanes in coupling via alkylcarbonic acid mixed anhydrides or the generation of *N*-acylureas and dehydration of asparagine side chains when DCC is used for peptide bond formation, have already been mentioned in this chapter. Yet, countless additional side reactions and by-products have been observed and reported, often only as a footnote. Thus, it would be difficult to give a historical account of their discovery. A review [50] of side reactions noted in peptide synthesis reveals that most of them are caused by strong acids and bases, by excessive protonation or deprotonation of the amino acid and peptide derivatives brought into reaction.

Strong acids play a role only in the removal of blocking groups (cf. Chap. 3), where they generate alkyl (benzyl, *tert*-butyl, etc.) cations which can alkylate certain side chains, for instance the phenol in tyrosine, the indole in tryptophan or the thioether sulfur atom in methionine. In most cases, however, not the cations themselves but the products formed in their reaction with the acidolytic reagent, e.g. benzyl bromide or *tert*-butyl trifluoroacetate, are the actual alkylating agents. In coupling, which is the subject of this chapter, side reactions are mostly base-induced and as a general rule it can be said that it is beneficial to perform the reaction without adding a tertiary amine or other auxiliary base to the reaction mixture. From a multitude of base induced side reactions, the ring closure of β-benzyl-aspartyl residue containing peptides

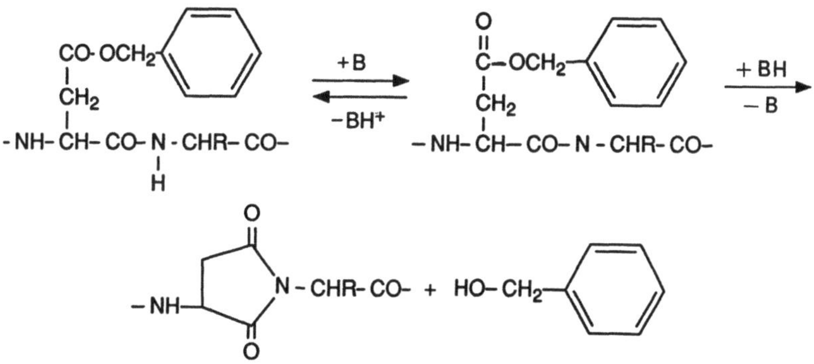


is mentioned here. A more general problem encountered in peptide synthesis, the risk of racemization, that is the partial or total loss of chiral purity in the activated residue during activation and coupling, requires a somewhat more detailed discussion.

While some concern about racemization has been voiced time and again, only around 1960 were systematic investigations started on this question. Several model compounds were proposed to serve in the comparison of coupling methods with respect to racemization, for instance the "Kenner model" [51], "Anderson-Callahan" model [52] or the "Young model" [53]. The latter, benzoyl-L-leucylglycine ethyl ester, formed the base of sustained studies by Geoffrey T. Young (Plate 53) and his associates at Oxford University in England. Their pioneering work included the examination of the influence of solvents the presence or absence of tertiary bases and various other factors on racemization and contributed in a major way to our understanding of the mechanisms participating in the process.

Activation of the carboxyl group in itself is conducive to racemization. The otherwise chirally stable acylamino acids and peptides can lose chiral homogeneity once their free carboxyl group is converted into a reactive derivative. The electron-withdrawing effect of the activating group (X) renders the proton on the α-carbon atom slightly acidic and hence abstractable by base and chirality is, of course, lost in the carbanion. This simple mechanism, however, is operative only

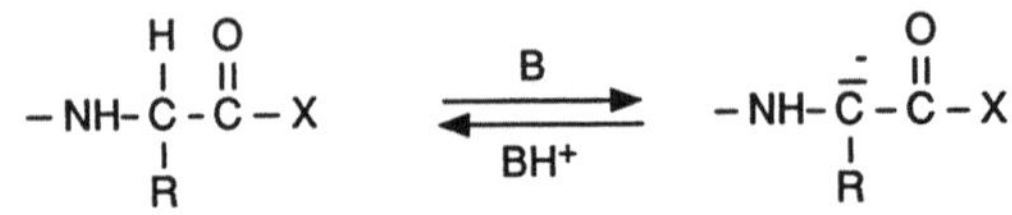

in some special cases. For instance the ready racemization of the phenylglycine residue, a constituent of microbial peptides but not of proteins, must be attributed to the direct abstraction of proton from the chiral center which is a benzylic carbon atom:

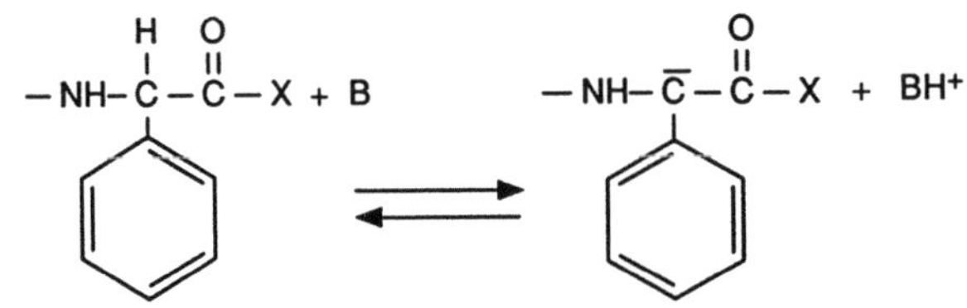

In the most commonly detected pathway of racemization azlactones (5(4*H*)-oxazolones, 5-oxazolinones) are implicated. Proton abstraction from these cyclic intermediates results in a resonance stabilized carbanion

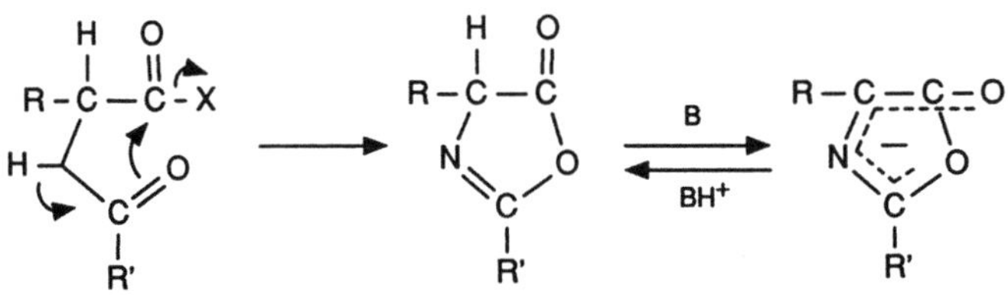

and its stability is the driving force of the process. The resistance of the (activated) proline residue to racemization was usually explained by the absence of azlactone formation, but the definite sensitivity of *N*-methylamino acids, in which the *N*—hydrogen of acylamino acids is similarly absent, suggests that the chiral stability of proline is due to its special geometry (L. Benoiton).

One of the most important practical results of these studies is the conclusion, drawn from theory and supported by experimental evidence, that racemization is greatly diminished if the amino component is acylated as such and not as a mixture of its salt with a tertiary amine.

Individual amino acids show considerable differences in their propensity for racemization. This is exceptionally pronounced in derivatives of *S*-benzyl-cysteine, *O*-benzyl-serine and $\beta$-cyanoalanine. The role of the substituent on the $\beta$-carbon atom is not obvious. Early assumptions of $\beta$-elimination, that is the reversible elimination-addition of benzyl mercaptane, benzyl alcohol or HCN, were not supported by the extensive studies of J. Kovacs. This example shows, however, that the azlactone mechanism, while it appears to be the most important pathway of racemization, is not the only process which leads to diminished chiral purity. For instance it is reasonable to consider that racemization in couplings carried out with the help of DCC involves enolization induced by the basic center within the *O*-acyl-isourea intermediate:

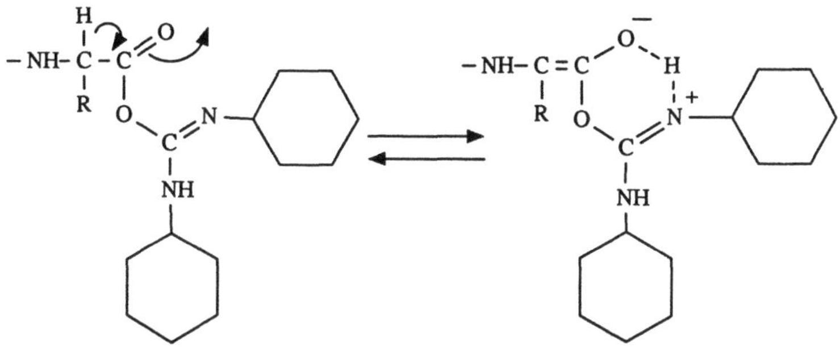

The mechanisms of racemization, the various models designed for their study and the methods proposed for the detection of undesired diastereosimers were treated in a profound review by D. Kemp [54]. In closing the discussion of this seemingly discouraging problem it seems to be necessary to stress that in order to maintain chiral homogeneity persistent vigilance is needed but only in the coupling of peptide segments. In syntheses which follow the stepwise strategy [55], that is chain elongation starting at the C-terminal residue and continued through the incorporation of single acylamino acid residues the risk of racemization is circumvented. As mentioned in connection of the benzyloxy-carbonyl group (p. 65) appropriately chosen blocking of the amine function prevents racemization in the activated amino acid derivative. This is true for the *tert*-butyloxycarbonyl, the 9-fluorenyl-methyloxycarbonyl and probably for other "urethane type" protecting groups

RO – CO – NH –

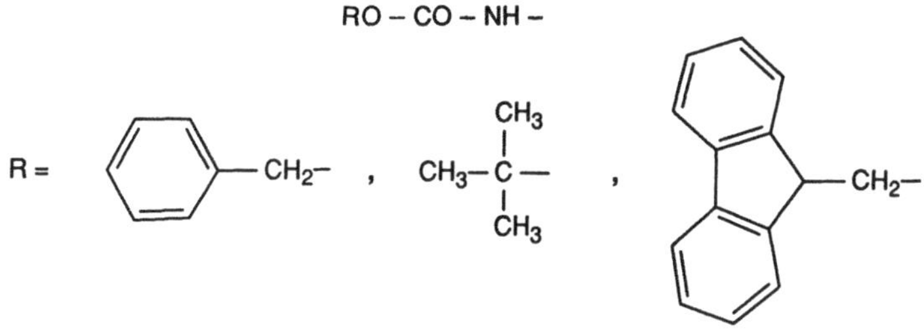

as well. In the case of a few noted exceptions, such as derivatives of *S*-benzyl-cysteine, the loss of chiral purity is greatly diminished if tertiary amines are absent from the reaction mixture. Therefore a real problem exists only in the condensation of segments, where instead of an acylamino acid a peptide has to be activated at its C-terminal residue. For such instances a major relief was provided by the discovery of racemization suppressing additives. Of these 1-hydroxy-benzotriazole, (HOBt) (I) [32] has already been mentioned. In several syntheses *N*-hydroxy-succinimide (II), proposed by Weygand and his associates [56] was found similarly useful. The promising 2-hydroximinocyanoacetic acid (III) [57] still awaits the test of major syntheses. Strangely the so far most effective

3-hydroxy-3,4-dihydro-1,2,3-benzotriazine-4-one (IV) [49] is less frequently used than HOBt

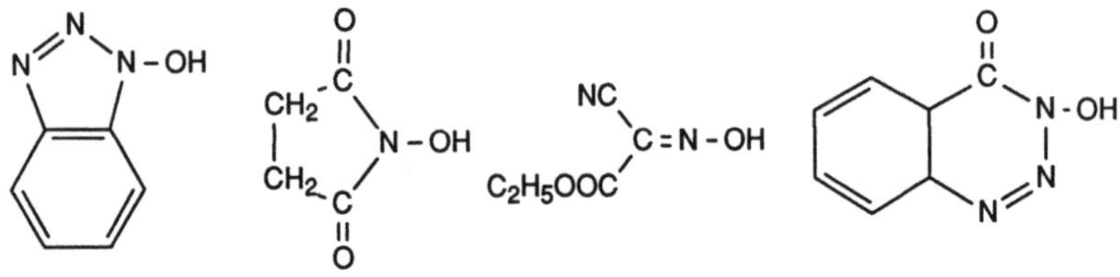


I II III IV

perhaps because the latter in addition to its ability to suppress racemization proved itself as a valuable remedy against several other side reactions as well.

Side reactions in peptide synthesis usually appear as an annoying difficulty. In the long run, however, they pose challenging problems and act as stimuli for further development of the field. For instance in order to prevent incomplete acylation in solid phase peptide synthesis (Chap. 5) active esters with exceptional reactivities were sought and found. Of course higher reactivity can also mean higher tendency for the production of undesired by-products. This conflict is characteristic for the methods that can be applied in the preparation of complex peptides. One can say (with M. Brenner, Plate 12) that the peptide chemist like Odysseus sails between Scylla and Charybdis, between overactivation and insufficient rates; of course he looks for calm waters. In these endeavors several non-conventional methods were designed and tried in practice. Such is the oxydation-reduction condensation [58] of Mukaiyama and his coworkers proposed in 1968. In one version of this versatile procedure the mixture of the two components to be coupled is treated with triphenylphosphine and 2,2′-dipyridyl disulfide. Oxidation of the phosphine and reduction of the disulfide afford the desired peptide bond formation (cf. p. 239 in Chap. 10).

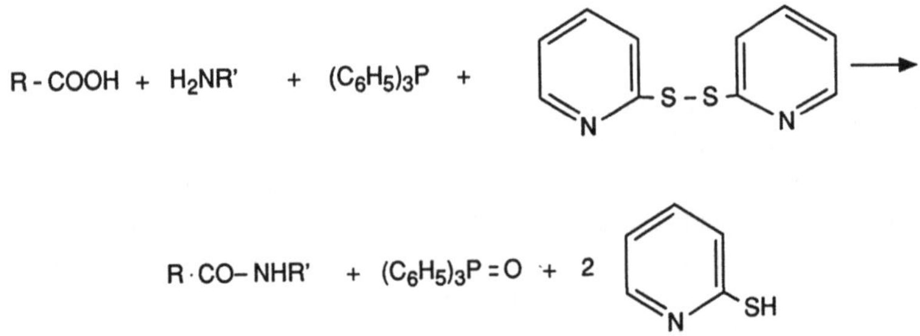


A novel type of coupling reagent was invented in 1976 in W. Steglich's laboratory in Bonn [59]. Opening of a cyclic carbonate yields a highly reactive enol ester smoothly and under mild conditions:

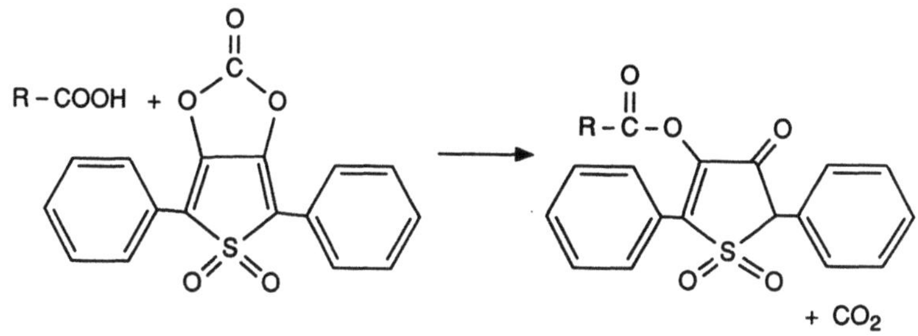

Our historical account would be incomplete without a few words about the surprisingly original four center condensation (4CC) method of Ivar Ugi [60]. The reaction of an acylamino acid, or peptide with an amine, an aldehyde and an isonitrile leads to a peptide derivative

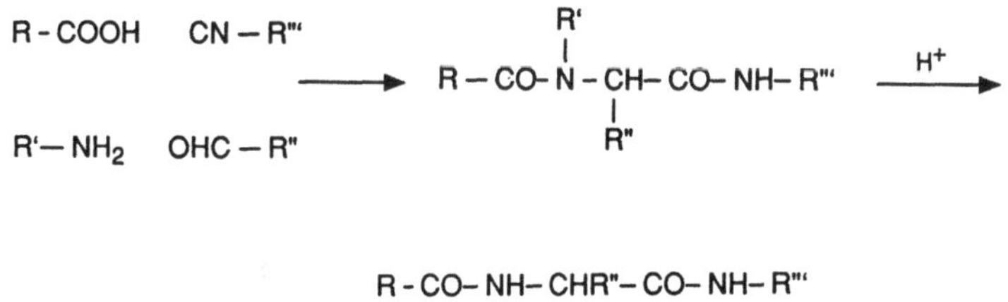


in which a newly formed residue (—NH—CHR″—CO—) appears. Through the use of certain optically active amines ($R'NH_2$) the new residue can be secured in near perfect chiral purity.

Biochemistry continues to have a major influence on the development of peptide synthesis. Peptide bond formation via catalysis with proteolytic enzymes has the promise of products with absolute chiral purity and should also be free from many side reactions encountered in synthesis by the methods of organic chemistry. Therefore coupling with the help of enzymes (cf. page 57) is receiving growing attention. Perhaps even more exciting is the exploitation of ribosomal protein synthesis for the production of selected target peptides, such as insulin. In recent years preparation of the necessary DNAs became almost routine and hence this avenue of peptide synthesis broadened to a major area that transcends the boundaries of this book.

## References

1. H. Chantrenne, Hippuric acid formation from glycine and dibenzoyl phosphate. Nature, 160, 603–604 (1947); Un modèle de synthèse peptidique. Propriétés due benzoylphosphate de phenyl. Biochim. Biophys. Acta 2: 286–293 (1948); A new method of peptide synthesis. Nature 164: 576–577 (1949); Peptide synthesis via glycyl phosphate. Biochim. Biophys. Acta 4: 482–492 (1950)
2. J.C. Sheehan, V.S. Frank, Peptide synthesis using energy-rich phosphorylated amino acid derivatives. J. Amer. Chem. Soc. 72: 1312–1316 (1950)
3. F. Lynen, E. Reichert, Zur chemischen Struktur der "activierten Essigsäure". Angew. Chem. 63: 47–48 (1951); F. Lynen, ibid. 63: 490 (1951)

4. Th. Wieland, W. Schäfer, E. Bokelmann, Über Peptidsynthesen V. Über eine bequeme Darstellungsweise von Acylthiophenolen und ihre Verwendung zu Amid-und Peptid-Synthesen. Liebigs Ann. Chem. 573: 99–104 (1951)
5. Th. Wieland, Sulfur in biomimetic peptide synthesis. In: Roots of Biochemistry, Fritz Lipmann-Meeting, Berlin 1987; H. Kleinkauf, H.v. Döhren, L. Jaenicke eds. De Gruyter Berlin, New York 1988, p. 213–223.
6. Th. Wieland, W. Kern, R. Sehring, Über Anhydride von acylierten Aminosäuren. Liebigs Ann. Chem. 569: 117–122 (1950); Th. Wieland, R. Sehring, Eine neue Peptid-Synthese. ibid. 569: 122–129 (1950)
7. K. Kraut, Fr. Hartmann, Über das Glycin. Liebigs Ann. Chem. 133: 99–108 (1865)
8. Th. Wieland, H. Bernhard, Über Peptid-Synthesen. 3. Mitteilung. Die Verwendung von Anhydriden aus N-acylierten Aminosäuren und Derivaten anorganischer Säuren. Liebigs Ann. Chem. 572: 190–194 (1951)
9. R.A. Boissonnas, Une nouvelle methode de synthèse peptidique. Helv. Chim. Acta 34: 874–879 (1951)
10. J.R. Vaughan Jr., R.L. Osato. The preparation of peptides using mixed carbonic-carboxylic acid anhydrides. J. Amer. Chem. Soc. 74: 676–678 (1952)
11. J. Meienhofer, The mixed carbonic anhydride method. In the Peptides, Vol. 1 (E. Gross, J. Meienhofer, eds.) Academic Press, New York 1979, pp. 241–314
12. G.W. Kenner, Synthesis of peptides. Chem. Ind. 1951, 15; G.W. Kenner, R.J. Stedman. Peptides. Part I. The synthesis of peptides through anhydrides of sulfuric acid. J. Chem. Soc. 1952, 2067–2076
13. G.W. Anderson, A.D. Welcher, R.W. Young, Diethyl chlorophosphite as reagent for peptide synthesis. J. Amer. Chem. Soc. 73: 501–502 (1951)
14. A.R. Emery, V. Gold, Quantitative studies of the reactivities of mixed carboxylic anhydrides. Part. I. The composition of the acylation products in the reaction between acetic chloroacetic anhydrides and primary aromatic amines. J. Chem. Soc. 1950: 1443–1447
15. J.R. Vaughan Jr, R.L. Osato, Preparation of peptides using mixed carboxylic acid anhydrides. J. Am. Chem. Soc. 73: 5553–5555 (1951)
16. M. Zaoral, Amino acids and peptides XXXVI. Pivaloyl chloride as a reagent in the mixed anhydride synthesis of peptides. Coll. Czechoslov. Chem. Comm. 27: 1273–1277 (1962)
17. N.F. Albertson, Synthesis of peptides with mixed anhydrides. Org. Reactions 12: 157–355 (1962)
18. E. Pacsu, E.J. Wilson Jr., Polycondensation of certain peptide esters I. Polyglycine esters. J. Org. Chem. 7: 117–125 (1942)
19. H. Brockmann, H. Musso, Versuche zur Synthese von Polypeptiden durch Kondensation von Aminosäure- und Peptidestern Chem. Ber. 87: 581–592 (1954)
20. R. Schwyzer, B. Iselin, M. Feurer, 8. Über aktivierte Ester der Hippursäure und ihre Umsetzungen mit Benzylamin. Helv. Chim. Acta 38: 69–79 (1955); R. Schwyzer, M. Feurer, B. Iselin, H. Kägi, 9. Über aktivierte Ester II. Synthese aktivierter Ester von Aminosäure Derivaten. ibid. 38: 80–83 (1955); R. Schwyzer, M. Feurer, B. Iselin, 10 Über aktivierte Ester III. Umsetzungen activierter Ester von Aminosäure- und Peptid-Derivaten mit Aminen und Aminosäureestern. Helv. Chim. Acta. 38: 83–89 (1955)
21. M. Bodanszky, Synthesis of peptides by aminolysis of nitrophenyl esters, Nature 175: 685–686 (1955)
22. M. Gordon, J.G. Miller, A.R. Day, Effect of structure on reactivity in ammonolysis of esters with special references to electron release effects of alkyl and aryl groups. J. Amer. Chem. Soc. 70: 1946–1953 (1948)
23. J.A. Farrington, G.W. Kenner, J.M. Turner, Preparation of p-nitrophenyl thiolesters and their application to peptide synthesis. Chem. Ind. (London) 1955, 601–602; J.A. Farrington, P.J. Hextall, G.W. Kenner, J.M. Turner, Peptides. Part VII. The preparation and use of p-nitrophenyl thiolesters. J. Chem. Soc. 1957, 1407–1413
24. M. Bodanszky, Active esters in peptide synthesis in The Peptides, Vol. I. (E. Gross, J. Meinhofer, eds.) Academic Press, New York 1979, pp. 105–196
25. J. Pless, R.A. Boissonnas, Über die Geschwindigkeit der Aminolyse von verschiedenen aktivierten, N-geschützten α-Aminosäure-p-nitrophenylestern, insbesondere 2,4,5-trichlorphenylestern. Helv. Chim. Acta. 46: 1609–1625 (1963)
26. J. Kovács, L. Kisfaludy, M.Q. Ceprini, On the optical purity of peptide active esters prepared by N,N′-dicyclohexylcarbodiimide and "complexes" of N,N′-dicyclohexylcarbodiimide-pentachlorophenol and N,N′-dicyclohexylcarbodiimide and pentafluorophenol. J. Amer. Chem. Soc. 89: 183–184 (1967)

27. G.H.L. Nefkens, G.I. Tesser, A Novel activated ester in peptide synthesis. J. Amer. Chem. Soc. 83: 1263 (1961)
28. G.W. Anderson, J.E. Zimmerman, N-Hydroxysuccinimide esters in peptide synthesis. J. Amer. Chem. Soc. 86: 1839–1842 (1964)
29. S.M. Beaumont, B.O. Handford, G.T. Young, The use of esters of NN-dialkylhydroxylamines in peptide synthesis and as selective acylating agents, J. Chem. Soc. Chem. Commun. 1965, 53–54; B.O. Handford, J.H. Jones, G.T. Young, T.F.N. Johnson, The use of esters of 1-hydroxypiperidine and other NN-dialkylhydroxylamines in peptide synthesis and as selective acylating agents. J. Chem. Soc. 1965, 6814–6827
30. H.D. Jakubke, A. Baumert, Vergleichende Studien über den Peptidknüpfungsschritt unter Verwendung verschiedener aktivierter Ester am Beispiel eines Modellpeptids. J. Prakt. Chem. 316: 241–248 (1974)
31. K. Lloyd, G.T. Young, The use of acylamino acid-esters of 2-mercaptopyridine in peptide synthesis. J. Chem. Soc. Chem. Commun. 1968, 1400–1401; Amino acids and peptides. Part XXXIV. Anchimerically assisted coupling reactions: the use of 2-pyridyl thiol esters. J. Chem. Soc. 1971, 2890–2896; A.S. Dutta, J.S. Morley, Polypeptides. Part XII. The preparation of 2-pyridyl esters and their use in peptide synthesis. J. Chem. Soc. C. 1971, 2896–2900
32. W. König, R. Geiger, Eine neue Methode zur Synthese von Peptiden: Aktivierung der Carboxylgruppe mit Dicyclohexylcarbodiimid unter Zusatz von 1-Hydroxybenzotriazolen. Chem. Ber. 103: 788–798 (1970)
33. D.F. Elliott, D.W. Russel, Peptide synthesis employing p-nitrophenyl esters prepared with the aid of N,N'-dicyclohexylcarbodiimide. Biochem. J. 66:49 P. (1957); M. Rothe, F.W. Kunitz, Synthese cyclischer Oligopeptide der α-Aminocapronsäure. Konstitutionsaufklärung der ring-förmigen Bestandteile von Polycaprolaktam. Liebigs Ann. Chem. 609: 88–102 (1957)
34. M. Fridkin, Polymeric reagents in peptide synthesis in The Peptides vol. 2 (E. Gross, J. Meienhofer eds.) Academic Press New York 1980, pp. 333–363
35. Th. Wieland, G. Schneider, N-Acylimidazole als energiereiche Acylverbindungen. Liebigs Ann. Chem. 580: 159–168 (1953); R.H. Mazur, Acceleration of p-nitrophenyl ester peptide synthesis by imidazole. J. Org. Chem. 28: 2498 (1963)
36. H.C. Beyermann, W. Maassen van den Brink, Use of bifunctional catalysts in peptide and other syntheses. Proc. Chem. Soc. (Lond) 1963, 266; H.C. Beyermann, W. Maasen van den Brink, F. Weygand, A. Prox, W. König, L. Schmidhammer, E. Nintz, Racemization and bifunctional catalysts in peptide synthesis. Rec. Trav. Chim. Pays. Bas. 84: 213–231 (1965)
37. W. König, R. Geiger, Racemisierung bei Peptidsynthesen, Chem. Ber. 103: 788–798; 2024–2034 (1970)
38. W. König, R. Geiger, N-Hydroxyverbindungen als Katalysatoren für die Aminolyse aktivierter Ester Chem. Ber. 106: 3626–3635 (1973)
39. J.F. Arens, The chemistry of acetylenic ethers XIII. Acetylenic ethers as reagents for the preparation of amides. Rev. Trav. Chim. Pays. Bas. 74: 769–770 (1955)
40. J.C. Sheehan, G.P. Hess, A new method of forming peptide bonds. J. Amer. Chem. Soc. 77: 1067–1068 (1955)
41. H.G. Khorana, Peptides. Part III. Selective degradation for the carboxyl end. The use of carbodiimides. J. Chem. Soc. 1952, 2081–2088; The chemistry of carbodiimides. Chem. Reviews 53: 145–166 (1953)
42. H.G. Khorana, The use of dicyclohexylcarbodiimide in the synthesis of peptides. Chem. Ind. (London) 1955, 1087–1088
43. R.B. Woodward, R.A. Olofson, The reaction of isoxazolium salts with bases. J. Amer. Chem. Soc. 83: 1007–1009 (1961); R.B. Woodward, R.A. Olofson, H. Mayer, A. new synthesis of peptides, J. Amer. Chem. Soc. 83: 1010–1012 (1961)
44. L. Claisen, Über α-Methyl-isoxazol. Ber. Dtsch. Chem. Ges. 42: 59–68 (1909); O. Mumm, G. Münchenmeyer, Überführung des Oxymethylenacetophenons in Benzoylbrenztraubensäure und einige neue Derivate. Ber. Dtsch. Chem. Ges. 43: 3335–3345 (1910)
45. H.A. Staab, Reaktionsfahige heterocyclische Diamide der Kohlensäure. Liebigs Ann. Chem. 609: 75–83 (1957)
46. B. Belleau, G. Malek, A new convenient reagent for peptide synthesis. J. Amer. Chem. Soc. 90: 1651–1652 (1968)
47. A.J. Bates, I.J. Galpin, A. Hallett, D. Hudson, G.W. Kenner, G.W. Ramage, R.C. Sheppard, A new reagent for peptide synthesis: μ-oxo-bis-[tris-(dimethylamino)phosphonium]-bis-tetrafluoroborate. Helv. Chim. Acta. 58: 688–696 (1975)

48. B. Castro, J.R. Dormoy, G. Evin, C. Selve, Reactifs de couplage peptidique IV. (1)-L-hexafluorophosphate de benzotriazolyl-N-oxitris-dimethylamino phosphonium (B.O.P). Tetrahedron Letters 1975, 1219–1222
49. W. König, R. Geiger, Eine neue Methode zur Synthese von Peptiden. Aktivierung der Carboxyl Gruppe mit Dicyclohexylcarbodiimid und 3-Hydroxy-4-oxo-3.4-dihydro-1.2.3-benzotriazin. Chem. Ber. 103: 2034–2040 (1970)
50. M. Bodanszky, J. Martinez, Side reactions in peptide synthesis. The Peptides, vol. 5 (E. Gross, J. Meienhofer, eds.) Acad. Press, New York 1983 pp. 111–216
51. D.W. Clayton, J.H. Farrington, G.W. Kenner, J.M. Turner, Peptides. Part VI. Further studies of the synthesis of peptides through anhydrides of sulfuric acid. J Chem Soc 1957, 1398–1407
52. G.W. Anderson, F.M. Callahan, Racemization by the dicyclohexylcarbodiimide method of peptide synthesis. J. Amer. Chem. Soc. 80: 2902–2903 (1958)
53. N.A. Smart, G.T. Young, M.W. Williams, Amino acids and peptides. Part XV. Racemization during peptide synthesis. J. Chem. Soc. 1960, 3902–3912; M.W. Williams, G.T. Young, Amino acids and peptides. Part XVI. Further studies of racemization during peptide synthesis. J Chem Soc 1963, 881–889
54. D.S. Kemp, Racemization in peptide synthesis. The Peptides, vol. 1. (E. Gross, J. Meienhofer eds.) Acad. Press New York 1979 pp. 315–383
55. M. Bodanszky, Stepwise synthesis of peptides by the nitrophenylester method. Ann N Y Acad Sci 88: 655–664 (1960)
56. F. Weygand, D. Hoffmann, E. Wünsch, Synthesis of peptides with dicyclohexylcarbodiimide by addition of N-hydroxysuccinimide. Z. Naturforschung 21 b: 426 (1966)
57. M Itoh, Racemization suppression by the use of ethyl hydroximino-2-cyanoacetate in Chemistry and Biology of Peptides (J. Meienhofer, ed.) Ann Arbor Science Pub. Ann Arbor, Michigan 1972 pp. 365–367
58. T. Mukaiyama, R. Matsueda, M. Ueki, The oxidation-reduction condensation. The Peptides, vol. 2 (E. Gross, J. Meienhofer eds.) Acad. Press New York 1979 pp. 383–416
59. O. Hollitzer, A. Seewald, W. Steglich, 4,6-Diphenylthieno [3,4d] [1,3]dioxol-2-one 5,5-dioxide. A novel activating agent in peptide synthesis. Angew. Chem. 15: 444–445 (1976)
60. I. Ugi, The four component synthesis. The Peptides vol. 2 (E. Gross, J. Meienhofer, eds.) Acad. Press New York 1979, pp. 365–381

# 5 A New Technology: Solid Phase Peptide Synthesis

In the nineteen-fifties peptide synthesis advanced by leaps and bounds but those who participated in the major endeavors of the period, such as building molecules of the size of angiotensin, oxytocin, gramicidin S or $\alpha$-MSH, keenly felt that the demands of the task in effort and time approached the limits acceptable by most investigators. While in the next decade chains of twenty or even more amino acid residues could be constructed, among them corticotropin, a 39-peptide, the synthesis of such large molecules was looked upon as a rather forbidding exercise. Hence new methods that would bring about some relief in the synthesis of long peptide chains appeared highly desirable.

The stepwise strategy (cf. p. 97), because of the repetitive character of the procedure, foreshadowed the possibility of mechanization and automation [1] but in itself did not absolve the practitioner from the drudgery of countless operations needed in the actual execution of peptide synthesis. Each coupling step had to be concluded with the removal of unreacted starting materials and by-products by extraction, precipitation, filtration and drying of the product. Similarly, removal of an amine-protecting group was usually completed by isolation of the partially deblocked material through evaporation, precipitation and filtration. An understandable, although not always warranted, concern about mechanical losses during these manipulation added to the aversion felt by many organic and biological chemists toward peptide synthesis. Therefore the solid phase method, which offered an answer to their problem, was greeted with enthusiasm.

Once an amino acid is linked to an insoluble polymer, it can be converted to a dipeptide derivative that remains attached to the support while excess starting materials and the by-products of acylation can be removed simply by washing the peptidyl polymer with appropriately selected solvents. Partial deprotection is then carried out in a similar fashion and the dipeptide transformed into a tripeptide while the link between peptide and resin remains intact. In this way the chain can be lengthened until complete, without a single concentration or precipitation or transfer from one container into another. Only addition of reactants and repeated washings are necessary. At the conclusion of the chain building process the bond between peptide and polymer is cleaved and the peptide purified.

This ingenious concept was developed simultaneously and independently by R.L. Letsinger and M.J. Kornet at Northwestern University in Evanston, Illinois,

Brue B. Merrifield

and by R.B. Merrifield at the Rockefeller University in New York City. The results of the two studies were published in the same year, 1963, both in The Journal of the American Chemical Society. Yet, while the work of Merrifield gained wide acceptance, the Letsinger–Kornet approach had no followers. The reason for this conspicuous difference lies in the choice of strategies adopted in the two laboratories.

In both methods styrene-divinylbenzene copolymers were used as the starting material for the preparation of functionalizmd resins suitable for linking ("anchoring") an amino acid to the insoluble polymeric support. Letsinger and Kornet [2] converted the commercially available "popcorn resin" to its hydroxymethyl-derivative, treated the latter with phosgene and applied the resulting benzyl chlorocarbonate derivative for the blocking of the amino group of an amino acid. The *N*-acylamino acid, a polymer-bound benzyloxycarbonyl derivative, was activated and used for the acylation of the second residue in the sequence and building of the chain continued in the same direction.

A major shortcoming of this stepwise procedure is that it starts with the N-terminal residue and continues through intermediates which have no protection against racemization. Therefore a grossly heterogeneous material has to be expected as the crude product, a complex mixture of diastereoisomers, from which the desired chirally homogeneous peptide could be secured only through the extensive application of separation methods, if at all. In contrast, the approach chosen by Merrifield [3] is based on stepwise chain-lengthening

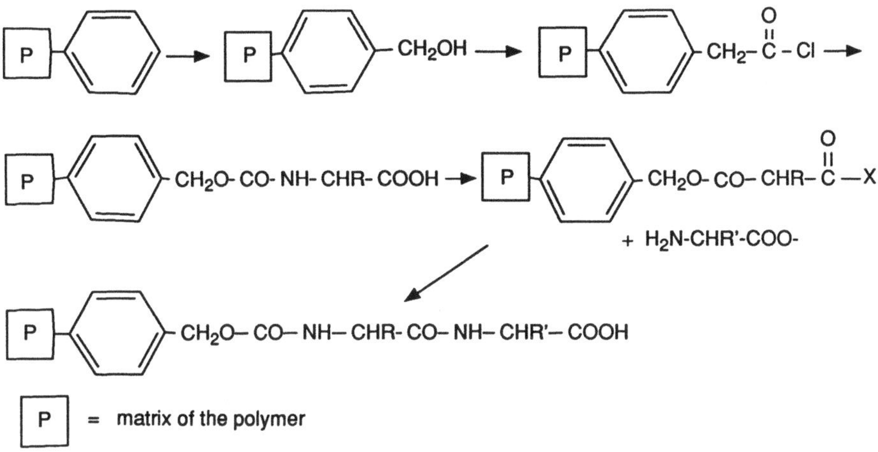


**Fig. 1.** Letsinger's approach to solid phase peptide synthesis

starting with the C-terminal residue. Since in this strategy, single amino acids, blocked by protecting groups which prevent racemization, are activated and coupled, the peptide cleaved from the resin should be chirally homogeneous. The styrene-divinylbenzene copolymer was chloromethylated and the benzyl chloride derivative, the Merrified resin, used for the anchoring of the first amino acid (the C-terminal residue in the sequence of the peptide to be synthesized) in the form of a benzyl ester derivative. After partial deprotection, i.e. removal of the blocking group from the $\alpha$-amino group, a second residue, the penultimate amino acid of the sequence, is added and the process continued until all residues are incorporated.

The anchoring ester bond, that should resist the various operations of the chain-building process, is cleaved at the completion of the procedure by alkaline hydrolysis[1] or better by acidolysis or, when a peptide amide is desired, by ammonolysis.

Not surprisingly, some difficulties were encountered in the early execution of solid phase peptide synthesis. For instance, the weak but not negligible acid sensitivity of the anchoring benzyl ester linkage led to a small loss of peptide from the resin in each deprotection step. Merrifield found a remedy for this shortcoming by nitrating the resin: nitrobenzyl esters are quite stable under the acidic conditions needed for the removal of the benzyloxycarbonyl group used in his first studies. A still better solution of the problem was the replacement of the benzyloxycarbonyl group with the more acid sensitive *tert*-butyloxycarbonyl (Boc) group. For acidolysis of N-protecting groups see Chap. 3.

[1] Instead of the not necessarily harmless saponification with alkali, in 1973 M.A. Barton and her associates proposed an elegant procedure: cleavage of the peptide from the supporting polymer by intramolecularly catalyzed transesterification of the (benzyl) ester bond with dimethylaminoethanol followed by the anchimerically assisted hydrolysis of the peptide dimethylaminoethyl ester.

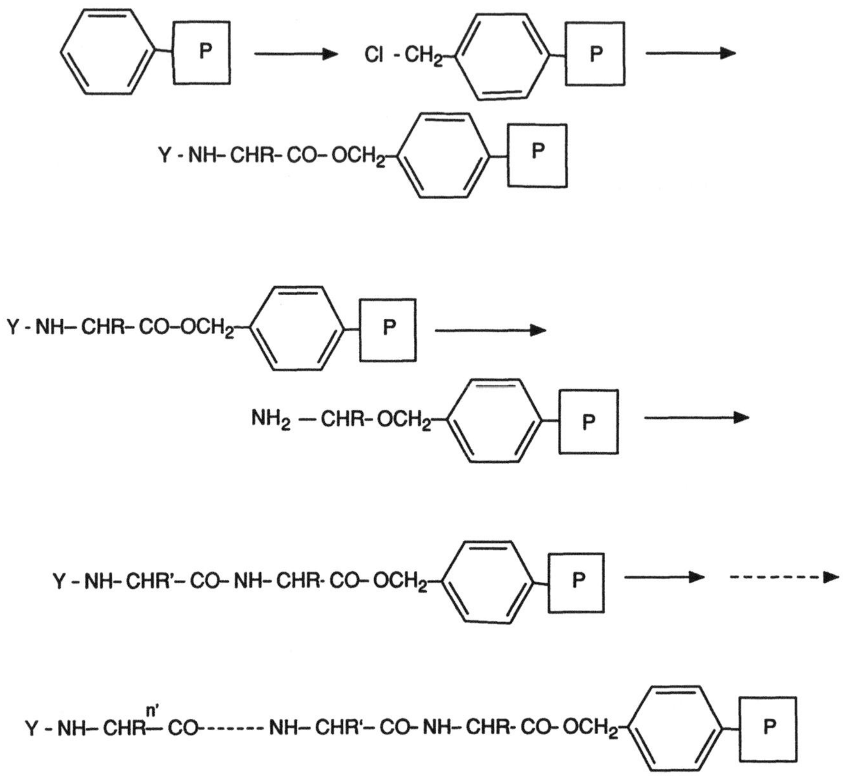


**Fig. 2.** Merrifield's approach to solid phase peptide synthesis

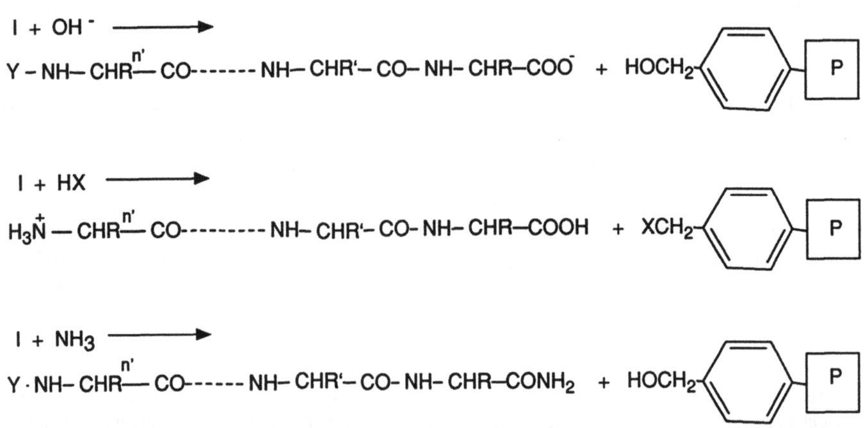


**Fig. 3.** Cleavage of the anchoring ester bond

Protection with Boc is still the most commonly used transient blocking of the α-amino function in solid phase synthesis. Its removal, mostly with a mixture of trifluoroacetic acid and dichloromethane, hardly affects the benzyl ester linkage at the C-terminus. For even greater selectivity the more acid labile *p*-methoxybenzyloxycarbonyl group was recommended recently. Suggestions involving the use of biphenylisoproyloxycarbonyl or the less acid sensitive phenylisopropyloxycarbonyl group were not heeded. A general shortcoming of schemes based on acid labile temporary blocking groups, which have to be removed at the end of each cycle of the chain-lengthening process, is the need for a less acid sensitive protection of the various side chain functions and for a rather acid-resistant anchoring bond since these must remain intact throughout the procedure. These relatively acid-stable groups, however, require very strong acids for their removal in the concluding step. With blocking groups based on the formation of benzyl cations HBr in trifluoroacetic acid was sufficient, but the subsequently introduced negatively substituted benzyl groups, such as in 2-chlorobenzyl esters, necessitated even stronger acids. In 1965 a remarkably efficient reagent, liquid hydrogen fluoride, was introduced into the practice of peptide synthesis by Sakakibara and Shimonishi (cf. p. 67). Soon it became extremely popular and the teflon HF-apparatus is now a standard piece of equipment in the laboratories engaged in solid phase peptide synthesis. Certain problems, however, related to liquid HF, must be kept in mind. For instance, the reagent can act as catalyst in Friedel-Crafts reactions between carboxyl groups, such as in the side chain of glutamyl residues, and aromatic nuclei in the polymeric support or in the molecule of anisole added as cation-scavenger to the reaction mixture (Fig. 4).

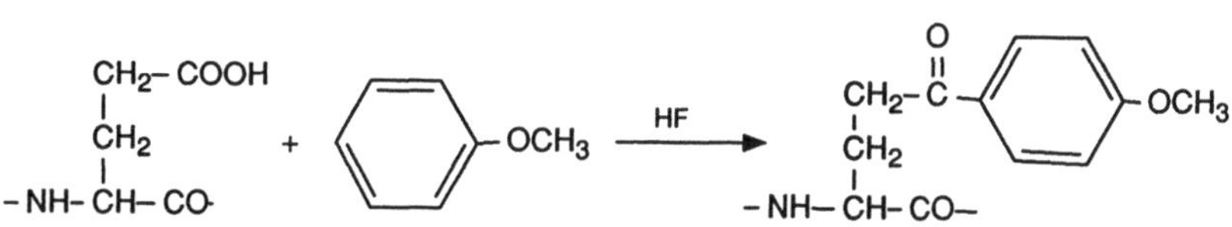


**Fig. 4.** Reaction with anisole

Some relief from similar difficulties could be obtained by the addition of diluents, such as dimethyl sulfide or cresols, to the liquid HF and also through its replacement by trifluoromethanesulfonic acid in trifluoroacetic acid. Serine containing peptides pose yet another problem: if instead of the usual mild conditions (0 °C, 20 min) because of the presence of acid resistant groups, for instance S-benzyl groups or nitro groups on arginine side chains, the cleavage must be carried out at room temperature rather than at 0 °C and for one hour instead of 20 minutes, then the well known N→O shift (or acyl migration) must be expected (Fig. 5). This side reaction can be reversed by the action of bicarbonate but not without some accompanying hydrolysis. Yet, in spite of all these concerns, liquid HF remains popular because it is a powerful reagent which smoothly cleaves the completed peptide from the resin and simultaneously removes all blocking groups including such acid resistant groups as the *N-p*-toluenesulfonyl group or the 2-chlorobenzyloxycarbonyl group.

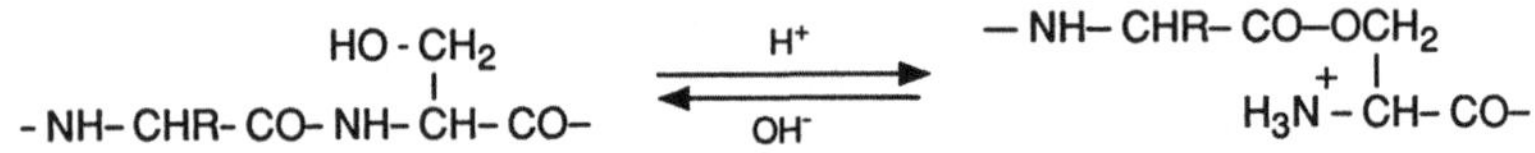


**Fig. 5.** N→O shift in serine side chain

A fundamentally different approach is based on the 9-fluorenylmethyloxycarbonyl (Fmoc) group proposed by L.A. Carpino and G.Y. Han in 1970. Its removal by secondary amines, usually by piperidine, from the α-amino group, leaves the *tert*-butyl based blocking groups on side chain functions intact. Accordingly final deblocking requires only moderately strong acids, for instance HCl in acetic acid or trifluoroacetic acid in dichloromethane. To reap the full benefits of this **orthogonal combination** (a term coined by Barany and Merrifield) also the anchoring benzyl ester grouping must be rendered more acid sensitive. This was readily achieved through the addition of electron-releasing substituents to the benzyl moiety as for instance in the Wang resin (S.S. Wang, 1973, Fig. 6). In syntheses performed with this type of polymeric support final deprotection and separation of the peptide-resin link are carried out by mild acid hydrolysis in a single operation.

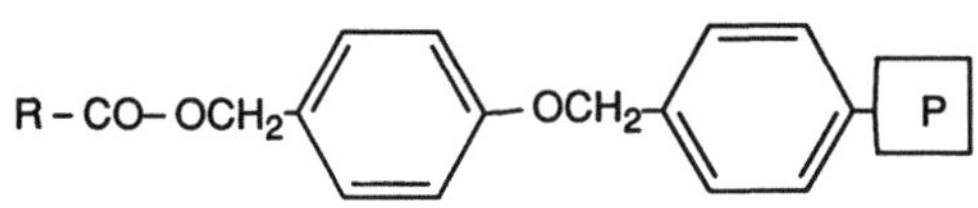


**Fig. 6.** Wang resin

In connection with different kinds of anchoring groups we should recall that numerous biologically active peptides have a carboxamide rather than a free carboxyl at their C-termini. As mentioned earlier, such peptide amides can be obtained by ammonolysis of the ester bond between peptide and polymer. A second possibility, to wit, acidolysis is equally useful and often practiced. In 1970 Pieta and Marshall constructed a new polymeric support in which the acid labile N—C bond in benzhydrylamine serves as the point of cleavage (Fig. 7). Of course, the bond's sensitivity to acids can be further increased by electron-releasing substituents such as methoxy groups.

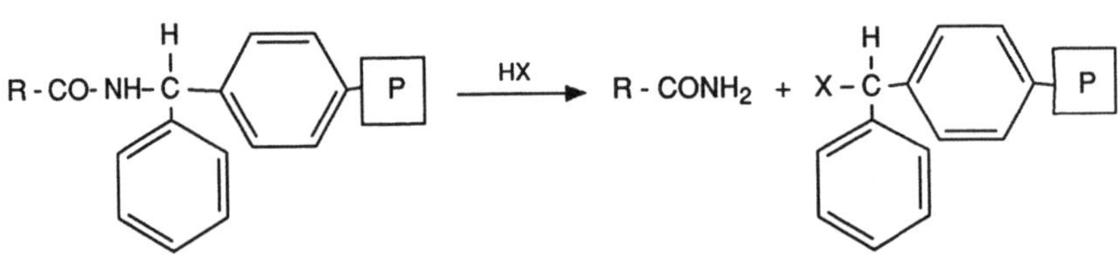


**Fig. 7.** Carboxamide formation from a benzhydryl resin

For the formation of the peptide bond in solid phase synthesis Merrifield applied the highly efficient dicyclohexylcarbodiimide (DCC) and it has remained the most widely used reagent ever since. The mode of application, however, has undergone a certain change. Instead of the traditional "coupling reagent mode" the "symmetrical anhydride mode" is practiced in most laboratories. Thus rather than adding the reagent to the mixture of carboxyl- and amine-components, two equivalents of the carboxyl component (an N-blocked amino acid) are treated with one equivalent of carbodiimide and the solution of the resulting symmetrical anhydride is used for acylation (cf. p. 83). Symmetrical anhydrides are rapidly generated in dichloromethane, more slowly in dimethylformamide. A part of the insoluble by-product, *N,N'*-dicyclohexylurea, might precipitate on or inside the beads of the polymer but is then readily removed in the following deprotection step since it is fairly soluble in the trifluoroacetic acid dichloromethane mixture used for cleaving the Boc group. Nevertheless, some practitioners prefer to use diisopropylcarbodiimide because the corresponding urea derivative is more soluble than the urea from DCC.

While symmetrical anhydrides, prepared with the help of carbodiimides and used without isolation, became the most commonly applied acylating agents in solid phase peptide synthesis, mixed anhydrides were only sporadically employed. The same is true for coupling reagents other than DCC. More attention was given to active esters. In fact, Merrifield in his often cited 1963 paper mentions that active esters would seem to be very much suited for his method, but in actual experiments they performed poorly. Not much later one of the authors of this volume, in collaboration with J.T. Sheehan, found that the failure experienced by Merrifield in the incorporation of blocked amino acids through their *p*-nitrophenyl esters was due to the solvent, benzene, in which both the swelling of the polymer and the aminolysis rates of the active esters are unsatisfactory and also that good results can be achieved in dimethylformamide. Subsequently several advantages of the active ester approach were demonstrated in various laboratories; some of them reported that with respect to homogeneity better products were obtained through the exclusive use of active esters than through coupling via carbodiimides. Nevertheless, the moderate rates observed when acylation was carried out in the hindered environment of the matrix of a polymeric support, discouraged most users of the new technology. In the late 1960s mechanization and automation of solid phase peptide synthesis revolutionized the field: it became quite simple to incorporate half a dozen or more amino acid residues in a 24-hour day. Such rapid chain-building is possible only with short coupling periods. Acylation with *p*-nitrophenyl esters or with 2,4,5-trichlorophenyl esters requires several hours for completion or even several days in the incorporation of hindered residues such as valine and isoleucine. Therefore, usually only asparagine and glutamine were introduced in the form of their active esters to circumvent the formation of $\beta$-cyanoalanine and $\gamma$-cyanobutyrine derivatives, a well known side reaction in coupling with carbodiimides. During the last decade a certain revival of the active ester concept can be noted. This became possible through the availability of very potent esters such as the esters of

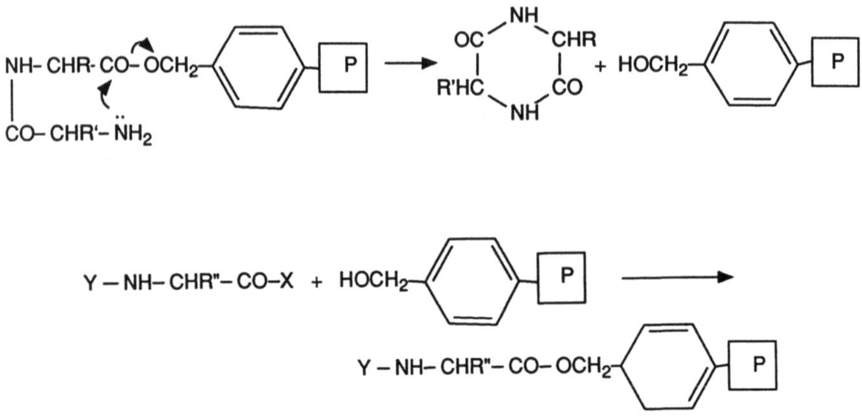


**Fig. 8.** Diketopiperazine formation and acylation of the released hydroxyl group

pentafluorophenol or 3-hydroxy-3,4-dihydrobenzotriazine-4-one. Let us add that less reactive esters perform equally well if catalyzed by 1-hydroxybenzotriazole or by the even more effective 3-hydroxy-3,4-dihydroquinazoline-4-one (König and Geiger, 1973) in the presence of a base. See the corresponding paragraphs in Chap. 4.

Peptide synthesis is quite regularly accompanied by undesired side reactions, for instance by the alkylation of side chains during deprotection. In addition to such general problems for which at least partial remedies have been found, also a special problem is encountered in solid phase syntheses: the formation of "deletion peptides" and "truncated sequences". Both incomplete acylation and incomplete deprotection result in peptide chains from which one or more amino acid residues are missing. A second kind of deletion, the loss of the C-terminal dipeptide sequence, is caused by diketopiperazine formation followed by acylation of the resulting hydroxymethyl polymer (Fig. 8). Premature chain termination (truncation) takes place if acetic acid or trifluoroacetic acid are not completely removed after deprotection and then co-activated in the following coupling step.

In spite of several shortcomings, solid phase peptide synthesis has reached an unprecedented degree of popularity and has become extremely productive. This is the consequence of studies of side reactions and to the introduction of counter-measures for their prevention. Improved polymeric supports, orthogonal protection schemes and in more recent years the systematic use of active esters, have all contributed to a more satisfactory methodology. Last but not least the availability of new, efficient methods of purification must be credited for the wide application of the solid phase method. With the help of high pressure liquid chromatography (HPLC, see p. 54) the target peptide can be isolated even from a complex mixture of accompanying related compounds. Affinity chromatography, although less frequently applied, adds a new dimension to separation techniques that can yield homogeneous peptides.

A glance through the catalogues of several research supply houses reveals a surprisingly large number of synthetic peptides that are available for studies in biochemistry, pharmacology, endocrinology, immunology, and various branches of medicine. Many of these compounds are peptides of considerable length and there can be little doubt about their origin: the majority of them was produced by solid phase peptide synthesis. Hence the question whether synthesis in solution will retain some of its earlier significance, is fully justified. Yet, it can be answered, at least at the time we are writing this book, in the affirmative. When unequivocal pathways and well defined intermediates are required, as in the preparation of biologically active peptides, on a large scale, for medical purposes, synthesis in solution is probably the method of choice, not lastly because it can be carried out in carefully documented good manufacturing practice. The economy of the process might also favor the older approach: if the reactions proceed in concentrated solutions, a lesser excess of the acylating agent is needed for practical rates. In solid phase synthesis the support itself acts as diluent and also as steric hindrance, hence the reactive intermediates (symmetrical anhydrides or active esters) are used in considerable excess. Thus solid phase synthesis plays a very important but not exclusive role in the production of peptides. The steadily increasing body of knowledge concerning the method was presented in two books [4, 5] and in a remarkably comprehensive review article [6].

The facilitation of peptide synthesis achieved through the application of polymers was really thought-provoking. Thus linking peptides via ion-pairs to ion-exchange resins was proposed as also the use of soluble polymers. Instead of binding the amine-component to a resin, acylation with polymer-bound active esters was recommended and even a technique in which both the acylating agent and the amino component are insoluble was invented. Yet, in spite of the originality of the underlying ideas and some obvious advantages of these methods, they have found no general acceptance in the peptide community. This was also the fate of several procedures which, in order to allow preparation of peptides in very short time, dispense with the isolation of intermediates. The reason behind the unique appeal of the Merrifield method is clearly its best feature: it is conducive to mechanization and automation.

More recent extensions of the solid phase technique permit the simultaneous preparation of several peptides. Graft polymers of different density (G.W. Tregear) provide analogs different from each other only with respect to certain residues and which can be separated from each other, prior to cleavage from the polymeric support, by sedimentation. Many studies requiring large numbers of discrete peptides can now be performed using systems of simultaneous multiple peptide synthesis (SMPS). One of them is the application of the resin in the form of small rods, another one is the "tea-bag method" of the Scripps Foundation group in La Jolla, Cal. (R.A. Houghten et al., 1985). The consecutive stepwise elongation reactions, as described above, are carried out in single packets of perforated polypropylene containing the aminoacylcopolymer particles in amounts of around 100 mg. These procedures enable one person to synthesize hundreds of peptides in 2–4 weeks in amounts large enough for virtually all chemical, biophysical and biochemical studies.

In closing this brief chapter on this ingenious procedure of peptide synthesis two additional aspects of the method should be mentioned. The solid phase principle is unusually stimulating and generated numerous studies aiming at improvements in methodology. The same principle could be extended also to other fields, particularly to the synthesis of oligo- and poly-nucleotides. Because of the difficulties inherent in the isolation of intermediates in the synthesis of DNA molecules, solid phase synthesis occupies here an almost exclusive position. Merrifield's work was recognized by the 1984 Nobel Prize in Chemistry.

## References

1. M. Bodanszky, Stepwise synthesis of peptides by the nitrophenyl ester method. Ann. N.Y. Acad. Sci. 88: 655–664 (1960)
2. R.L. Letsinger, M.J. Kornet, Popcorn polymer as support in multistep syntheses. J. Amer. Chem. Soc. 85: 3045–3046 (1963)
3. R.B. Merrifield, Solid phase peptide synthesis I. Synthesis of a tetrapeptide. J. Amer. Chem. Soc. 85: 2149–2154 (1963)
4. C. Birr, Aspects of the Merrifield synthesis. Springer Verlag, Berlin Heidelberg New York 1978
5. J.M. Stewart, J.D. Young, Solid phase peptide synthesis. 2nd edn, Pierce, Rockford, Illinois, 1984
6. G. Barany, R.B. Merrifield, Solid phase peptide synthesis. In: The Peptides, vol 2 (E. Gross, J. Meienhofer, eds.) Academic Press, New York, 1979

# 6 Structure Elucidation

The classical approach to determining the structure of peptides has been through analytical methods, including purification and derivatization, hydrolysis and identification of the amino acid constituents, partial hydrolysis and stepwise degradation of fragments. The analytical arsenal has been enlarged by the application of gas-chromatography combined with mass-spectrometry and by sophisticated homonuclear and heteronuclear magnetic resonance (NMR) measurements.

In more recent years, these approaches have lost some of their significance because the speedy and precise sequencing of nucleic acids [1, 2] allows the prediction of the order of amino acids directly from the nucleotide sequences of the messenger-RNA-s (transcribed from the corresponding DNA sequences). This method, however, is limited to deriving the linear order of common amino acids in polypeptides; yet, peptides are often modified after their biosynthesis in the cell. Some have their ends trimmed and others arise by cleavage of longer chains. Cysteine residues are oxidized to form disulfide linkages or changed into lanthionine moieties (cf. p. 223). Specific side chains can be altered, for instance by methylation or glycosylation. Such posttranslational modifications cannot be recognized merely from the primary structure of the gene. Furthermore, since for small peptides, containing up to about 20 amino acids, the application of recombinant DNA technology is not necessarily the most practical approach, the classical methods still occupy a strong position.

Several modern microanalytical methods have their origin in peptide chemistry. Paper chromatography, its precursor partition chromatography, and ion-exchange chromatography have already been mentioned in Chap. 3. Similarly, the electrophoretic techniques, indispensable in biochemical and molecular-biological laboratories, also grew out of early peptide research. These methods were developed, shortly after World War II, for separation of peptides and amino acids in order to elucidate the composition, and ultimately, the structure and molecular shape of biologically important peptides: their conformation by which the binding to specific target molecules is controlled.

The analytical tools available in the early history of peptide research were rather poor. The starting point for analytical, at first qualitative, studies was probably the introduction of ninhydrin by S. Ruhemann (cf. p. 51) in 1910. For the determination of small amounts of $\alpha$-amino acids ninhydrin is far superior to the combustion methods in use until then.

The classical method for the quantitation of amino-nitrogen came from D.D. van Slyke who, in 1911, measured the nitrogen gas generated from R-$NH_2$ by nitrous acid first through volumetry, then, from 1929, manometrically. For the determination of individual amino acids more or less specific precipitating reagents, such as calcium, copper, silver, mercury and cadmium salts, were applied. These were followed later by strong organic acids, extensively studied by Max Bergmann and his associates, whose work, including the solubility product method, was discussed in Chap. 3.

In 1940 D. Rittenberg and G.L. Foster described the determination of amino acids in mixtures, for instance glycine, glutamic acid and aspartic acid in the hydrolysate of cattle fibrin [3], by the so called *isotope dilution* method. In this process an amino acid containing an excess of a particular isotope, in this case $^{15}N$, is added to the mixture to be analyzed and the quantity of the amino acid is inferred from the amount of the isotope in the substance after isolation and purification. The error in the determination is independent of the method of isolation of the pure substance or of the yield. In spite of its accuracy and elegance, the method did not gain wider application because of the limited availability of labelled amino acids and of the primitive methods for the determination of isotopes at that time. Later, when isotopes became routinely used in chemical laboratories, the simpler chromatographic methods of amino acid analysis had already conquered the field. This is similarly true for the specific colorimetric and spectrometric methods described in detail by Martin and Synge in 1945 [4] and for the microbiological assay of amino acids summarized by E.E. Snell [5].

## 6.1 Analysis of Amino Acid Sequences

In order to determine the sequence of amino acids in a peptide, initially it was subjected to partial hydrolysis. Thereby fragments were formed, smaller peptides the sequence of which could be more easily established. After micropreparative separation, e.g. by paper chromatography, the fragments were hydrolyzed and analyzed for amino acid composition both before and after treatment with nitrous acid. The amino acid missing in the treated, desaminated, sample, was its N-terminal residue. The sequence of the whole peptide was then reconstructed through the appropriate combination of the structurally identified peptide fragments. A typical example, the elucidation of the structure of gramicidin S, is described on p. 205.

### 6.1.1 End Groups Determination

With longer peptides, however, this kind of sequence determination does not always lead to unambiguous results. Therefore, controlled methods of degradation were needed, starting from a defined point in the chain, preferentially from the amino terminus. In any case, the N-terminal amino acid had to be identified. As a reagent tagging free amino groups, 2,4-dinitrochlorobenzene was introduced by E. Abderhalden (with P. Blumberg) as early as 1910. Since

2,4-dinitrophenyl (DNP) amino acids, under the conditions used for the total hydrolysis of peptides, are not decomposed by boiling hydrochloric acid, Abderhalden (with W. Stix), in 1923, tried to determine amino end groups in a mixture of peptides in a partial hydrolysate of silk fibroin. The experiment, however, gave no encouraging results, because, with the methods available at that time, it was not possible to separate the DNP amino acids formed from the bulk of free amino acids or from each other.

Decisive progress in the method was achieved in 1945 by Frederick Sanger [6] (Plate 38), who found that 2,4-dinitrofluorobenzene reacts with amino

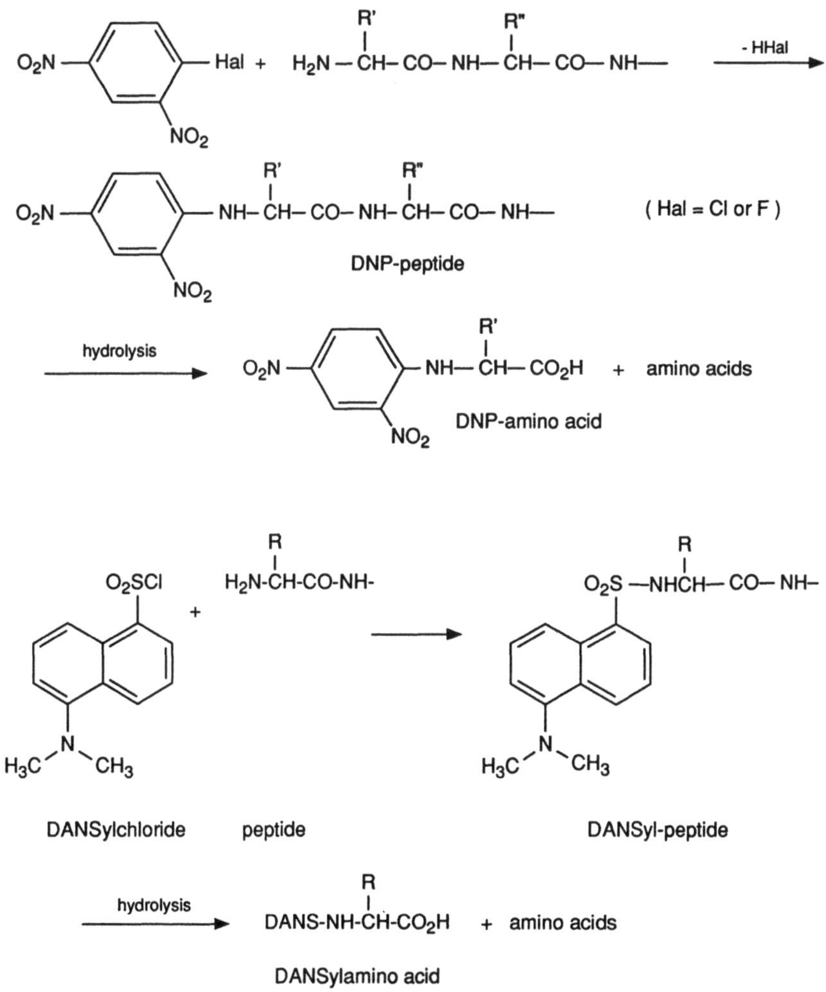


**Fig. 1.** Dinitrophenyl (DNP) and 5-dimethylaminonaphthalene-1-sulfonic acid (DANS) peptides and amino acids

groups much more readily than the chloro-compound, and that the DNP-amino acids (Fig. 1) can easily be separated from each other, and also characterized, by paper chromatography. This was the beginning of Sanger's famous research, soon leading to the elucidation of the structure of insulin (p. 158)

A second reagent, 5-dimethylamino-naphthalene-1-sulfonic acid chloride (DANSyl-chloride), that allows a very sensitive identification of the N-terminal amino acid, was introduced by Gray and Heartley in 1963 [7]. It reacts with amino groups under mild conditions and yields DANSyl-peptides in which the sulfonamide bond is more resistant to acid hydrolysis than the peptide bonds are. Hence DANSyl-amino acids (Fig. 1) are released intact on hydrolysis and can be visualized, after chromatography on paper or on thin layer plates, even in minute amounts by their fluorescence under UV-light.

The identification of the carboxyl terminal amino acids of peptides seemed to be of less interest and was experimentally somewhat more difficult. In 1952, S. Akabori (Plate 9) showed that on heating with anhydrous hydrazine to 100 °C for several hours, all peptide bonds of a peptide chain are cleaved by hydrazinolysis. Hereby the amino acid constituents are released as the respective hydrazides, $R\text{-}CONHNH_2$, except the C-terminal amino acid which, sometimes with difficulty, is isolated and characterized as such. Shortly before the

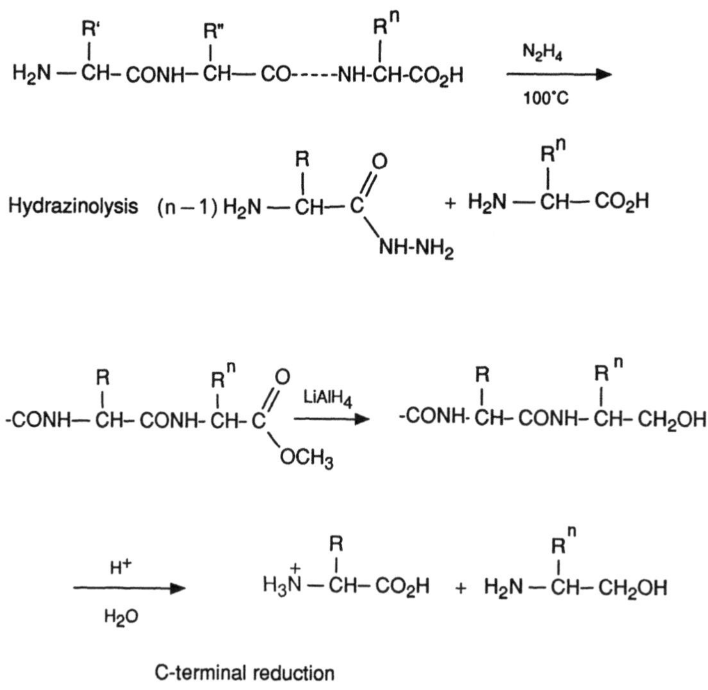


**Fig. 2.** Analysis of C-terminal amino acids by hydrazinolysis or by reduction

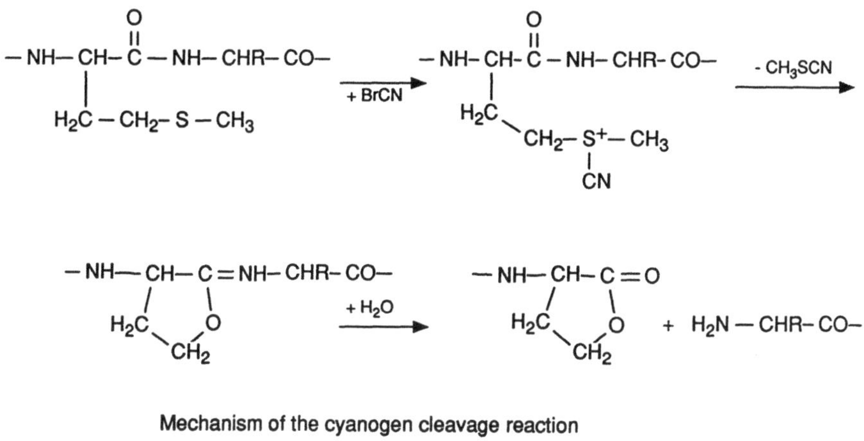


**Fig. 3.** Mechanism of cleavage of a Met-peptide bond by BrCN

introduction of the Akabori reaction, Claude Fromageot and his associates demonstrated that the carboxyl end group, or better its methyl ester, can be reduced, without attacking the peptide bonds, to the primary alcohol by a brief reaction with complex hydrides. After total hydrolysis that follows the reduction, one of the amino acids present in the original peptide is missing: this is the C-terminal residue (Fig. 2). The newly formed amino alcohol can be identified, best as the DNP or DANSy1-derivative, by paper or thin layer chromatography.

Here we have to mention briefly the partial cleavage or fragmentation of longer peptide chains. This can be achieved enzymatically by peptidases (trypsin, chymotrypsin, etc. cf. p. 57) of different specificity or by specific chemical means [8]. Well known and frequently applied is the cyanogen bromide method of Gross and Witkop (Plate 49) [9]. The methionine side chain in a peptide is degraded by reaction with BrCN forming a $\gamma$-lactone, which facilitates the hydrolytic splitting of the adjacent peptide bond (Fig. 3).

This reaction can also serve cleaving (methionine-free) polypeptides, produced by gene-technology in bacterial or other cells, from their larger propeptides at a Met residue incorporated in an appropriate position.

### 6.1.2 Stepwise Degradation

Since complete information on the structure of a peptide can be obtained only by the determination of the sequence of its amino acid residues, stepwise degradation of the peptide chain, starting either from the amino terminus or from the carboxyl end, was a long-standing aim of peptide chemists. Hans Brockmann, while working on his doctoral thesis under E. Abderhalden in the late twenties, resumed Max Bermann's experiments with phenylisocyanate. In 1927, with Kann and Miekeley, Bergmann observed that the N-terminal amino group reacts with phenylisocyanate to form a phenylureido derivative which, on boiling with hydrochloric acid, is transformed, via ring-closure, to a phenylhydantoin. Under the strong reaction conditions used, peptide bonds were also cleaved. The authors, apparently, were not aware of the relative easiness of ring formation by acids. Later, Abderhalden and Brockmann [10] were able to demonstrate that from *N*-phenylcarbamoyl peptides the terminal acid is cleaved as phenylhydation by treatment with HCl-saturated methanol already at 60–65 °C and within 30 minutes, and that the intact shortened peptide (glycyl-leucine) can, in turn, be subjected to the same reaction (Fig. 4).

Decisive progress in stepwise degradation was brought about by Pehr Edman (Plate 15) in 1950 [11], who found that phenylisothiocyanate, the sulfur analog of Bergmann's reagent, reacts with the N-terminal amino group at pH 9 even at room temperature, and that the thioureido compound hereby formed is readily split by weak acids, e.g. trifluoroacetic acid, to yield a 2-anilinothiazolin-5-one.

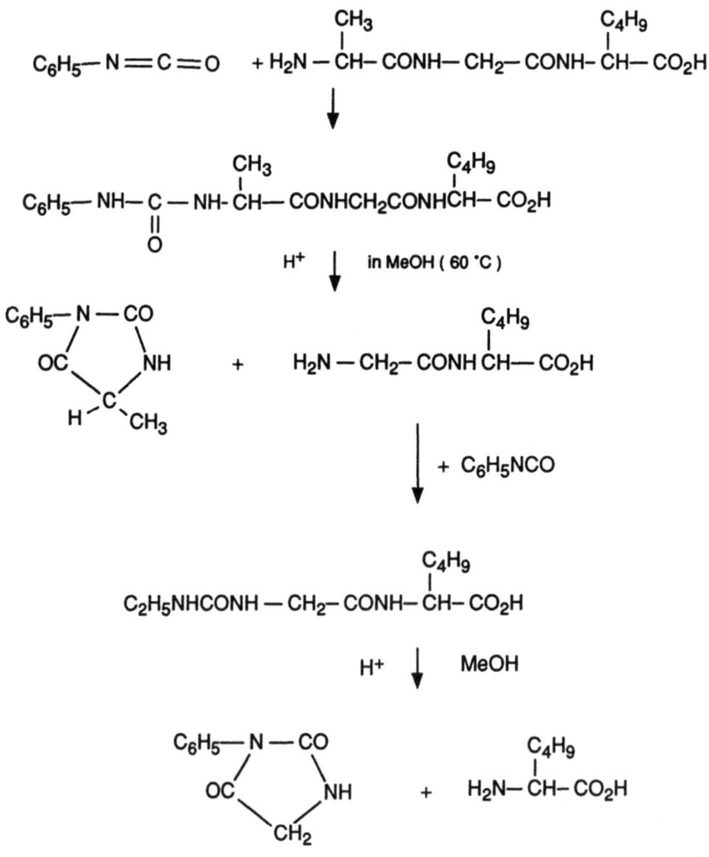


Stepwise degradation of a tripeptide via phenylhydantion

**Fig. 4.** Peptide degradation via phenylhydantoines [10]

The peptide, shortened by the corresponding amino acid, is then subjected to a new cycle of degradation. The thiazolinones are treated with stronger acids or heated and thus converted to the more stable 3-phenyl-2-thiohydatoins (PTH-s) which can be identified by chromatographic methods or by mass spectrometry (Fig. 5).

The introduction of an automatic sequenator by P. Edman and C. Begg [12] in 1967 brought a great saving of time and labor. The degradation takes place in a rotating glass beaker. The solution of the peptide adheres to the wall of the beaker as a film and is subjected to the degradation reactions in an automatically controlled sequencer. The thiazolinones are collected and separately converted to the PTH-s. At this time the method allows the analysis of nanomolar amounts through 50 or more cycles of degradation.

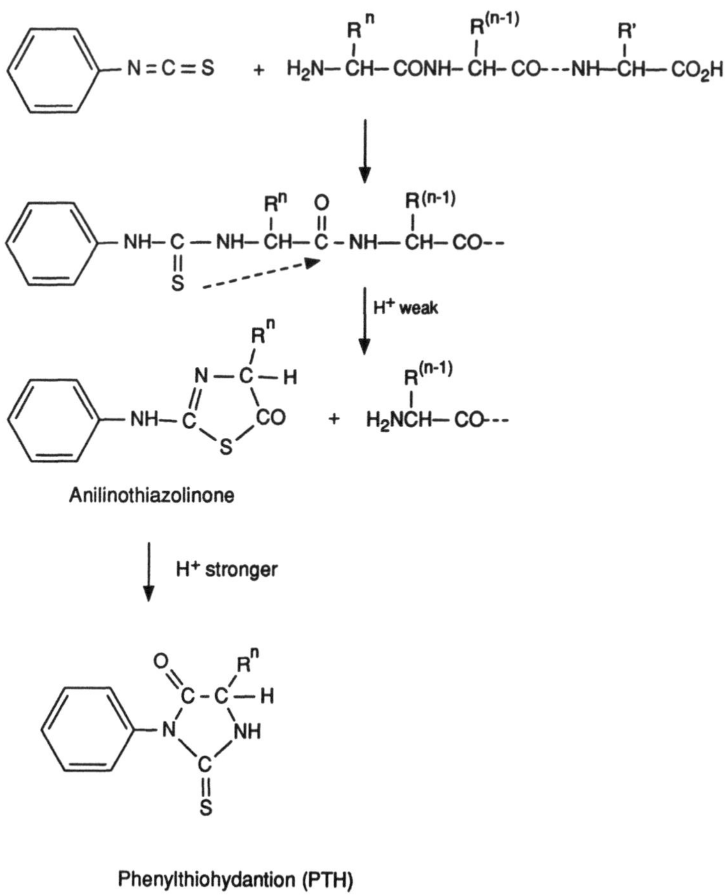


**Fig. 5.** Scheme of Edman degradation [11]

Several years later R.A. Laursen [13] presented a sequenator based on the Edman degradation of peptides anchored to a solid phase via their carboxyl ends (cf. Chap. 5). This approach was further developed by the introduction of various supports and methods of fixation [14]. The picomole range was achieved in the beginning of the eighties by the most recent technique, gas-phase Edman-degradation (M.W. Hunkapiller and associates, Pasadena). Here the sample is spread on a tiny fiberglass disc and treated with gaseous reagents in high vacuum.

## 6.2 Spectroscopic Methods

The spectroscopic methods most frequently used in peptide chemistry involve the measurement of absorption (or transmission) of electromagnetic radiation due to excitation of electron transitions (200–400 nm, ultraviolet, and 400–800 nm, visible region), and to intramolecular vibrations of bonds between C, O, N and H atoms (3000–10 000 nm, infrared region), to the excitation of magnetic transitions of atomic nuclei (0.1–20 cm, nuclear magnetic resonance, NMR) and, less frequently, of unpaired electrons (electron spin resonance, ESR). A characteristic property of dissymmetric substances, like amino acids and peptides, is their optical activity or optical rotatory power, a consequence of their ability to refract and absorb right and left circular-polarized light to different extents. These differences are manifested in the optical rotatory dispersion (ORD) spectra and the circular dichroism (CD) spectra in which they are recorded as a function of the wave-length. Finally, in mass spectroscopy (MS) the molecular mass ions of peptides and their fragments, produced by bombardment with electrons, fast atoms (FAB) or ultraviolet light (laser), are monitored after resolution in an electric and magnetic field.

### 6.2.1 Ultraviolet Absorption Spectroscopy

The naturally occurring amino acids and peptides are colorless, they do not absorb visible light. Rare examples containing a chromophore, for instance the actinomycins, or colored derivatives obtained by reactions with chromogenic reagents, will not be discussed here.

Absorption studies of amino acids in the UV region go back to the mid-thirties [15]. An overview on this subject appeared as early as 1952 [16]. Only the aromatic amino acids, phenylalanine, tyrosine and tryptophan, absorb in a conveniently observed region (Fig. 6). The weakest of them is phenylalanine, revealing at least 6 absorption bands; it has a molecular absorption coefficient of $\varepsilon_{mol}$ $0.195 \times 10^3$ at 257 nm. The UV-spectrum of tyrosine is strongly dependent on pH: in the presence of 0.1 N HCl (protonated hydroxyl group) $\varepsilon_{mol} = 1.34 \times 10^3$ at the maximum, ($\lambda_{max}$ 275 nm); in 0.1 N NaOH (phenolate) $\varepsilon_{mol} = 2.33 \times 10^3$ at $\lambda_{max}$ 293 nm. The absorption of tryptophan, is nearly independent of the pH; it shows a maximum at 280 nm with $\varepsilon_{mol} = 5.55 \times 10^3$ and a second maximum (shoulder) at 288 nm of $\varepsilon_{mol} = 4.55 \times 10^3$. These parameters have often been used in the quantitative analysis of peptides. The sulfur-containing side chains show weak absorption below 250 nm and are of interest only in rare special cases.

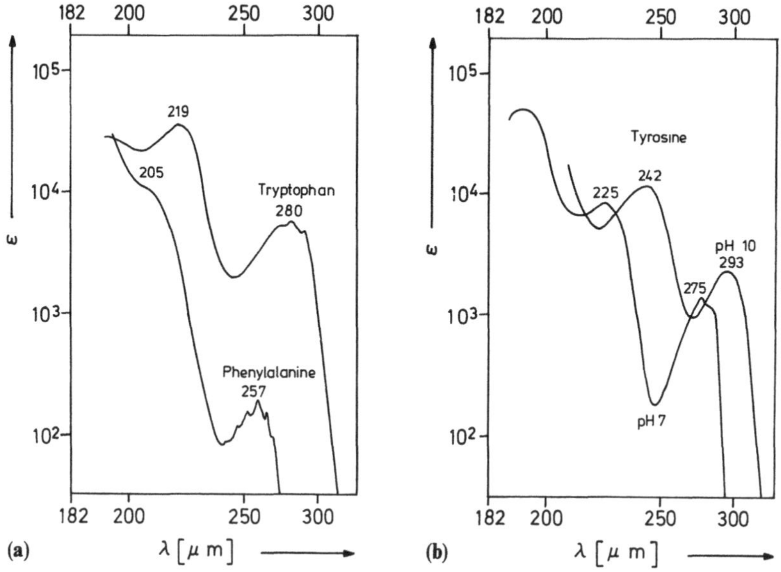


**Fig. 6.** Ultraviolet absorption spectra of phenylalanine and tryptophan (**a**) and tyrosine (pH 7 and pH 10) (**b**)

The general estimation of the amino group, in the form of copper salts, by UV-absorption at 230 nm, was suggested by Spies and Chambers [17].

The UV absorption of the peptide bond, —NH—CO—, is centered around 210 nm. Its molar absorption coefficient, approx. $3 \times 10^3$, is much higher than that of the amide group of saturated fatty acids at the same wave length, a consequence of the mesomeric electron-distribution (partial double bond character) of the —NH—CO— moiety.

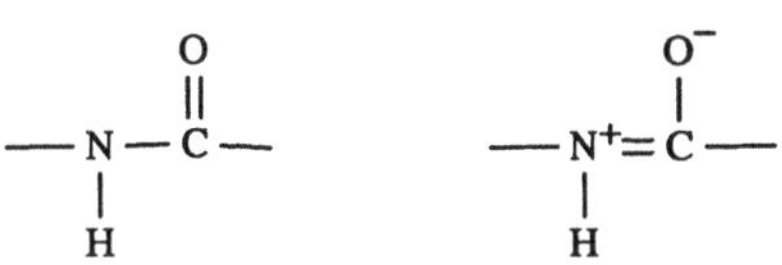

*Circular dichroism* (CD) spectroscopy is based on the difference of absorption of right and left circularly polarized radiation by chiral (dissymmetric) molecules. In peptide research CD spectra can give valuable information on the conformation of peptide molecules in solution. The differences in absorption (ellipticities, $\theta$) are pronounced in the 200 to 220 nm region. The positive or negative Cotton effects allow empirical statements on the arrangement of peptide chains, through comparisons with the CD spectra of peptides that have helical, pleated sheet or random coil conformations, established by other means, e.g. by X-ray diffraction studies. Optical rotatory dispersion (ORD) spectroscopy, which is based on the different refraction of circularly polarized light by chiral compounds, is the older

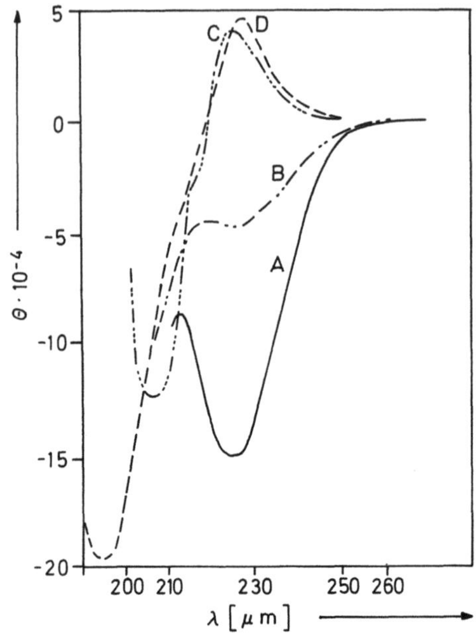


**Fig. 7.** CD-spectra of free antamanide, and with successively added Na-ions

of the two "chiroptical" methods, it has been applied in peptide research for about 30 years.

As mentioned in Chap. 9 (p. 215) the cyclic decapeptide antamanide exists in several conformations depending on the presence or absence of alkali metal ions and on the polarity of the solvents. The transition of the free form to a complexed one can be easily followed by CD spectra (Fig. 7). On the addition of Na ions or water the large negative ellipticity (Cotton effect) observed in apolar solvents, such as dioxane, disappears and changes into a positive Cotton effect in the same wave length region.

According to numerous studies, for instance by G.T. Fasman, the arrangement of peptide chains in $\alpha$-helices is distinctly reflected by two negative Cotton effects around 205 and 220 nm, whereas $\beta$- or pleated sheet structures give rise to one negative effect at 215 nm. Random coil conformations are revealed by a strong negative band at about 195 nm [18] (Fig. 8). Based on the study of helical synthetic polypeptides G. Holzwart and P. Doty [19] assigned the ultraviolet CD data in the region between 190 and 240 nm to $n \rightarrow \pi$ and $\pi \rightarrow \pi^*$ electron transitions of the peptide bond.

A continued and successful application of CD spectroscopy for the study of biologically active peptides was carried out by Karel Blaha (Plate 11).

### 6.2.2 Vibrational Spectroscopy. Infrared Absorption. Raman Spectra

In the infrared (IR) and in Raman spectra of a molecule, bands are observed that correspond to normal modes of vibration of its particular structure. Vibration occurs by oscillation of two atoms in one direction (stretching

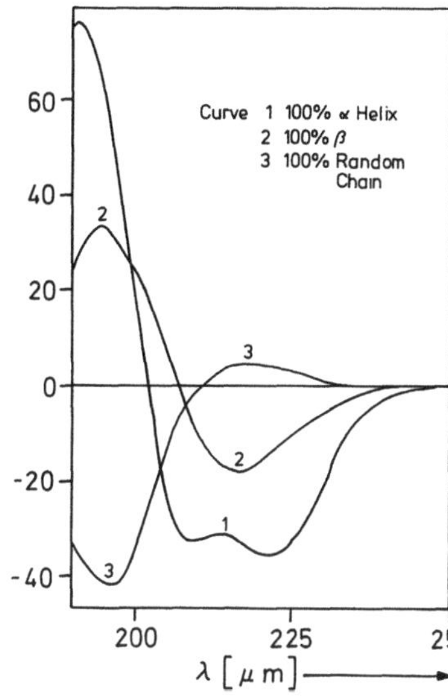


**Fig. 8.** CD spectra of poly-L-lysine in different chain conformations [18]

frequency), in three dimensions ("deformation" frequencies) or in a torsional way. The most studied IR bands can be assigned to the different modes of vibration. These appear in the wavelength range of 2.5–50 μm, or, expressed in frequency numbers $\nu(\nu = 10^4/\lambda)$ between 4000 and 200 $cm^{-1}$. In investigations of peptides the bands related to the peptide bond are obviously of profound interest.

The history of vibrational spectroscopy begins in 1936, with J.T. Edsall's studies [20] of the Raman spectra of amino acids. In Raman spectroscopy a strong monochromatic beam of any wavelength, today from a laser, is transmitted through the probe. It excites the molecules which then emit, in addition to Raleigh-scattering at the same frequency, discrete frequencies, Raman bands. Fundamentally, these are produced by the same vibrational modes which absorb specific frequencies of continuous IR light thus generating an IR absorption spectrum. Later work on peptides and proteins was pursued mainly with the help of IR spectroscopy, yet Raman spectra have gained considerable ground in the last decade [21].

In the infrared field a very thorough study of amides and of the —CO—NH— peptide link was made in 1943 to 1945 in several laboratories in Great Britain and in the USA as part of the IR analysis of penicillin (p. 198). In this work it became well established that all compounds of the type RCONHR′ exhibit strong absorption near 3, 6 and 6.4 μm (3330, 1670 and 1560 $cm^{-1}$). The first of these bands was assigned to the stretching vibration of the NH group, the second to the stretching vibration of the CO group and the third, with reservations, to the deformation vibration of the NH group. Further important advances came only 15 years later with the determination, by Myazawa [22], of the normal modes of vibration of the peptide grouping in the simplest model containing a trans-peptide group, *N*-methyl-acetamide.

By calculation 12 bands were assigned to different modes of vibrations, in part to their combinations. Here only the strong bands are mentioned. The band at 3236 $cm^{-1}$ (amide A, cf. 3330 $cm^{-1}$ above) is completely due to NH stretching, the one at 1653 $cm^{-1}$, amide I to CO and CN stretchings and to a small contribution from CCN deformation. Amide II at 1567 $cm^{-1}$ is mainly due to the N—H in-the-plane bending and to CN stretching vibration. A fourth strong band that follows the medium strong amide V at 725 $cm^{-1}$ is assigned to CN torsion.

In peptides, principally, the band frequencies are positioned at the same region, but can be shifted according to the close environment of the moving atoms. Thus they depend on the conformation of the chain, whether it is helical or pleated sheet or folded. Hydrogen bonding plays a similarly important role. Although NH stretching is not likely to be sensitive to chain conformation, it frequently depends on the strength of the N—H·O=C hydrogen bond. In some polypeptides of known structure it was noted that the NH frequency of the free peptide NH is considerably higher than that of the hydrogen-bonded one. There is, therefore, a correlation between frequency and the length of the N—H··O distance.

Shifts and certain splitting by resonance interactions of amide I and amide II bands in polypeptides in folded and extended conformation are discussed in Ref. 22 and in the early investigations of Miyazawa and of Elkan R. Blout [23], who used IR spectra for the characterization of the conformation of poly-amino acids.

IR spectra are generally recorded as transmittance bands with 100 at the top and zero at the bottom, (Fig 9).

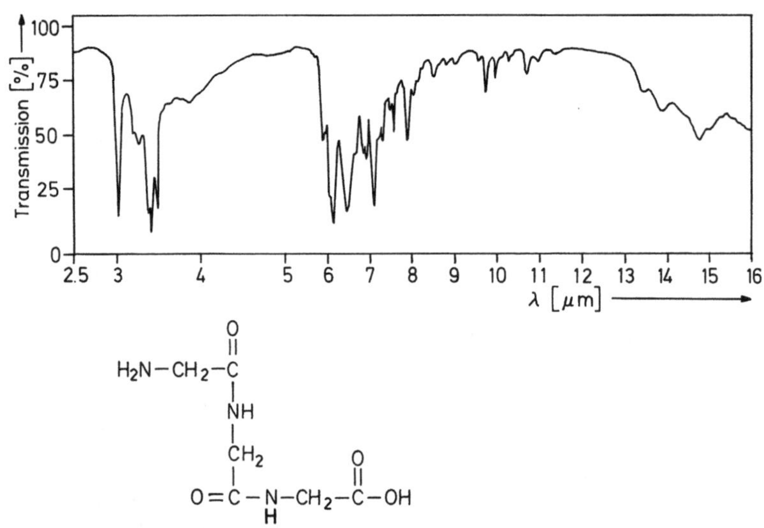


**Fig. 9.** IR-spectrum of triglycine (nujol)

### 6.2.3 Nuclear Magnetic Resonance Spectroscopy (NMR)

NMR spectroscopy has developed to such an extent that it has become the most useful method for the elucidation of the conformation of peptides in solution. It can also identify the amino acid sequence in medium size peptides. The method is based on the magnetic properties of the nuclei of isotopes with spin numbers I = 1/2. Of these the proton ($^1H$), the carbon isotop $^{13}C$, and to a lesser extent $^{15}N$ are of practical importance.

In a strong magnetic field the magnetic nuclear moments are oriented to the lines of the magnetic field in two directions different in energy. Energy input by a tunable high-frequency electric field lifts the low energy state to the higher one at a discrete field-strength which depends on the nature of the atom or its binding partner and its close environment. The absorption of energy by resonance is registered as narrow peaks, signals in the NMR spectrum. Their intensity, which is proportional to the number of nuclei, their position and their fine structure (multiple lines due to the mutual influence of neighboring nuclear spins, "spin–spin coupling") provide reliable information on the arrangement of the atoms and allow conclusions on the structure of the molecules.

The NMR methods was introduced in 1946 by Bloch and Purcell (Nobel Prize 1952). Ten years later NMR spectrometers were available, but only with weak magnetic field, yielding spectra of poor resolution.

The resolving power of an NMR instrument depends on the strength of the magnetic field. It is expressed by the proportional frequency of the electric field generator, in megahertz (MHz). The first instruments worked with 50 MHz, today expensive spectrometers with up to 600 MHz are in use and they can discriminate between two signals in a distance of only 0.01 ppm (on the scale ranging from zero to eight ppm; see Fig. 10).

As early as 1957, American researchers described the $^1H$-NMR spectra of amino acids, dipeptides [24] and even of the enzyme ribonuclease [25]. At that time only a few peaks in the spectrum (Fig. 11) could be assigned to structural elements such as —$CH_2$—, to methyl protons (of alanine, valine, leucine, isoleucine), to aromatic protons (in phenylalanine, tyrosine, tryptophan) or to the proton in the peptide bond. The corresponding signals are distinguishable by their different "chemical shifts". Not much later it was found that in heavy water the protons bound to N, O or S are exchanged by the magnetically inactive $^2H$-ions. The rate of exchange, that is the rate of fading of the respective proton signals, is a measure of the strength of their linkage. In hydrogen bridges, for instance in helical or $\beta$-sheet regions, the exchange can be extremely slow. The intensity of NMR absorption of one nucleus may be altered by a close neighboured second nucleus. The closer the nuclei the greater the "nuclear Overhauser enhancement" (NOE effect). Since the relevant distance is that through space, so molecular conformations can be determined.

The main field of application of NMR spectroscopy in peptide research is the analysis of peptides in solution. Although the most accurate picture is obtained by X-ray analysis of single crystals (p. 131), this is restricted to crystalline samples and provides information only about the conformation in the solid state. It has

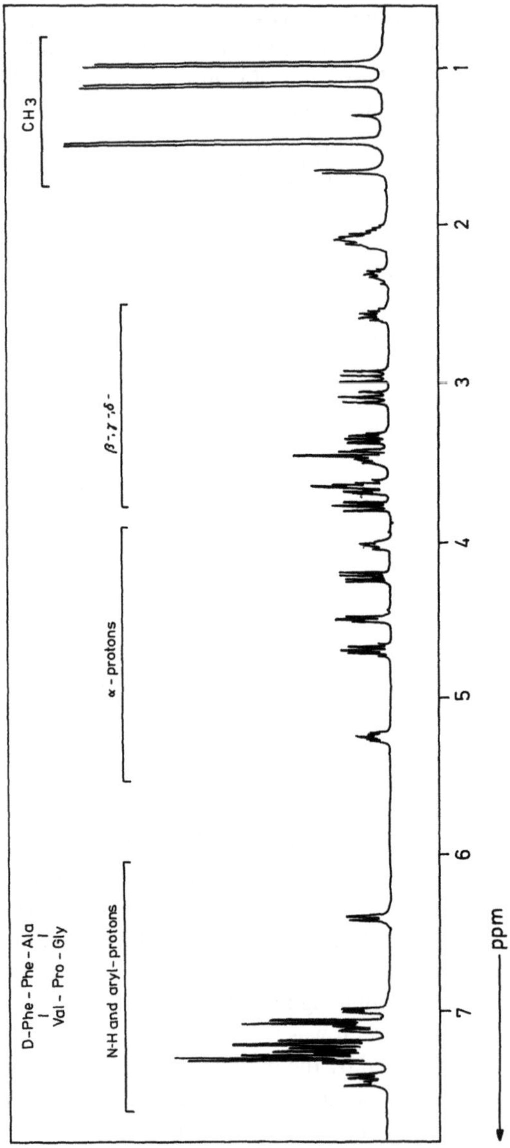


**Fig. 10.** NMR spectrum of a cyclic hexapeptide at 500 MHz (Courtesy of Dr. K.-H. Pook, Ingelheim)

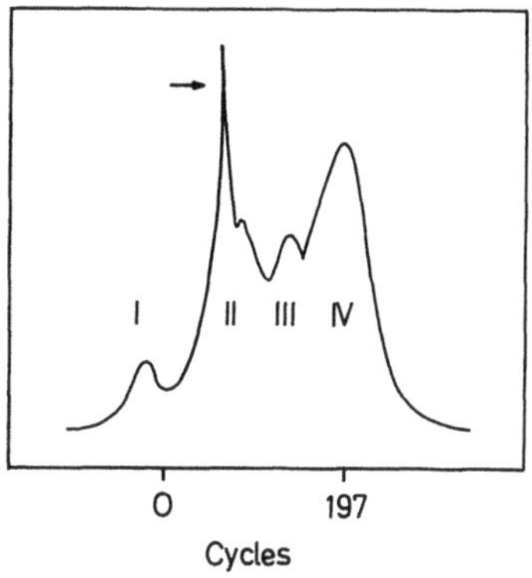


**Fig. 11.** First $^1$H-NMR spectrum (50 MHz, in 1957), of ribonuclease in $D_2O$ [25]. The assignment of the fused peaks corresponds in principle to those completely resolved in the spectrum of Fig. 10

been shown that the conformation of smaller peptides in solution may greatly differ from that found in the crystals. The crystal structure is strongly determined by hydrogen bonds between the molecules, whereas the conformation in solution is determined mainly by hydrogen bonds within a single molecule or to solvent molecules. The relevance of NMR spectroscopy is due to its ability to determine the spatial arrangements of individual $^1$H, $^{13}$C and $^{15}$N atoms in the molecule. In recent years this capability has been revolutionized by the introduction of two-dimensional techniques (2D NMR). Examples of modern NMR analysis of peptides can be found in a review article by H. Kessler et al. [26].

### 6.2.4 Electron Spin Resonance

An alternative analytical method, electron spin resonance (ESR) spectroscopy, functions only with paramagnetic substances, in organic chemistry with molecules which have an odd number of electrons, that is free radicals. In a homogenic magnetic field the magnetic moments of the electrons are ordered in states with parallel and antiparallel spins relative to the field lines, states that have different energies. If energy is added by electromagnetic irradiation, the higher energy level will be more heavily populated because of specific absorption. In practice, the sample is irradiated with a microwave of constant frequency (e.g. $10^{10}$ Hz) and the absorption line registered while the magnetic field strength is varied with time.

Radicals normally do not occur in peptides but they can be generated by irradiation at low temperature with UV, electrons or X-rays, usually at the $\alpha$-carbon atom, and then can be recognized by ESR. There are some radicals that are stable at ambient temperature, for instance N-oxides with the structural element C–NO–C. In biochemistry, e.g. derivatives of 2,2,5,5-tetramethyl-pyrrolidine-1-oxide are being used as spin labels, conjugated to bioactive compounds. In this way ESR can localize the position of possible receptor sites. For instance, in hormone research the distances in the neurophysin complex between spin-labeled small peptides, models of oxytocin, have been determined [27].

### 6.2.5 Mass Spectroscopy

The method of mass spectroscopy (MS) was introduced in 1920 by F.W. Aston for the exact determination of atomic masses but routine application of mass spectrometers for the analysis of organic molecules began in only about 1960. A minimal amount of a substance is vaporized in high vacuum, at ca. $10^{-6}$ Torr, and ionized by bombardment with a beam of electrons of sufficient energy, e.g. 70 eV. These electrons shoot out one electron from each molecule thus generating positive radical molecule ions, $R^{+\bullet}$, which by the elimination of neutral parts form a series of fragment ions. In this way a spectrum of different masses is obtained. Separation of masses is achieved through acceleration in an electric field followed by deviation in a magnetic field perpendicular to the direction of the ion beam. The magnetic field forces the ions into circular pathways with different radii and thus the ions with different masses are dispersed as polychromatic light is diffracted by a prism. The separated ions are recorded in the spectrum according to their masses and intensities.

At first MS did not seem adequate for peptide research because of the poor volatility of the compounds which are zwitter-ionic and held together by hydrogen bridges. Therefore derivatization was necessary. The pioneering work of Klaus Biemann circumvented these difficulties, as early as 1959, through acetylation of the terminal amino group and reduction of the peptide bonds to poly-aminoalcohols [28] by boiling the acetylated peptide with $LiAlH_4$, or with $LiAlD_4$, for several hours (P. Karrer, 1952). In these compounds, which have a succession of —NH—CH(R)—$CH_2$—NH— group, the carbon–carbon bonds are easily cleaved on electron impact. Since the positive charge is better stabilized on the more substituted carbon atom, the fragment pertaining to the N-terminal part of the molecule, —NH—CHR—, will be somewhat more pronounced. Thus, (in an idealized instance) from reduced *N*-acetyl-leucyl-proline a mass spectrum is obtained in which ions of masses 114, 171 and 254 are prominent (Fig. 12).

An advantage of this method is the good volatility and hence ready separation of medium sized poly-aminoalcohols by gas chromatography. A mixture of small peptides, obtained by enzymatic or chemical partial hydrolysis of a polypeptide, can be reduced and separated in this way before each fraction is submitted to

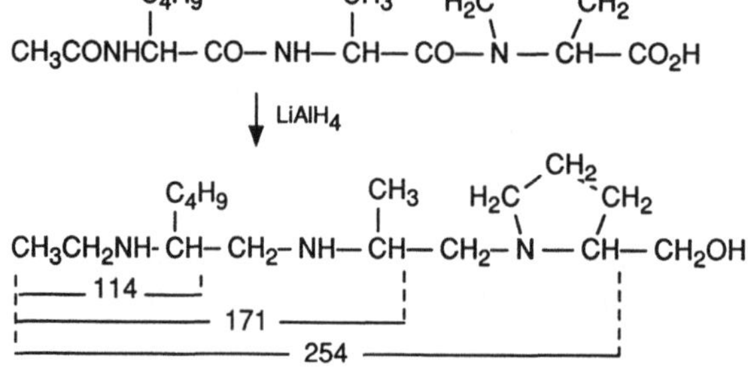


**Fig. 12.** Mass spectroscopic fragmentation scheme of reduced *N*-acetylleucylalanylproline

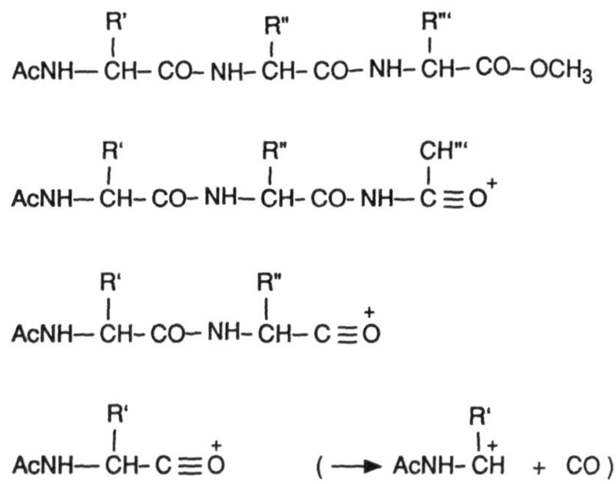

**Fig. 13.** Molecular ion and sequence peak ions of an acyl tripeptide methylester

sequence analysis via mass spectroscopy. The approach was later successfully applied to other volatile peptide derivatives as well.

At about the same time, in 1963, it was reported that simple *N*-acetyl-peptides (Heyns and Grützmacher [29]) or *N*-trifluoroacetyl peptide esters (F. Weygand et al. [30]) are cleaved under electron impact at the peptide bonds, giving rise mainly to ions corresponding to N-terminal parts of the chain. If such fission occurs at each peptide bond, a series of ions are formed all having the same N-acyl group. This is equal to a stepwise degradation of the chain from its carboxyl end (Fig. 13).

Since *N*-trifluoroacetyl peptide methyl esters are suitable for separation by gas chromatography, the procedure just described can also be applied to peptide mixtures obtained through partial methanolysis of larger peptides. This method permitted the structure elucidation of cyclic peptides such as antamanide (p. 213) and cyclolinopeptide (p. 216) by Prox and Weygand [31].

In 1967 E. Lederer (Plate 28) and B.C. Das reported [32] that the volatility of acetyl-peptide esters can be increased by replacing all dissociable protons with methyl groups. This permethylation can be performed with methyl iodide in anhydrous dimethylsulfoxide in the presence of a strong base. In *N*-methyl peptides the coherence caused by hydrogen bonds is overcome. Moreover, the fragmentation patterns are much simpler than the patterns found in *N*-acetyl peptide esters without permethylation. Fragmentation occurs almost exclusively at the peptide bonds and the mass spectra consist mainly of sequence peaks of high intensity (Fig. 14).

In the early days of the application of MS in peptide chemistry masses of at most several hundred mass-units could be investigated and the molecular ion peak $(m/e)^+$ could be weak or even missing. Chemical ionization, a technique introduced later, yields stronger peaks. It is achieved through the reaction of the molecules of the vaporized sample with ions generated in a reagent gas by electron bombardment. For instance, with methane $CH_5{}^+$ ions are generated

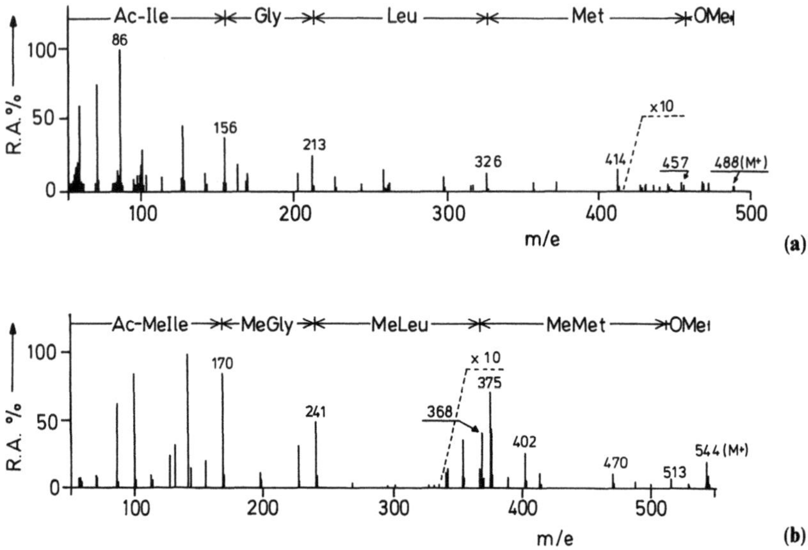


**Fig. 14a, b.** Mass spectra of acetylated and acetylated-permethylated (**a**); AlaMet AlaLeuPheGly (**b**) [32]

which protonate the molecules to give positively charged molecular ions containing an additional hydrogen atom: $(M+H)^+$.

In the field desorption technique (FD) the sample is placed on a specially prepared emitter wire and ionized in a high voltage electrostatic field. In this way little or no fragmentation occurs and mass peaks are obtained even from practically non-volatile peptides [34].

A quite novel ionization technique, fast atom bombardment (FAB) was introduced in 1980 by M. Barber et al. in Manchester. Their procedure is rather similar to secondary ion ionization (SI). In both techniques the sample in a surface layer is bombarded with accelerated particles to produce molecular ions. In SI the particles are ions, such as argon $Ar^+$, while in the FAB ionization method a beam of fast moving neutral atoms, mostly argon atoms, is used instead of ions. This beam is generated by charge exchange between accelerated argon ions, produced in an ion gun, and argon gas in a collision chamber.

$$Ar^+_{fast} + Ar \longrightarrow Ar^+ + Ar_{fast}$$

The neutral fast atom beam ionizes the sample in a soft way. Generally the efficiency of the ionization is further improved if the sample is mixed with a liquid matrix, mostly glycerol, sometimes containing NaCl. By FAB, protonated mass ions $(M+H)^+$ are produced together with negatively charged $(M-H)^-$ ions. Molecular ions and fragment ions can easily be obtained from non-volatile, thermally labile compounds of molecular masses up to several thousand, including peptides [35], with little damage to the samples. In Fig. 15 the marked

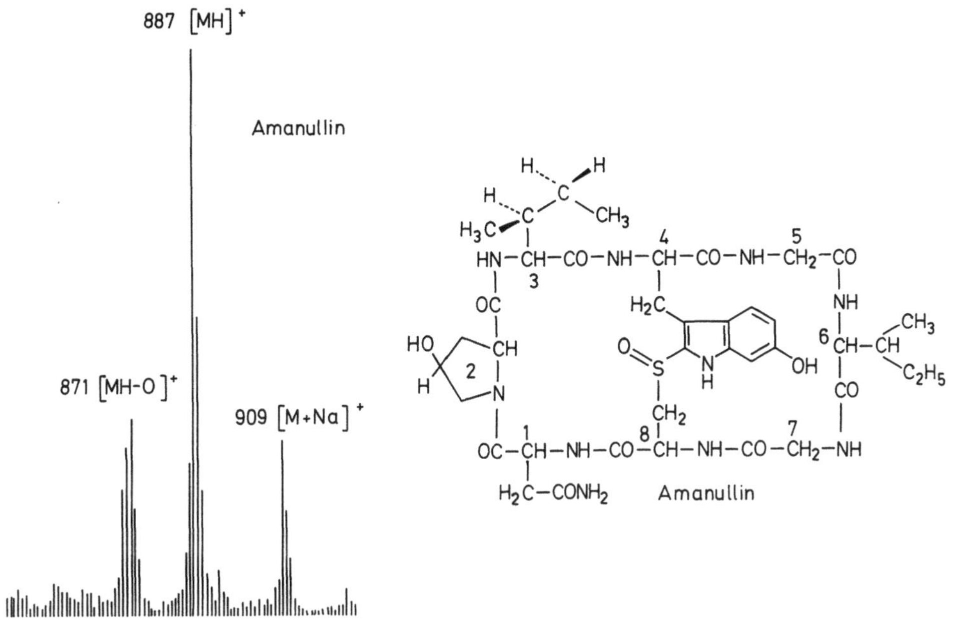


**Fig. 15.** FAB mass spectrum of amanullin (courtesy of W.A. König) and formula

$(M + H)^+$ ion of the mushroom peptide amanullin is depicted, accompanied by the S-desoxo compound $(M + H - 16)^+$ and $(M + Na)^+$ ions. The multiplets are due to natural isotopes.

The most recent methods use desorption ionization of high molecular weight samples, up to 100000 dalton, by UV laser irradiation of a matrix surface.

## 6.3 X-ray Crystallography

In 1912, Max von Laue, Walther Friedrich and Paul Knipping detected the interference of X-rays in crystals, proving thereby both the regular lattice-like arrangement of ions and the wave-nature of X rays. As early as 1913 the English physicists W.H. and W.L. Bragg used this observation for determining the structure of NaCl and ZnS. Subsequently the new method was further developed to an extremely valuable tool in chemical and biological research. A beam of X rays passing through a crystal interacts with the electrons of the atoms thus generating a spherical wave from every atom. The waves then form a pattern of diffraction maxima, reflexes of various intensities. The arrangement of the reflexes yields the dimensions of the elementary cell, while their intensities provide information on the structure of the substance. Evaluation of the diagram requires very complicated and extensive calculations, but these are now greatly facilitated by modern computer techniques. For instance, 20 years ago determination of the mathematical relation called "Fourier transform", that yields an electron-density wave for each spot, required weeks or months, yet can be done in minutes at present.

A pioneer in X-ray studies of amino acids and peptides was Robert B. Corey of the California Institute of Technology in Pasadena. In the early 1930s, with R.W.G. Wyckoff, he investigated the amide bond in simple amides such as urea and thiourea and, in 1938, determined the dimensions, interatomic distances and bond angles in the more peptide-like simplest diketopiperazine, cyclo$(Gly)_2$. The state of the peptide field, at a time when a more detailed knowledge of the structure of proteins started to emerge, has been reviewed by Corey [37]. In addition to exact data on crystals of glycine, DL-alanine and glycylglycine, conclusions were reached and extrapolations made on the probable arrangement of carbon, oxygen and nitrogen atoms along the peptide chains in proteins. Further considerations, with Linus Pauling (Plate 34) resulted in the formulation of stable configurations of polypeptide chains, including the $\alpha$-helix and the pleated $\beta$-sheet. In the next two decades a large body of more accurate structural information became available by single crystal X-ray investigations of all naturally occurring amino acids and several peptides including glutathione. These are described in a review article by R.E. Marsch and J. Donohue [38] which appeared in 1967 and which shows how little the picture had changed (since Ref. 37) with respect to the geometry of the planar trans peptide bond and the $\phi$ and $\psi$ angles in a fully extended peptide chain (cf. p. 12). In the meantime, however, crystal structure analysis was extended also to biologically interesting macromolecules: for the elucidation of the structures of myoglobin and hemoglobin J.C. Kendrew and M. Perutz were awarded the 1962 Nobel Prize. The culmination of this development, thus far, was the publication [39] of the photosynthetic reaction center of the red bacterium *Rhodopseudomonas viridis* by Robert Huber, Johann Deisenhofer and Hartmut Michel (Nobel Prize 1988). This membrane protein system consists of four protein subunits bearing the cooperative heme-, chlorophyll-, quinone-factors and the non-heme iron by which the conversion of light energy into electron shifting is directed. Eventually this brings about the synthesis of adenosine triphosphate (ATP). More than 200 protein structures have been "solved" up to now. Knowledge of their structure has provided insight into their biological functions, mainly enzymic activities. In protein crystals the spatial arrangement of the atoms is fundamentally the same as in solution and this allows conclusions on the mechanisms of interactions with other biomolecules.

With many biologically active peptides conformational identity in solid and dissolved states is not the rule. Crystal structure analysis is informative only if the molecular shapes are the same in both states. This had been assumed for insulin which posed a challenge for X-ray crystallography for many years. The structure of the hexameric Zn-complex of pig insulin was solved in Oxford, England by Dorothy Hodgkin's group (Nobel Prize 1964, Plate 21) in the years 1969 to 1972 [40]. From 1967 crystal analysis of Zn-insulin was also going on in China, and led, in the following years, to results corresponding to those reached in Oxford. In Fig. 16 a view

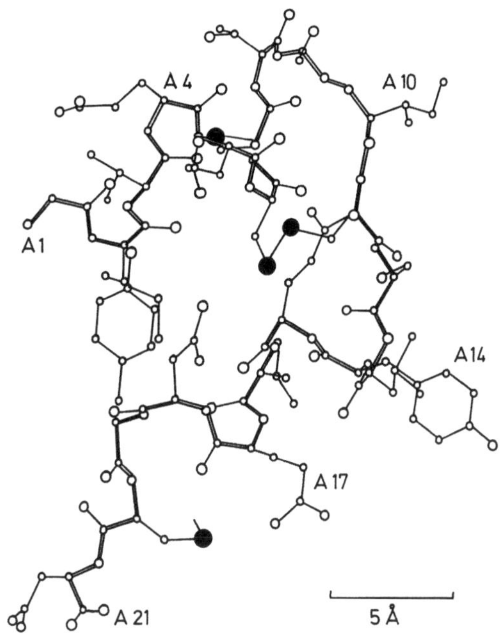


**Fig. 16.** View to the A-chain of Zn-pig insulin perpendicular to the 3-fold axis

of the A chain (21 amino acid residues), perpendicular to the 3-fold axis, is reproduced from Ref. 40.

One or more stable conformations can also be assumed for cyclic peptides which can adopt more or less rigid structures stabilized by hydrogen or other bridges or, strongly, by complexing various metal ions. As an example of many published structures [41], there is the Li-complex of the antitoxic mushroom peptide antamanide (see p. 215), a structure which was determined in 1973 by Isabella L. Karle [42] using the "direct" method of Jerome Karle (Nobel Prize 1987, shared with J.A. Hauptmann).

## References

1. F. Sanger, A.R. Coulson, The use of thin acrylamide gels for DNA sequencing FEBS Lett. 87: 107–110 (1978)
2. A.M. Maxam, W. Gilbert, A new method for sequencing DNA Proc. Natl. Acad. Sci. USA 74: 560–564 (1978)
3. D. Rittenberg, G.L. Foster, A new procedure for quantitative analysis by isotope dilution with application to the determmation of aminoacids and fatty acids. J. Biol. Chem. 133: 737–744 (1940)
4. A.J.P. Martin, R.L.M. Synge, Analytical chemistry of the proteins, Advan. Prot. Chem. 2: 1–83 (1945)
5. E.E. Snell, The microbiological assay of amino acids, Advan. Prot. Chem. 2: 85–118 (1945)
6. F. Sanger, The free amino groups of insulin, Biochem. J. 39: 507–515 (1945)
7. W. Gray, B.S. Hartley, Fluorescent end-group reagent for proteins and peptides Biochem. J. 89: 59 P (1963)
8. B. Witkop, Nonenzymatic methods for the preferential and selective cleavage and modification of proteins Advan. Protein. Chem. 16: 221–321 (1961)

9. E. Gross, B. Witkop, Selective cleavage of the methionyl peptide bonds in ribonuclease by cyanogen bromide J. Amer. Chem. Soc. 83: 1510–1511 (1961)
10. E. Abderhalden, H. Brockmann, Beitrag zur Konstitutionsermittlung von Proteinen bzw. Polypeptiden Biochem. Z. 225: 386–425 (1930)
11. P. Edman, Method for determination of the amino acid sequence in polypeptides, Acta Chem. Scand. 4: 283–293 (1930)
12. P. Edman, C. Begg, A. protein sequenator, Eur. J. Biochem. 1: 80–91 (1967)
13. R.A. Laursen, Solid-phase Edman degradation. An automatic peptide sequencer, Eur. J. Biochem. 20: 89–102 (1971)
14. see e.g. Methods in peptide and protein sequence analysis, Chr. Birr ed. Elsevier/North Holland, Biomedical Press, Amsterdam New York Oxford, 1980
15. E.R. Holiday, Spectrophotometry of proteins. I. Absorption spectra of tyrosine, tryptophan and their mixtures, Biochem. J. 30: 1795–1803 (1936)
16. G.H. Beaven, E.R. Holiday, Ultraviolet absorption spectra of proteins and amino acids, Advan. Protein Chem. 7: 320–386 (1952)
17. J.R. Spies, D.C. Chambers, Spectrophotometric analysis of amino acids and peptides with their copper salts, J. Biol. Chem. 191: 787–797 (1951)
18. N. Greenfield, G.D. Fasman, Computed circular dichroism spectra for the evaluation of protein conformation, Biochemistry 8: 4108–4116 (1969)
19. G. Holzwart, P. Doty, The ultraviolet circular dichroism of polypeptides, J. Amer. Chem. Soc. 87: 218–228 (1965)
20. For an early review see G.B.B.M. Sutherland, Infrared analysis of the structure of amino acids, polypeptides and proteins. Advan. Protein Chem. 7: 291–318 (1952)
21. Recent review: S. Krimm, J. Bandekar, Vibrational spectroscopy and conformation of peptides, polypeptides and proteins, Advan. Protein Chem. 38: 181–364 (1986)
22. T. Miyazawa, Perturbation treatment of the characteristic vibrations of polypeptide chains in various conformations. J. Chem. Phys. 32: 1647–1652 (1960)
23. T. Miyazawa, E.R. Blout, The infrared spectra of polypeptides in various conformations. J. Amer. Chem. Soc. 83: 712–719 (1961)
24. M. Takeda, O. Jardetzky, Proton magnetic resonance of single amino acids and dipeptides in aqueous solution. J. Chem. Phys. 26: 1346–1347 (1957)
25. M. Saunders, A. Wishnia, J.G. Kirkwood, The nuclear magnetic resonance spectrum of ribonuclease, J. Amer. Chem. Soc. 79: 3289–3290 (1957); O. Jardetzky, Ch.D. Jardetzky, An interpretation of the proton magnetic resonance spectrum of ribonuclease, J. Amer. Chem. Soc. 79: 5322–5323 (1957)
26. H. Kessler, W. Bermel, A. Müller, K.-H. Pook, Modern nuclear magnetic resonance spectroscopy of peptides in The Peptides Vol. 7 Acad. Press 437–473 (1985)
27. S.T. Lord, E. Breslow, Synthesis of peptide spinlabels that bind to neurophysin and their application to distance measurements within neurophysin complexes. Biochemistry 19: 5593–5602 (1980)
28. K. Biemann, F. Gapp, J. Seibl, Application of mass spectrometry to structure problems I, Amino acid sequence in peptides. J. Amer. Chem. Soc. 81: 2274–2275 (1959)
29. K. Heyns, H.F. Grützmacher, Massenspektren von N-Acetyl-peptiden einfacher Monoamino-carbonsäuren, Liebigs Ann. Chem. 669: 189–201 (1963)
30. F. Weygand, A. Prox, W. König, H.M. Fessel, Massenspektrometrische und gaschromatographische Sequenzanalyse von Peptiden, Angew. Chem. 75: 724–724 (1963)
31. A. Prox, F. Weygand, Sequenzanalyse von Peptiden durch Kombination von Gaschromatographie und Massenspektrometrie, Peptides North-Holland Publishing Company, Amsterdam (1967) 158–172
32. B.C. Das, S.D. Gero, E. Lederer, N-Methylation of N-acyl oligopeptides, Biochem. Biophys. Res. Commun. 29: 211–215 (1967)
33. P. Roepstorff, K. Norris, S. Severinsen, K. Brunfeldt, Mass spectrometry of peptide derivatives. Temporary protection of methionine as sulfoxide during permethylation FEBS Lett. 9: 235–238(1970)
34. H.U. Winkler, H.D. Beckey, Field desorption mass spectroscopy of peptides, Biochem. Biophys. Res. Commun. 46: 391–398 (1972)
35. M. Barber, R.S. Bordoli, R.D. Sedswick, A.N. Tyler, E.T. Whalley, Fast atom bombardment mass spectrometry of bradykinin and related oligopeptides, Biomed Mass Spectrometry of bradykinin and related oligopeptides, Biomed. Mass Spectrom. 8: 337–340 (198!)

36. M. Karas, U. Bah, A. Ingendoh, F. Hillenkamp, Laserdesorptions und Massenspektrometrie von Proteinen mit Molmassen zwischen 100 000–250 000 Dalton, Angew. Chem. 101: 805–806 (1989)
37. R.B. Corey, X-ray studies of amino acids and peptides, Advan. Protein Chem. 4: 385–406 (1948)
38. R.E. Marsh, J. Donohue, Crystal structure studies of amino acids and peptides, Advan. Protein Chem. 22: 235–256 (1967)
39. J. Deisenhofer, H. Michel, Das photosynthetische Reaktionszentrum des Purpurbakteriums *Rhodopseudomonas viridis* (Nobel-Vortrag), Angew. Chem. 101: 872–892 (1989)
40. T. Blundell, G. Dodson, D. Hodgkin, D. Mercola, Insulin: the structure in the crystal and its reflection in chemistry and biology, Advan. Protein Chem. 26: 279–402 (1972)
41. W. Burgermeister, R. Winkler-Oswatitsch, Complex formation of monovalent cations with biofunctional ligands, Topics in Current Chemistry 69: 91–196 (1977)
42. I. Karle, K. Karle, Th. Wieland, W. Burgermeister, H. Faulstich, B. Witkop, Conformation of the Li-antamanide complex and Na- ($Phe^4$, $Val^6$)-antamanide complex in the crystalline state, Proc. Natl. Acad. Sci. USA 70: 1836–1840 (1973)

# 7 Peptide Hormones

The development of peptide chemistry in the second half of this century is so closely related to hormone research that it appears necessary to dedicate a chapter to some historically significant accomplishments in this area.

When in 1902 W.M. Bayliss and E.H. Starling reported [1] that an extract from the jejunum of a dog stimulates the secretion of pancreatic juice, this raised many eyebrows. In the last century control of physiological action in animals was attributed to reflexes transmitted by electrical impulses through the nervous system. Thus, it was dramatically new that a *substance* can play a major role in this process. The active principle in the intestinal tissues was promptly named *secretin* (from secretion) and a few years later the word *hormone* (from the Greek *ʽορμαω* = to set in motion) was coined for the yet to be discovered messengers carrying instructions from organ to organ. In spite of arduous effort in many laboratories, it took about six decades to isolate secretin in pure form. A most interesting account of the history of secretin research was written by the investigators, J. Erik Jorpes and Viktor Mutt (Plate 23) [2], who accomplished both isolation and sequence determination of the hormone.

A few years after the discovery of secretin, a second gastro-intestinal hormone, *gastrin*, was identified by J.S. Edkins [3] in extracts of the antral part of the stomach, a substance that stimulates the secretion of the acidic digestive mixture in the same organ. Isolation of gastrin in pure form by R.A. Gregory and H.J. Tracy [4], determination of its structure [5] and synthesis of the peptide [6] concluded a long period, in fact more than half a century, of erratic research during which even the existence of gastrin has repeatedly been questioned. The related intestinal hormone cholecystokinin-pancreozymin, which carries messages to the pancreas and to the gall bladder, and the important pancreatic hormones insulin and glucagon are all participants in the interplay between peptide chemistry and hormone research. Yet, before turning our attention to these fairly complex substances, we should rather deal with two smaller and hence simpler peptide hormones, *oxytocin* and *vasopressin*, because they were the targets of efforts that led to a major breakthrough in peptide chemistry and particularly in peptide synthesis.

## 7.1 Oxytocin

### 7.1.1 Isolation

A whole series of important physiological impulses originate from the pituitary gland or neurohypophysis, a small organ attached to the base of the brain. Administration of extracts from the posterior lobe of this gland causes the release of milk, raises blood pressure and stimulates the reabsorption of water in the kidneys. A quite conspicuous effect, namely rhythmic contractions of the uterus recognized by H.H. Dale in 1906, led to the name *oxytocin* (from the Greek for quick childbirth) for the responsible active principle in the pituitary extracts. The self-explanatory designation *vasopressin* belongs to a blood pressure raising substance that turned out to be the origin of the antidiuretic effect as well. Purification of the crude extracts of dried and powdered posterior pituitary lobes of animals by Kamm [7] in 1928 produced at one point two fractions, one mainly uterus contracting (oxytocic) the other blood pressure raising and antidiuretic. The pharmaceutical company Parke-Davis, interested in the usefulness of the two hormones in medicine, provided Vincent du Vigneaud, at that time professor of biochemistry at George Washington University School of Medicine, with a substancial quantity of dry pituitary powder in the expectation that the isolated hormones will be more suitable for clinical application than the already available preparations. These were applied in childbirth and in the control of diabetes insipidus, a fairly rare condition, in which the affected people excrete large volumes of water, sometimes more than ten liters a day, and have to drink the same amount of fluids. The efforts toward the isolation of the pure hormones, started in 1933, were interrupted during the Second World War, because du Vigneaud's group, by then at Cornell University Medical College in New York City, became involved in the structure elucidation and synthesis of penicillin, a project that had to be given priority because of the war. Fortunately, the pituitary powder survived the years of storage (in dry ice) and, when the project was resuscitated, yielded, through purification by chromatography and electrophoresis, very potent preparations of oxytocin and vasopressin. The most striking results were achieved, however, with a new method of separation: countercurrent distribution. This sophisticated process of liquid-liquid extraction developed by L.C. Craig at the Rockefeller Institute for the purification of antimalarial drugs (see p. 55) was made known to the Cornell researchers (in the next city block) prior to publication. They immediately adopted the method for the isolation of oxytocin. The material obtained by countercurrent distribution in a system of secondary butanol-acetic acid-water was homogeneous as shown by the good agreement between calculated and experimentally found distribution curves [8]. Moreover, the method afforded the hormone in amounts unheard of until then. Several hundred milligrams of pure oxytocin became available, more than enough for the elucidation of its structure. It is noteworthy that the hormone was also secured in the form of a crystalline salt with flavianic acid [9].

### 7.1.2 Structure

Hydrolysis of a small sample of the peptide with hydrochloric acid at elevated temperature and examination of the hydrolysate by means of paper chromatography revealed that the molecule of oxytocin is built of eight amino acids: cystine, aspartic acid, glutamic acid, glycine, isoleucine, leucine, proline and tyrosine. It was necessary, however, to complete this information with the determination of the molar ratios of the constituents. Fortunately during the years when the oxytocin studies were in progress at Cornell, a major effort was directed toward the quantitative determination of the amino acid composition of peptides and proteins. This endeavor, started by Max Bergmann at the Rockefeller Institute in the nineteen forties (see Chap. 3) was first based on the selective precipitation of amino acid salts with carefully chosen acidic reagents. Bergmann's associates, Stanford Moore and William H. Stein, further advanced the methods of analysis and applied chromatography, initially with a starch column, later on ion-exchange materials as described on p. 50 in Chap. 3. Chromatography showed that oxytocin is built of two half cysteine residues and one each of aspartic acid, glutamic acid, glycine, isoleucine, leucine, proline and tyrosine. Three moles of ammonia were also released during hydrolysis indicating that both aspartic acid and glutamic acid were present not as such as in the parent molecule but rather in the form of their amides, asparagine and glutamine and also that the chain ends in a carboxamide and not in a free carboxyl group.

The N-terminal residue, the amino acid with a free, unacylated amino group, was determined with the help of dinitrophenylation method of Sanger [10] (see p. 115) that was applied in the study of insulin with considerable success. In several additional aspects the oxytocin study was helped by the methods used in the structure elucidation of insulin. Thus oxytocin was oxidized with performic acid in order to diminish the complication caused by the disulfide bridge. A single compound resulted from the oxidation, a chain of nine amino acids. Dinitrophenylation with 2,4-dinitrofluorobenzene followed by hydrolysis yielded only one amino acid that has been tagged with the dinitrophenyl group, cysteic acid

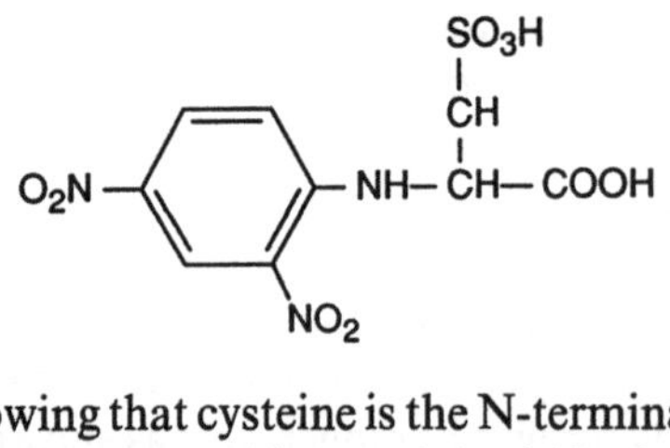


showing that cysteine is the N-terminal residue of the nonapeptide. The chain was then fragmented by partial acid hydrolysis, the fragments dinitrophenylated and examined. Reconstruction of the parent sequence from the information gained through the analysis of fragments was aided by an alternative method of cleavage. Treatment with bromine water caused the fission of the chain of the oxidized molecule. Of the two products formed one contained only cysteic acid and a transformation product of tyrosine; the other segment, a heptapeptide; was analyzed through partial hydrolysis and dinitrophenylation of its smaller fragments. Similar degradations were carried out on a sample of oxytocin desulfurized with Raney nickel. In this material the cysteine residues were converted to alanine residues; therefore a different set of peptides were secured by partial acid hydrolysis. The partial sequences were combined (on paper, in the manner of jigsaw puzzles) to form the structure of oxytocin. Last but not least the, at that time, new method of sequential degradation, the Edman procedure [11], (p. 119) was applied to a sample of the performic-acid-oxidized material. The results were in harmony with the conclusion drawn from the analysis of fragments produced by partial acid hydrolysis. Hence, in 1953, du Vigneaud, Ressler and Trippet [12] proposed the following structure for oxytocin:

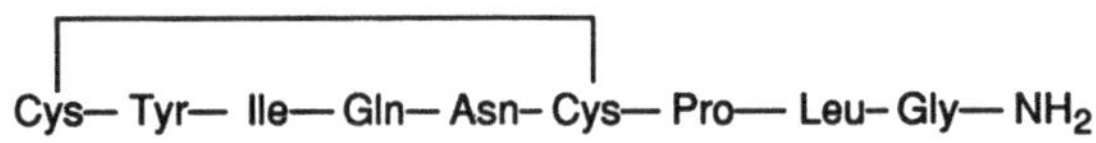


The disulfide bridge between the two cysteine residues appears more realistically if the structure is drawn in the manner of organic chemists:

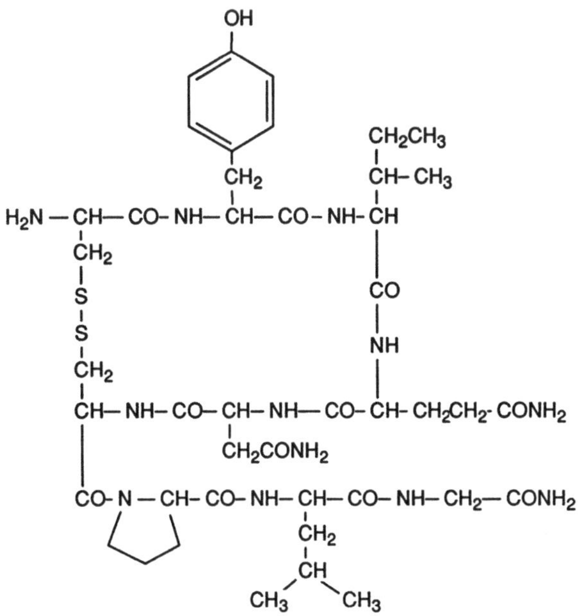

**Fig. 1.** Structure of oxytocin

At the time of the conclusion of this study the elucidation of the structure of a nonapeptide was regarded as a major achievement. It could be realized only because of the advances made in the methodology of peptide research during the preceding years, for instance purification by Craig distribution or sequential degradation by the Edman method. The approach that yielded the even more complex structure of insulin [13], namely partial acid hydrolysis followed by determination of the sequence in the hydrolytic fragments through dinitrophenylation, was instrumental in the elucidation of the structure of oxytocin as well. It is not too surprising, that H. Tuppy, who participated in the insulin structure determination in Sanger's laboratory in Cambridge, England, on his return to Vienna, attacked the problem of oxytocin. Simultaneously with the du Vigneaud publication he proposed the same structure for the hormone [14]. His work was based on partial acid hydrolysis and on selective hydrolysis with a bacterial enzyme. These studies were carried out by remarkably simple means, with paper chromatograms and electropherograms as the sole analytical tools. Yet, the conclusions were based, in part, on some details of the chemistry of the hormone molecule elaborated and published by the investigators at Cornell.

### 7.1.3 Synthesis

The traditions of organic chemistry more or less require that a structure derived through degradation be confirmed by synthesis. The art of peptide synthesis was, however, rather limited in 1953 and it was far from obvious how a ring containing nine amino acid residues should be constructed. The task appeared fairly formidable at that time. The inherent difficulties were, fortunately, greatly reduced by the results of a series of experiments [15] in which the ring of the molecule was opened up through reduction, the thiol groups benzylated, the biologically inactive dibenzylderivative debenzylated and oxytocin regenerated by oxidation:

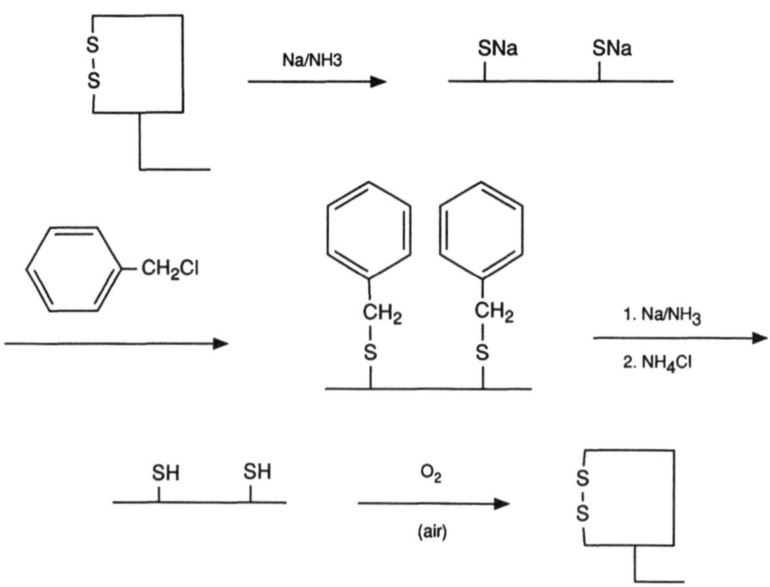


Reduction, S-benzylation, de-benzylation and oxydative recyclization of oxytoxin

Scheme 1.

The recovered material had full hormonal activity and was shown to be identical with the sample of oxytocin in all respects, including its ability to form a crystalline flavianate. Although the yield in terms of recovered biological activity was rather modest, the results provided the encouragement needed for undertaking the synthesis. It became clear that, if the open chain **S,S'**-dibenzyl derivative (dibenzyloxytocein) can be secured, the road will be free for the concluding steps leading to the cyclic molecule of the hormone.

At this point it is interesting to note, that the method of reducing disulfides with sodium in liquid ammonia [16], benzylation of the thiol groups in situ and their debenzylation in the same solvent [17] had been elaborated by du Vigneaud many years earlier (cf, p. 71) as if some intuition would have directed his efforts toward the chemistry of oxytocin. An account of this continued interest in sulfur containing natural products can be found in his book "A Trail of Research" [18].

In the existing know-how of the field of peptide synthesis only limited guidance could be found and designing a scheme for the synthesis of oxytocin was a major challenge. No similar undertaking had been published up till then. Peptides obtained by synthesis were either much smaller or of fairly simple amino acid composition. Yet, the partially blocked dipeptide *N*-benzyloxycarbonyl-*S*-benzyl-L-cysteinyl-L-tyrosine, the N-terminal sequence, had already been pre-

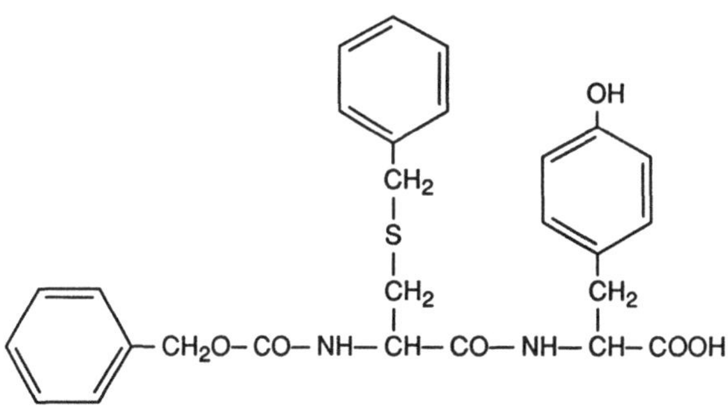


pared by Harington and Pitt Rivers [19] and their procedure was applied in the synthesis of oxytocin. The C-terminal dipeptide portion of the nonapeptide chain was assembled with the help of a novel method of peptide bond formation, the mixed (or unsymmetrical) anhydride process, discussed in detail in Chap. 4. Synthesis of the dipeptide derivative benzyloxycarbonyl-L-leucyl-glycine ethyl ester

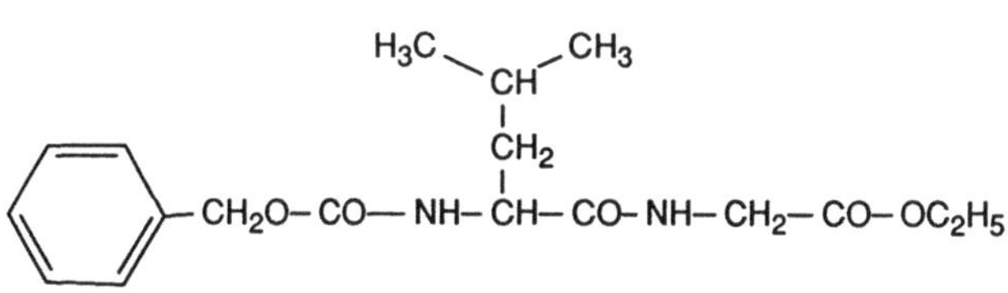


had just been reported by Vaughan and Osato [20] and the Cornell groups applied the same approach for the first synthesis of oxytocin; activation in the form of isovaleric acid mixed anhydridide. The product was partially deblocked by catalytic hydrogenation and the dipeptide ester acylated with benzyloxycarbonyl-L-proline, again by the isovaleric acid mixed anhydride method. Unmasking the amino group of the tripeptide derivative by catalytic hydrogenation opened the way for the synthesis of the partially blocked tetrapeptide, *S*-benzyl-L-cysteinyl-L-prolyl-L-leucylglycinamide, the C-terminal segment:

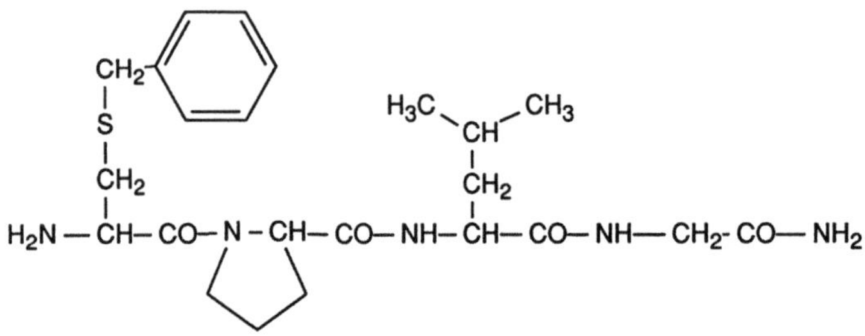

The actual preparation of the tetrapeptide amide required several steps, among them acylation with di-benzyloxycarbonyl-L-cystine, activated in the form of acid chloride, removal of the benzyloxycarbonyl group by reduction with sodium in liquid ammonia, rebenzylation of the thiol group, conversion to benzyl ester and then to the amide. The tetrapeptide amide was secured in crystalline form.

The central part of the molecule, comprising isoleucine, glutamine and asparagine, posed a yet unsolved problem: the incorporation of glutamine into the peptide chain. This was solved by a novel approach elaborated in connection with the oxytocin project. *p*-Toluenesulfonyl-L-glutamic acid was converted to the acid chloride of *p*-toluenesulfonyl-pyroglutamic acid and the latter used for the acylation of asparagine. Treatment of *p*-toluenesulfonyl-L-pyroglutamyl-L-asparagine with ammonia yielded the blocked derivative of glutaminyl-asparagine, which was acylated, in turn, with the similarly protected acid chloride of L-isoleucine. This afforded the central segment of the molecule, *p*-toluenesulfonyl-L-isoleucyl-L-glutaminyl-L-asparagine:

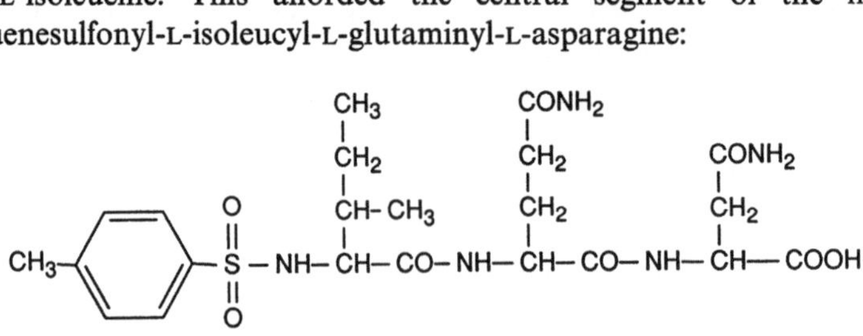

The initially planned 2 + 3 combination (linking the N-terminal dipeptide segment with the central tripeptide) ran into technical difficulties and therefore the plan was changed and the central portion of the molecule attached to the C-terminal tetrapeptide segment to give a heptapeptide derivative. The heptapeptide amide was partially deblocked and acylated with the N-terminal dipeptide.

For the combination of segments according to this 3 + 4 = 7; 2 + 7 = 9 scheme, the tetraethyl pyrophosphite method, which had just been developed, (G.W. Anderson et al.) [21] was applied. The blocked nonapeptide was secured in low

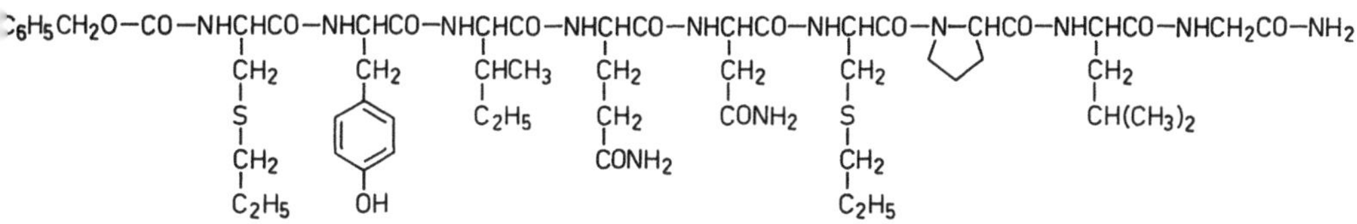

yield, but after removal of the protecting groups in a single step of reduction with sodium in liquid ammonia, oxidation with air afforded the desired cyclic disulfide. Countercurrent distribution in the solvent system used for the isolation of the natural product proved itself once again to be of great practical value: the material recovered from the tubes corresponding to the peak in the distribution curve was fully active and could not be distinguished from natural oxytocin.

In retrospect it might appear that the first synthesis of oxytocin, which was also the first synthesis of a peptide hormone, was not too efficient. The overall yield in the chain building process and conversion of the blocked nonapeptide intermediate to the biologically active peptide was well below one percent of the calculated. Certain shortcomings in the various operations were evident from the amount of biological activity generated by synthesis. Thus, if the blocked nonapeptide intermediate had been a single compound and deprotection followed by cyclization had been quantitative, then the 360 mg blocked material used in these steps should have yielded 275 mg oxytocin. Since the best preparations of the hormone had about 500 units per mg potency in the avian depressor assay, a total of 137,500 units of activity should have been obtained, but only 29,000 units were found in the crude product. This lack of perfection, however, could not diminish the enthusiasm felt by the readers of the report of du Vigneaud and his associates [22] on "The Synthesis of an Octapeptide Amide with the Hormonal Activity of Oxytocin". (At that time the two half-cysteine residues were regarded as one amino acid, cystine; today a nonapeptide would be mentioned.) The synthesis of oxytocin was a real breakthrough in peptide chemistry. It was impressive to note that no effort was spared in the purification and characterization of the synthetic material. Beyond the determination of potency, specific rotation and chromatographic-electrophoretic mobilities, the experiments extended to quantitative amino acid analysis of the final product and to the determination of the molecular weight in the ultracentrifuge. A sample was oxidized with performic acid and then treated with bromine water: it showed the fragmentation observed with the natural product. Even the ability of the synthetic peptide to form a crystalline flavianate was checked. The melting point of the flavianate was the same as that of the salt obtained from natural oxytocin and no melting point depression occurred when the two flavianate samples were mixed. This careful comparison of the synthetic material with the isolated product was a characteristic feature of the research carried out in du Vigneaud's laboratory: no stone was left unturned. A year after the publication of the oxytocin synthesis in detailed form [23] this epochmaking contribution, a real milestone in the history of peptide synthesis, was recognized by the Nobel Prize [24].

Synthesis of oxytocin was soon attempted in several laboratories. In 1955, Boissonnas and his group [25] in Basel, Switzerland, and in 1956, Rudinger and his associates [26] in Prague, Czechoslovakia, reported overall yields of about 5%. This was substantially higher than the yield achieved in New York. In both instances mixed anhydrides were used for the preparation of segments and the classical azide procedure of Curtius [27] (p. 26) for their combination into the molecule of the hormone. Further improvement both in yield (about 10%) and in the quality of the blocked nonapeptide intermediate was reported in a modified

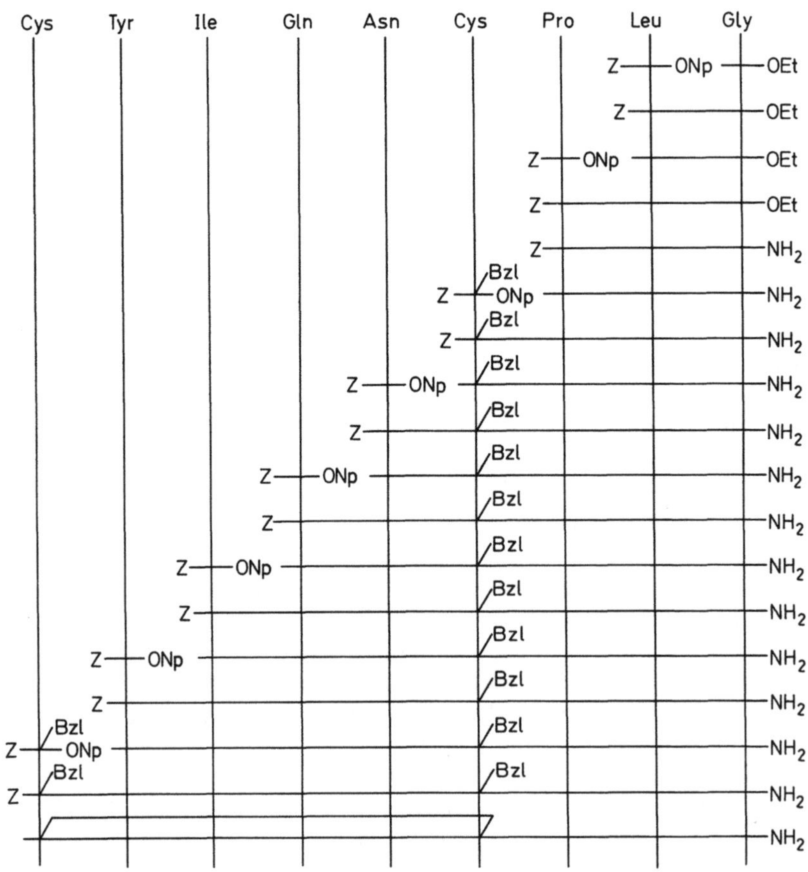


**Scheme 2.** Stepwise synthesis of oxytocin [29] (ONp = *p*-nitrophenyl ester)

synthesis of the hormone by Bodanszky and du Vigneaud [28], an approach in which the originally planned segment condensation scheme ($2 + 3 = 5; 5 + 4 = 9$) was put to test. In the third oxytocin synthesis [29] in New York, by Bodanszky and du Vigneaud in 1959, a more significant change in methodology was tried out. The protected nonapeptide was built by stepwise chain-lengthening starting at the C-terminal residue and proceeding through the incorporation of single amino acid residues in the form of their active esters. The new strategy [30] (Scheme 2) afforded oxytocin in 38% overall yield.

It might be unrewarding to enumerate the long series of synthetic variations leading to oxytocin. The molecule of the hormone became the testing ground of new methods, for instance of the final removal of blocking groups with liquid hydrogen fluoride by Sakakibara and Shimonishi [31]. Also, one of the early applications of solid phase peptide synthesis (cf. Chap. 5) was the preparation of oxytocin through the strategy of stepwise chain-lengthening starting with the C-terminal residue that was attached to an insoluble polymeric support [32].

### 7.1.4 Studies of Relationships between Structure and Biological Activity

Once the synthesis of a biologically active peptide is well in hand it becomes possible to explore the role of the individual amino acid constituents in the interaction between the agonist and its receptor. Replacement of one of the residues by a judiciously selected other amino acid provides an analog of the hormone, which can be more active or less potent than the parent compound. From such changes in potency important conclusions can be drawn on the contribution of certain structural features to biological activity. It seems impractical to enumerate the analogs of oxytocin synthesized for such studies. We must limit our account to a few examples and will try, therefore, to describe compounds which provided information transcending the immediate area of oxytocin.

An inspection of the structure of oxytocin reveals only two functional groups: a free amino group in the N-terminal cysteine residue and a phenolic hydroxyl in the side chain of tyrosine, the second residue in the sequence. The latter was the first target of structure-activity studies because replacement of the hydroxyl group by hydrogen could readily be accomplished by repeating the synthesis of oxytocin but incorporating phenylalanine instead of tyrosine. The thus modified hormone [33, 34] was biologically active, yet its potency, both on isolated uteri of guinea pigs and in the avian depressor assay, was only about one sixth of the potency of oxytocin. It is clear from these results that the phenolic hydroxyl is not essential for activity, but it contributes to the affinity between hormone and receptor. Of course it became even more intriguing to examine the role of the free amino group. Synthesis of desamino-oxytocin was achieved [35] by incorporation of (*S*-benzyl) 3-mercaptopropionic acid rather than cysteine at the conclusion of the chain-lengthening procedure. The new analog turned out to be *more* potent than oxytocin itself: the free amino group plays no role in the hormonal activity of the peptide. The increase in potency is probably due to the enhanced stability of the analog under physiological conditions. In contrast to the parent compound it can not be degraded by aminopeptidases. It is interesting to note that desamino-oxytocin can be dried at elevated temperature without loss in potency and that it is readily crystalized from acetic acid [36]. In view of the results of earlier experiments in which methylation of oxytocin caused inactivation, the activity of the desoxy and desamino derivatives was quite surprising. Clearly, synthesis of analogs is distinctly superior to modifications performed on the natural product: the role of a functional group is better determined if it is replaced by hydrogen rather than by an alkyl or acyl group.

The remarkable finding that neither the phenolic hydroxyl nor the free amino group is essential for hormonal activity, directed the attention to the general architecture of the molecule of oxytocin, to its cyclic structure and to the disulfide bridge. The importance of the size of the ring was tested with the help of analogs in which asparagine or glutamine was replaced by isoasparagine [37] or isoglutamine [38] respectively and also through the replacement of the cysteine residues with homocysteine [39]. The virtual absence of activity in these compounds showed that the ring formed of twenty atoms is a rigid requirement. Yet, the cyclic portion of the molecule in itself was also devoid of biological activity [40]. Thus, a further intriguing possibility had to be investigated, the role of the disulfide in the interaction of the hormone with its (hypothetical) receptors. It was conceivable that the disulfide is reduced in the tissues and the resulting sulfhydryl groups are involved in binding. Such speculations found support in prior experiments by Sealock and du Vigneaud [41] in which oxytocin was reduced by treatment with excess cysteine, in aqueous solution at pH 7.5, without any loss in activity. The activity remained unchanged on reoxidation, but was lost if the sulfhydryl groups were benzylated. In later studies, aimed at the examination of the role of the sulfhydryl groups in binding, oxytocin labeled with a radioactive sulfur isotope was applied [42]; radioactivity was incorporated into the target tissues and could be displaced from them by excess thiols. Thus, participation of the disulfide in the triggereing of hormonal action seemed confirmed. Yet, the conclusions drawn from these studies were quite erroneous. This was clearly demonstrated in the definitive work of Rudinger and Jost [43] who synthesized an oxytocin analog in which one of the sulfur atoms of the disulfide bridge was replaced by a methylene group. The thioether group in the new analog could not participate in binding in the way proposed, that is in formation of new disulfides between hormone and receptor. Nevertheless, the carba-analog had high potency in the various assays. This unambiguous evidence allowed only one interpretation: the disulfide grouping in oxytocin is an architectural component, that provides for a well defined geometry, but does not participate directly in the biological effects. The dicarba-analog, in which both sulfur atoms are replaced by $CH_2$ group has also been synthesized [43] and its high potency further corroborated the results obtained with the carba analog.

At this point it might be appropriate to mention that it was not fully justified to attribute a direct biological role to the disulfide. The biological activity of reduced oxytocin is obviously due to reoxidation in the blood or in the target tissues. When the experiments with the reduced form (oxytocein) were carried out on uterus strips suspended in glucose containing solution and nitrogen rather than air was bubbled through the bath, then only a fraction of the original activity could be observed. The misleading results obtained with radioactive labeling had no lasting effect: peptide chemists turned their attention to the significance of geometry. It might be worthwhile to recall a remark of du Vigneaud, made many years earlier in his book "A Trail of Research" (p. 21): "Some investigators have been inclined to conclude that one or two disulfide bridges have a special function in insulin. Our own tendency was, and still is, to regard the architecture of the molecule as a whole as the important factor with regard to its hypoglycemic action".

The experiments of Rudinger and his associates and the results obtained with the help of the synthetic carba-analogs of oxytocin shed light on an important aspect of peptide and protein chemistry. Disulfide bridges, in general, are structural elements that give final definition to the geometry of biologically active peptide chains. The fundamental determination of the conformation lies in the sequence of the constituent amino acids. For instance, reduction of the four disulfide bridges in ribonuclease followed by destabilization of the secondary structure through the addition of urea, removal of the urea by dialysis and reoxidation resulted in the reconstitution of the active enzyme molecule [44]. Thus, the amino acid sequence, the primary structure, defines the geometry of the chain, at least to a considerable extent; otherwise a large number of different molecules would have been generated by various combinations of the sulfhydryl groups. In the smaller molecule of oxytocin, definition of the geometry is less pronounced and unless oxidation is carried out in very dilute solution appreciable amounts of dimers (a parallel and an antiparallel dimer) are produced and some oligomers and polymers as well. Factors which affect the conformation have a certain influence on the rate and outcome of the process. For example, it requires more time or oxidants which are more powerful than air to generate the ring of desamino-oxytocin. On the other hand cyclization occurs readily in lysine vasopressin by oxidation with air and the hormone is obtained in fairly pure form, practically without dimers.

Our understanding of the role of disulfide bridges in peptides and proteins considerably improved through the studies of the "baby protein" oxytocin, but neither the rigidity provided by the disulfide bridge, nor the conformation determining effect of the sequence were sufficient for final conclusions on the 3-dimensional geometry of the hormone. Some details of its architecture were revealed by the temperature dependence of NMR spectra [45] (cf. p. 125). The intramolecular hydrogen bonds recognized in these studies were confirmed by X-ray crystal structure analysis [46] (cf. p. 131). These investigations, however, performed on desamino oxytocin rather than on the natural hormone, might not be valid for unaltered oxytocin itself. Furthermore, two different conformation were recognized in the molecule of desamino-oxytocin even in its crystalline form and this leaves doubts about the geometry of the molecule in dilute solution or on the surface of its receptor. Thus, a full clarification of the biologically active conformation of oxytocin remains an unmet challenge.

It can not be our purpose to give an exhaustive account of the analogs of oxytocin synthesized so far. They are too numerous for this volume and were prepared to answer too many questions. As interesting examples the analogs, in which the carboxamide groups in the asparagine and glutamine side chains were replaced by hydrogen, can be mentioned [47]. Pharmacological tests on these compounds proved that only one of these carboxamide groups is essential for hormonal activity: in the analog in which glutamine was replaced by $\alpha$-aminobutyric acid, a major portion of the potency was retained, but replacement of asparagine by alanine resulted in a virtually complete loss of efficacy. Most analogs are less potent than the parent hormone, yet a notable exception was found in desamino-hydroxy-oxytocin [48].

Certain alterations in the sequence resulted in oxytocin derivatives which inhibit rather than stimulate the characteristic hormonal activities. Thus, glycyl-oxytocin, a compound that contains a glycyl residue attached to the N-terminal amino group of the parent molecule, is an antagonist, albeit a weak one [49]. More pronounced inhibition was shown by the analog in which the tyrosine residue is replaced by its *O*-methyl derivative [50]. The most conspicuous inhibition found so far was in analogs with 2-alkylcysteine residues, such as penicillamine (2,2-dimethylcysteine) [51] or 3-mercapto-3,3-pentamethylene-propionic acid [52] at their N-termini. Also, analogs could be designed with enhanced selectivity. For instance the uterus contracting effect is retained by 4-threonine-7-glycine oxytocin [53] while it has practically no pressor or antidiuretic effect. Oxytocin itself has a weak but measurable affinity to the vasopressin receptors.

The analogs discussed so far reveal certain features of the various oxytocin receptors. The molecule binds, however, not merely to the receptors in the target organs, but also to a carrier protein, originally known as the van Dyke protein [54] that was renamed neurophysin after its isolation in pure form. It is rather remarkable that for binding of oxytocin to this protein the free N-terminal amino group, which is not needed for activity, is quite essential [55].

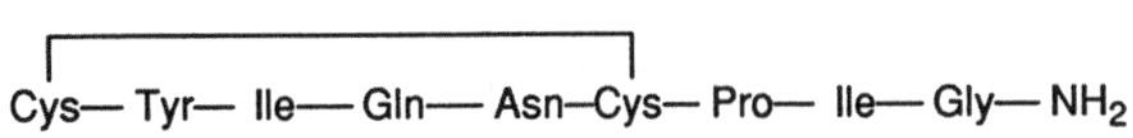


Mesotocin

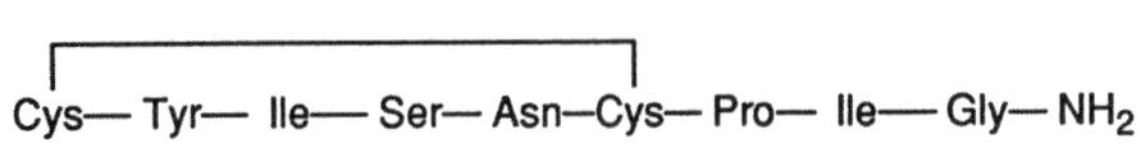


Isotocin

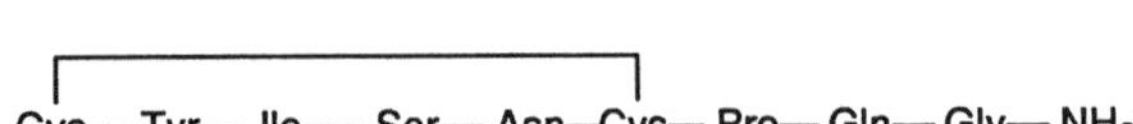


Glumitocin

Even this brief discussion of oxytocin analogs can not be concluded without pointing out that Nature experimented with hormone analogs well before organic chemists undertook the synthesis of such compounds. Thus, 8-isoleucine oxytocin (mesotocin) was discovered in the frog [56], 4 serine-8-isoleucine oxytocin (isotocin) in teleost fishes [57] and 4-serine-8-glutamine oxytocin (glumitocin) in the ray [58].

It appears that oxytocin might become the subject of many more investigations in the future. This should certainly be the case if some of its interesting physiological effects turn out to be useful in medicine. The significance of the hormone might indeed transcend its application in childbirth. In this connection it is noteworthy that oxytocin initiates the spawning reflec in killfish [59], prevents digitalis-induced arrhythmia in dogs [60], increases the life span of male rats [61] and has antiarrhythmic action in anesthetized man [62].

## 7.2 Vasopressins

In the course of the isolation of oxytocin, the second principal hormone of the pituitary gland, vasopressin, a peptide with pressor and antidiuretic effects, was similarly secured in pure form. Porcine glands yielded a vasopressin that contained lysine, while an arginine containing peptide was obtained from the neurohypophysis of cattle. Structure elucidation [63, 64] revealed that the two vasopressins are quite analogous to oxytocin, except that isoleucin in position 3 of oxytocin is replaced by phenylalanine and position 8 is occupied by one of the basic amino acids, lysine or arginine, rather than by leucine:

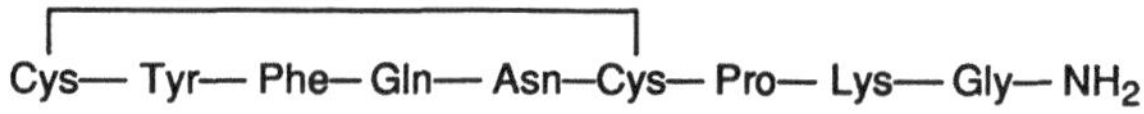


Lysine vasopressin

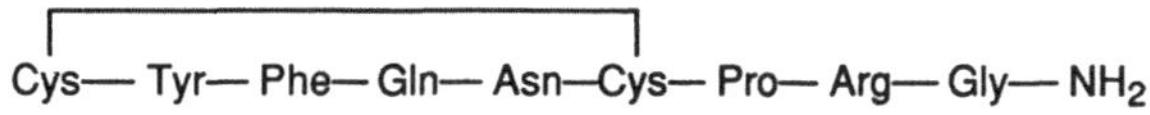


Arginine vasopressin

Lysine vasopressin is the less commonly found of the two peptides, but it does occur in species other than the hog, for instance in the hippopotamus. Arginine vasopressin is the hormone of many species, including man.

The physiological significance of vasopressin is beyond doubt. Unlike in the case of oxytocin for which no deficiency syndrom is known, the lack of vasopressin is the cause of a rather unpleasant condition, *diabetes insipidus*. In persons afflicted by it, the excessive excretion of water was controlled earlier by the administration of dried pituitary powder as a snuff, because the hormone is absorbed through the nasal mucosa. With the availability of pure vasopressins this therapy became more convenient: a solution of the peptide is applied in the form of nasal spray. Synthesis of vasopressins in several laboratories [65–72] opened the way to large scale manufacturing of preparations for the control of *diabetes insipidus* in patients and made the pharmaceutical industry independent of the availability of the glands of animals. A not less important consequence of the development of advanced methods by synthesis was the preparation of numerous vasopressin analogs.

As in the case of oxytocin, the analogs of vasopressin were designed to provide information on the relations between structure and hormonal activities. The presence of a basic amino acid in position 8 in both vasopressins suggested that electrostatic interaction might play a role in binding the hormone to its specific receptors. The amino group of the lysine side chain or the guanidino group in arginine should provide the cation of an ion pair while the anion might be the deprotonated carboxyl group of an aspartyl or a glutamyl residue at the active site of the receptor molecule. This assumption was tested with the help of vasopressin analogs in which the side chain amino group of lysine was replaced by a methyl or by a hydroxyl group. Only a small fraction of the original pressor and antidiuretic effects was found in 8-homonorleucine vasopressin [73] and only a modest potency was exhibited by 8-ε-hydroxynorleucine vasopressin [74]. These results support the ion-pair concept. The methyl group in homonorleucine is a poor substitute for the amino group in the lysine side chain. The hydroxyl group in the hydroxynorleucine analog appears to contribute to binding, probably through a hydrogen bond, but can not fully compete with the ion-forming amino group. In a similar vein the guanidino group in arginine vasopressin was replaced by the ureido group. The latter is similar to the guanidine moiety both in size and shape, but does not form salts with weak acids. The citrulline analog [75] retained the hormonal activities of the parent compound but its potency was a magnitude lower. Binding through an ion-pair is once again indicated. On the other hand, the length of the side chain to which the cation forming group is attached to the peptide backbone is less important: 8-ornithine vasopressin [70, 76] and 8-homolysine vasopressin [77] are nearly as potent as lysine vasopressin.

In the vasopressins, like in oxytocin, the N-terminal amino group is not required for hormonal activity. In fact, 1-desamino-8-D-arginine vasopressin is applied in medicine because of its long-lasting effect. In the absence of a free amino group at the N-terminus this peptide is not degraded by aminopeptidases and the replacement of L-arginine by the same amino acid with D-configuration renders the chain resistant to the action of trypsin as well. Hence, the analog remains intact in the tissues longer than the parent hormone. Selective agonists which show only pressor or only antidiuretic activity are challenging objectives, while antagonists of the antidiuretic effect should provide medicine with a valuable diuretic agent. A systematic and continued study aiming at the design of

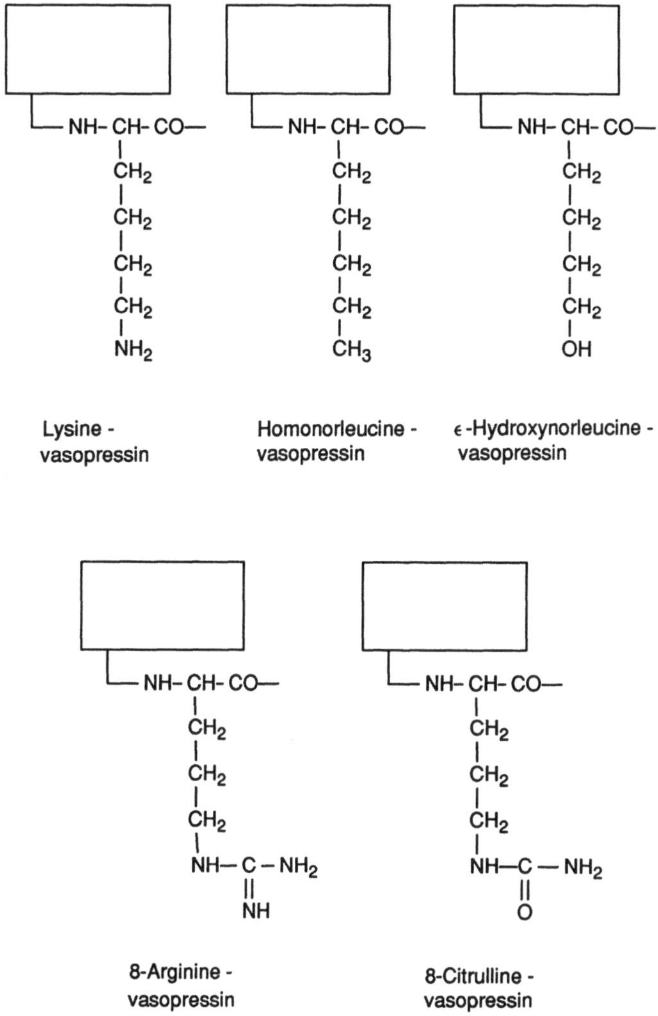


**Fig. 2.** Some vasopressin analogs

such selective analogs of the pituitary hormones in M. Manning's laboratory in Toledo, Ohio, led to sophisticated inhibitors, such as the vasopressin analog in which the N-terminal cysteine residue was replaced by 2-(1-mercaptocyclohexyl)-acetic acid, tyrosine in position 2 by D-phenylalanine, glutamine by isoleucine and the C-terminal glycinamide by D-alaninamide [28].

Equally exciting observations revealed that vasopressin acts on the central nervous system: it has a beneficial effect on memory. It is perhaps even more striking that some metabolic products of arginine vasopressin, such as

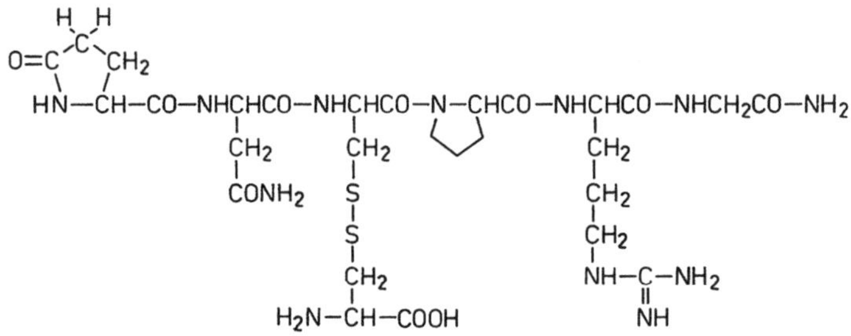

show a similarly selective memory effect [79].

Prompted by curiosity, Katsoyannis and du Vigneaud synthetized [80] in 1958, a peptide with the ring of oxytocin and the side chain of vasopressin. The hybrid molecule, *vasotocin*, had the biological properties of both parent hormones. Soon after its synthesis the peptide was detected in extracts from the pituitary glands of birds and reptiles and later in the pineal gland of mammals, including man.

Lysine-vasotocin

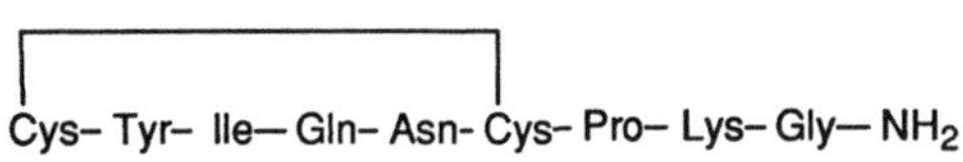


Arginine-vasotocin

It is conceivable, that vasotocin is the ancestor of the two principal pituitary hormones.

## 7.3 Two Exponents

As described at the beginning of this chapter, progress in the art of peptide synthesis in the second half of this century was very much stimulated by the recognition of the peptide nature of several hormones. An outstanding pioneer in this area was Vincent du Vigneaud, whose contributions and personal style justify a few retrospective paragraphs. Another scientist of great influence on the development of peptide chemistry in this new epoch was Joseph Rudinger. His remarkable role in the history of peptide chemistry should also be remembered.

### 7.3.1 Vincent du Vigneaud (1901–1978)

The son of a machine designer and inventor, was born in Chicago, in 1901. At the age of 17 he entered the University of Illinois in Urbana to study chemical engineering, but impressed by the lectures of C.S. Marvel he changed to organic chemistry and stimulated by the research of H.B. Lewis he decided to work in the area of sulfur containing compounds. When W.C. Rose, returning from a visit in Toronto, gave an account of the recent discovery of insulin (Banting and Best, 1922), the young student turned his interest to the study of this uniquely important molecule. Thus, after graduation he joined Professor J.R. Murlin for graduate studies at the University of Rochester, New York, to work on the chemistry of insulin. His investigations revealed that the alkali-labile sulfur in insulin belongs to cystine residues and in his dissertation entitled "The sulfur in insulin" there is a suggestion that insulin is a peptide.

With his fresh Ph.D. degree, du Vigneaud moved to Baltimore to work with J.J. Abel as a National Research Fellow. It was Abel's laboratory where insulin was first obtained in crystalline form. Here the young researcher found the opportunity he sought; he could finally study the chemistry of the hormone. In collaboration with Hans Jensen and Oskar Wintersteiner, he showed that insulin contains several amino acids, that it is a protein or large peptide. However, when his fellowship was extended for a second year, du Vigneaud traveled to Dresden,

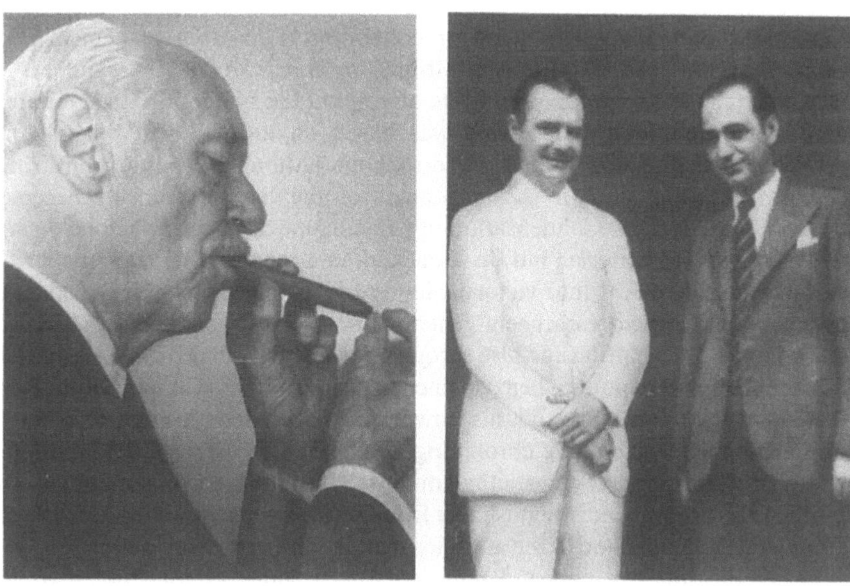

**Plate 6.** V. du Vigneaud (1901–1978) and with L. Zervas in 1934

Germany, to work with Max Bergmann, the most distinguished peptide chemist at that period. There he collaborated with a young researcher from Greece, Leonidas Zervas (cf. Chap. 3, p. 44). Before returning to the U.S. he spent shorter periods with G. Barger in Edinburgh, Scotland and Charles Harington in London.

In the U.S. du Vigneaud joined his alma mater, the University of Illinois in Urbana, with an appointment in the Department of Physiology, but a few years later, in 1932, he accepted an invitation to head the Department of Biochemistry at the George Washington School of Medicine in Washington D.C. and remained in this position until 1938, when a still more tempting invitation called him to Cornell University Medical College in New York City as Professor and Chairman of the Department of Biochemistry. In 1967 he retired from Cornell in New York and joined the Department of Chemistry of Cornell in Ithaca, N.Y., where he remained vigorously active for almost a decade. Professor du Vigneaud died in 1978 in White Plains, N.Y.

Early in his career du Vigneaud witnessed the experiments of L.F. Audrieth who carried out organic reaction in liquid ammonia. To his great satisfaction he found that insulin is soluble in this remarkable solvent. Soon he tried to reduce cystine to cysteine with sodium in liquid ammonia and found that the sulfhydryl group can be benzylated in situ. It was similarly possible to use reduction with sodium in ammonia for the removal of the S-benzyl group and to regenerate thereby the sulfhydryl function. The process turned out to be applicable for the cleavage of the benzyloxycarbonyl group as well (p. 63). These achievements, while valuable in themselves, paved the way toward his later studies on oxytocin and vasopressin.

During the Second World War these investigations were interrupted by du Vigneaud's participation in a major collaborative research of penicillin (cf. Chap. 9, p. 198). An important contribution from Cornell, the synthesis of penicillamine, was soon followed by an attempt at the synthesis of the antibiotic itself. This ambitious undertaking was based on the so-called thiazolidine-azlactone structure, now regarded as erroneous, although, as pointed out by Sir Robert Robinson [85], it is a protomer of the later universally accepted thiazolidine-$\beta$-lactam. Condensation of penicillamin with the azlactone derivative gave inactive material, but du Vigneaud, in a manner characteristic for his research, turned defeat into victory. Simply by heating the inactive product in pyridine, the Cornell researchers generated some biological activity, that was shown by isotope dilution and also through its sensitivity to penicillinase, to stem from the target compound, benzylpenicillin. Instead of concluding the study at this point, du Vigneaud and his coworkers proceeded through a series of purification steps, including chromatography, countercurrent distribution and finally crystallization of the triethylamine salt, to homogeneous material. The purified sample was indistinguishable from natural penicillin in battery of test, among them comparison of the antimicrobial spectra and X-ray diffraction [86].

After the war du Vigneaud returned to the endeavor interrupted by the penicillin problem, to the isolation of the pituitary hormones oxytocin and vasopressin. In a thorough and unrelenting effort his group utilized several contemporary separation techniques among which countercurrent distribution, developed next door in the Rockefeller Institute, by L.C. Craig, just about that time, played an important role (see p. 55). Similarly, once the peptide hormones were secured in homogeneous form, the most up-to-date methods were applied for their study. Determination of the amino acid composition by an early version of the Stein-Moore method and of the amino acid sequence through fragmentation by partial acid hydrolysis and analysis of the fragments by the DNP-process of F. Sanger (cf. p. 115) were completed by sequential degradation via the Edman procedure (p. 119).

Both the structure proposed for oxytocin and the synthesis corroborating that structure (cf. pp. 138–145) were published in 1953 and two years later the Nobel Prize in Chemistry was awarded to Professor du Vigneaud. While recognition of his contributions was expressed in membership of foreign academies in the US and abroad, in honorary doctorates, medals etc., du Vigneaud, unaffected by acclaim, continued the in-depth study of the neurohypophyseal hormones for over two more decades. From the many important contributions made in this period let us mention the design and synthesis of highly active analogs of the two hormones, such as desamino-oxytocin. As before, no effort was spared, no stone was left unturned by the investigators led by du Vigneaud. For instance, a new version of partition chromatography was utilized to secure oxytocin, the reduced form of oxytocin [87].

Chemistry of the pitutary hormones provided an excellent training ground for a generation of peptide chemists: the list of du Vigneaud's associates (members of the "VduV Club") makes impressive reading. It contains the names of many distinguished scientist, including Nobel laureates (Fritz Lipmann, Robert W. Holley). Those, who like one of the authors, were fortunate enough to participate in du Vigneaud's endeavors, guard a priceless legacy, a special spirit of research. They learned to look upon their experiments not merely as a source of new knowledge, but also as a great adventure with discoveries just around the corner.

### 7.3.2 Josef Rudinger (1924–1975)

Josef Rudinger was born in Jerusalem, in 1924. Three years later his parents with their small child returned to Czechoslovakia. Thus it was in Prague where the young boy received his primary education. In 1939, however, they emigrated to England, where he became a student at King's College of the University of Durham, in Newcastle upon Tyne. After a year his studies were interrupted by service in the Royal Air Force. Following a brief training in Canada he served as a radio operator on airplanes searching for German U-boats in the Atlantic. With the end of the war he returned to Newcastle, to his studies in chemistry and after graduation remained there to work on his doctoral dissertation under G.R. Clemo. In 1949 the Rudinger family left England and Josef was once again a resident of Prague. He became a member of the peptide research group at the Institute of Chemical Technology, led by F. Sorm, which was then reorganized to form the Institute of Organic Chemistry and Biochemistry of the Czechoslovak

**Plate 7.** Joseph Rudinger (1924–1975)

Academy of Science. Rudinger headed the Laboratory of Peptide Chemistry until 1968, when with his wife Edita Adler, herself a chemist, and their 9 year old daughter, he emigrated to Switzerland. They found a new home in Zurich and Rudinger was appointed Professor in the Institute of Molecular Biology and Biophysics of the ETH, an appointment that was terminated by his untimely death in 1975.

Rudinger's early work, in Prague, on the chemistry of glutamic acid was already remarkable. It showed his ability to go in depth and to attack a problem from many sides. A whole range of reactions, all starting with glutamic acid, was envisaged and realized with admirable thoroughness. In 1955, impressed by the work of du Vigneaud, he embarked on a new synthesis of oxytocin and with the help of an excellent research team assembled the molecule of the hormone [26]. A study of structure-biological activity relationships followed. The most spectacular result of this effort was the synthesis of carba and dicarba analogs of oxytocin (cf. also p. 146), compounds in which one or both sulfur atoms of the molecule have been replaced by the $CH_2$ group. The high biological activity of the carba analogs proved that the disulfide bridge is an architectural feature of oxytocin, but not a functional group that participates in the interaction between the hormone and its receptor. The structure stabilizing role of disulfide bridges turned out to be generally valid for other peptides and proteins as well.

We can not enumerate here Rudinger's many valuable contributions to the methodology of peptide synthesis or to the design of molecules for specific pharmacological studies. It seems to be more important to recall one of his ideas, that had and still has a major impact on peptide chemistry. In 1958 he invited the handful of European investigators active in peptide research to participate in a symposium in Prague. This first European Peptide Symposium was followed, first at yearly intervals, then in every second year, by meetings held in various locations in Europe (cf. Ref. 42 in Chap. 3). Their excellence acted as a stimulus or perhaps as catalyst and led to an unprecedented growth in research and understanding of peptide chemistry. Not less important are the friendships and collaborations formed at the Symposia. Their example was followed in the United States and also in Japan, where recently also an international symposium was held.

While he was among us, Rudinger represented a spirit of dedication, cooperation and integrity. His legacy remains a major asset for all whose life is enriched by a sincere interest in peptides.

## 7.4 Insulin

By discussing insulin after the section dealing with the neurohypophyseal hormones we do not wish to suggest that insulin is less important than oxytocin or vasopressin. In fact, there can be little doubt about the exceptional significance of insulin in medicine: it controls the unfortunately rather common deficiency disease *diabetes mellitus.* While *diabetes insipidus,* caused by the absence of vasopressin, affects relatively few people, perhaps about 100 000 in the United States, millions lack insulin, either because their organism is unable to produce the hormone or because they have circulating antibodies that combine with their own insulin. Yet, over and above the obvious attention received by insulin, it became the objective of broad and continued studies because its isolation, crystallization, the elucidation of its structure and three dimensional architecture, synthesis and synthetic modification appeared as a major challenge for a large number of investigators, tantalizingly difficult problems for inquisitive minds.

### 7.4.1 Discovery

About a hundred years ago J. von Mering and O. Minkowski extirpated the pancreas of dogs and produced artificial diabetes in them. It soon became known that the so called Langerhans islet cells of the gland are the site of hormone production and in 1909 J. de Meyer proposed the name *insulin* for the active principle. It took, however, many more years until this physiological concept turned into hard chemical reality. In 1922 in the laboratory of J.J.R. McLeod in Toronto, Canada, a remarkable success was achieved. Frederick (later Sir Frederick) G. Banting and Charles H. Best isolated insulin in solid, albeit amorphous, form. The crucial innovation in their process was the use of *acidified* aqueous ethanol for the extraction of the glands. In acidic media the proteolytic enzymes trypsin and chymotrypsin, present in the same tissue, can not exert their hydrolytic action and hence do not cleave peptide bonds in insulin. A few years later crystallization of insulin was reported from Baltimore, Maryland, by Geiling and Abel. It is interesting to point out that only the insulin preparation obtained from one pharmaceutical manufacturer E.R. Squibb could be crystallized while the not less pure material from Ely Lilly yielded no crystals. Subsequently the cause of this discrepancy could be traced to an impurity in the Squibb preparations: zinc salts. The crytalline material was a Zn complex of insulin.

Crystalline insulin was well suited for chemical investigations and it soon became clear [88–90] that the hormone is built of amino acids, that it is a peptide. The importance of the disulfide bridges was clarified not much later [91] and the role of molecular architecture in the hormonal activity of insulin postulated in the late thirties [92].

### 7.4.2 Structure elucidation

Probably the most important achievement in insulin research was the determination of its primary structure by Frederick Sanger and his associates in Cambridge, England. In order to elucidate the amino acid sequence of the hormone it was necessary to separate the two chains constituting the molecule. This was accomplished by oxidation with performic acid. This operation cleaved the three disulfide bridges by converting each cystine to two cysteic acid residues:

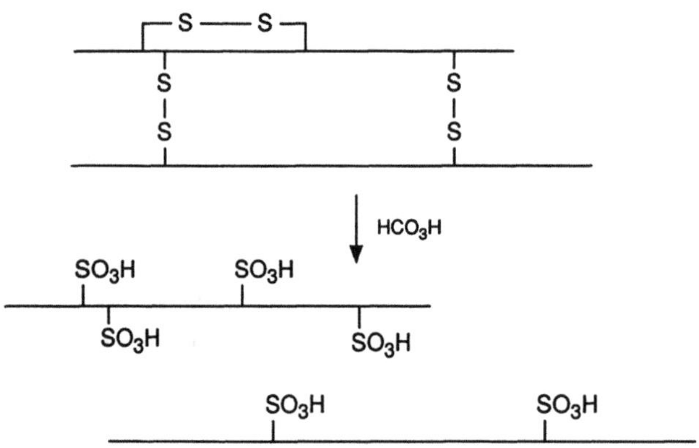

Disulfide splitting of insulin by oxidation with performic acid

**Scheme 3.**

The separated chains were subjected to the action of concentrated hydrochloric acid at moderate temperature to generate fragments, small peptides, which were then analyzed by Sanger's dinitrophenylation method [10]. Skillful combination of the data secured through the study of the fragments yielded the amino acid sequence of each individual chain [93, 94]. The study was then completed with the determination of the position of the disulfide bridges connecting the two chains and that of the intra-chain bridge in the A-chain. Of course this concluding work had to be carried out on the intact insulin molecule. The necessary fragmentation was achieved not by acid catalyzed hydrolysis but with the help of

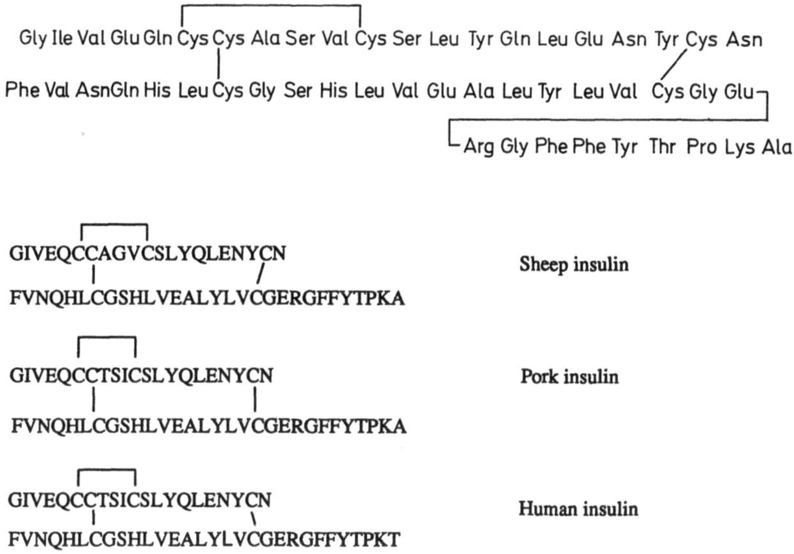


**Fig. 3.** The amino acid sequence of bovine insulin and other mammalian insulins

proteolytic enzymes. Finally, in 1955, the entire structure of insulin was proposed [95].

An account of this historical study can be found in the Nobel lecture [13] of (later Sir Frederick) Sanger (Plate 38).

The 3-dimensional structure of insulin became an important target for research and it was attacked by various physicochemical means in many laboratories. A definitive picture emerged, however, only through the application of X-ray crystallography by Dorothy Hodgkin and her associates [96] (p. 132). An impressive, highly helical architecture was revealed by the diffraction patterns obtained with single crystals of metal-complexes of insulin, but the biological relevance of the geometry thus determined remains open to questions. The structure is that of a hexamer and it appears that the high helix content of the molecule is due to stabilization of the secondary structure through *intermolecular* interactions. Other studies, for instance by optical rotatory dispersion (p. 121) or by circular dichroism (p. 122) indicate rather low helicicity in dilute solutions. Therefore the geometry of insulin in contact with its specific receptors might or might not be the one determined in the solid state by crystallography. The complexity of this problem is further enhanced by the discovery in recent years of several different crystalline forms of Zn-insulin.

### 7.4.3 Synthesis

The synthesis of insulin required two separate achievements, both rather problematic at the time when the structure of the hormone became known. In the late 1950s construction of the two chains, the A chain (or glycine chain) with 20 and the B chain (or phenylalanine chain) with 30 residues appeared as challenging objectives. Only a few years earlier the synthesis of the 9-residue peptide oxytocin was hailed as a major accomplishment. Yet, even more formidable was the problem of combining the two chains into a single molecule by the formation of two disulfide bridges. Reductive cleavage of insulin results in two chains that can be separated. The method of oxidative sulfitolysis can be applied for this purpose:

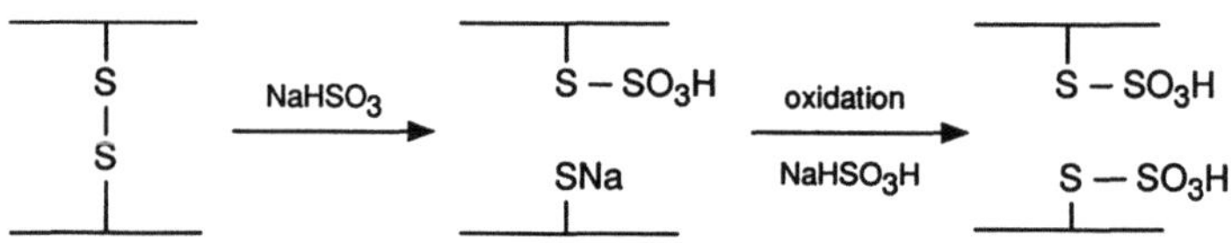


**Scheme 4.**

The products of the reaction are the two chains in the readily separable form of S-sulfonates. Treatment of the latter with thiols converts them to peptides with free thiol groups which then are smoothly oxidized to disulfides. A random formation of disulfides, however, should yield a mixture of a vast number of different molecules because the sulfhydryl groups can combine in various ways. Calculations [97] showed that only an impractically small amount of insulin can be expected in such experiments, but in 1960 a low but definitely higher than the calculated yield was observed by Dixon and Wardlaw [98] and therefore the way to synthetic insulin appeared to be open. Still higher yields were reported a year later from China [99] where in the recombination experiments the solution containing the two separated chains was incubated prior to oxidation. This suggests that the conformation of the two peptides undergoes a gradual change toward geometries in which they are held together by secondary forces in an arrangement that leads to the formation of the desired disulfide bridges.

While these recombination studies were in progress, synthesis of the individual chains was started in two laboratories, one in Aachen, Germany, the other in Pittsburgh, Pennsylvania. By mid 1963 the American team, Katsoyannis, Fukuda and Tometsko, reported the successful combination of their synthetic A-chain with natural B-chain and in the same year Helmut Zahn's group in Aachen (Plate 54), achieved [100] the combination of two synthetic chains and thus secured fully synthetic sheep insulin. Within a short time the total synthesis of insulin was accomplished in Pittsburgh as well [101]. The schemes for the construction of the insulin molecule were quite similar in the two syntheses. The chains were built from almost identically selected segments and the methods applied were also not too different. In both laboratories fairly large peptides were prepared by stepwise chain-elongation, mostly with the help of active esters [30] and then coupled by the azide procedure or by the application of dicyclohexylcarbodiimide (cf. Chap. 4). The design of the schemes was dictated by retrosynthetic analysis that led to the selection of segments that can be linked to each other without racemization, for instance because their C-terminal residue is glycine, the only amino acid without a chiral center. Thus the presence of a glycine residue in a strategically favorable position in the A-chain of sheep insulin led to its selection as the target compound in both laboratories. The methods of blocking and coupling corresponded to the know-how of the period. This explains the many points of similarity between the two independent endeavors. In China, the already mentioned important studies of recombination were completed by the combination of the synthetic chains of beef insulin prepared in laboratories in Shanghai and Peking [102]. See also page 231.

From the long series of studies aimed at synthesizing insulin and its analogs we point here only to two: the chains of the molecule were assembled through stepwise incorporation of the amino acid residues on an insoluble polymeric support (cf. Chap. 5) by Marglin and Merrifield [103] at the Rockefeller Institute and an exceptionally elegant synthesis of human insulin, in which large segments already containing the disulfide bridges were constructed and combined, was reported [104] by the Ciba-Geigy group in Basel, Switzerland. The solid phase approach has been applied [105] in the preparation of analogs for the determination of structure-activity relationships as well. The two laboratories which were first in the synthesis of insulin continued to work in this intriguing area. The Deutsche Wollforschungsinstitut in Aachen, led by Professor H. Zahn, broadened the scope of insulin studies by including proinsulin, the biological precursor of the hormone, as one of its objectives. The group guided by Professor P.G. Katsoyannis in New York (Plate 26) designed and prepared a wide range of potent analogs, among them some superagonists.

The extensive homology between the sequences of human and porcine insulins (Fig. 3) prompted several attempts to convert the peptide from the animal to the human hormone. Only a single amino acid residue, alanine at the C-terminus of the B chain in the insulin from hogs (and also from whales) had to be changed: in human insulin this position is occupied by threonine. It had been known before that carboxypeptidase A preferentially cleaves the bond between this alanine and the preceding lysine residue without significantly attacking the A-chain. Fortuitously, the lysine residue that becomes C-terminal in the process is not removed by the enzyme because it is preceded by proline. The resulting desalanino pork insulin has full biological activity and is a potentially valuable tool in the control of diabetes, but so far it has been exploited mainly as starting material for the production of human insulin. The same enzyme, carboxypeptidase A can catalyze not only the cleavage but also the synthesis of the peptide bond in question, since it merely accelerates the establishment of the equilibrium (cf. Chap. 3). Thus by adding threonine in large excess to the reaction mixture the B-chain of the desalanino compound is lengthened and human insulin forms [106]. In an efficient version of this approach the acid labile *tert*-butyl ester of threonine was used [107] Acidolysis under mild conditions removes the *tert*-

butyl group and thus completes the semisynthesis of human insulin. In the practical implementation of the method carboxypeptidase Y and trypsin were also applied.

In the development of gene-technology, preparation of human insulin with the help of the protein synthesizing machinery of microorganisms was one of the first major objectives. Two alternative methods emerged from the broad and intensive efforts. In the first approach the two individual chains were generated separately and then combined. In the second procedure proinsulin [108], a 86 residue single-chain peptide is produced and converted through the action of a specific enzyme or by trypsin to the hormonally active two-chain molecule. In both processes new desoxyribonucleic acids (DNA-s) are incorporated into the genetic make-up of a microbial cell, usually Escherichia coli, to provide the necessary information for the production of the desired peptide. In the two-chain procedure synthetic DNA-s were used, while a natural DNA was applied, although synthetically modified, for the purpose of producing proinsulin in the bacterial cell. The new genetic information was inserted into plasmids, small rings of cytoplasmic DNA, that were opened up by restriction enzymes. In addition to the insulin or proinsulin sequence the genetic information incorporated into the plasmids also contained the DNA sequence of a promoter protein, such as beta-galactosidase or tryptophane-synthetase, followed by the tetranucleotide code of methionine that was intended to link the protein to the insulin sequence. The ring of the plasmid is then closed with a ligase enzyme and the plasmid incorporated into the cell by a process of "transformation". The large scale culture of the modified cells yields a "chimeric protein" which is isolated from the fermentation mixture. Treatment of the chimeric protein with cyanogen bromide cleaves the bond between methionine and the insulin (or proinsulin) sequence. In the process where the two chains were the target compounds these were combined by the methods already discussed but with improved efficiency. The use of the A-chain in 100% excess was one of the major factors contributing to high yield. The excess A-chain is recycled in the process. In the second approach the proinsulin obtained through fermentation is treated with trypsin and then with

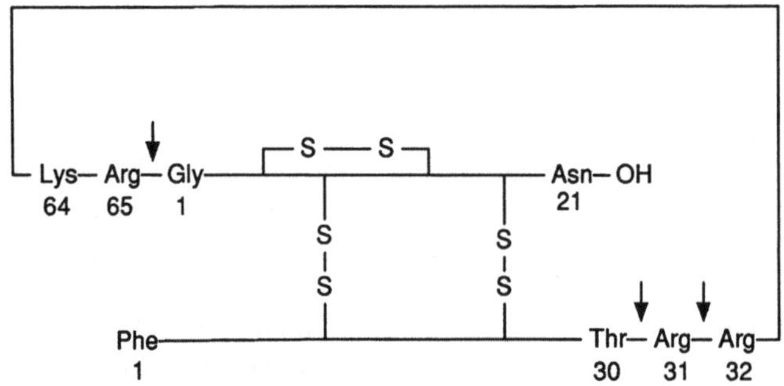


**Fig. 4.** Conversion of proinsulin to insulin

carboxypeptidase B. The first hydrolysis cleaves proinsulin at the two neighboring basic residues which are the signal sequences for the conversion of proinsulin to insulin in the pancreas. At the same time also one of the arginine residues is removed while the remaining arginine is cleaved by carboxypeptidase B, an enzyme with preference for basic residues [109].

Synthesis of peptides by methods of biogenetic engineering transcends the boundaries of peptide chemistry: the method is generally applied for the production of proteins. Nevertheless, it seemed to be proper to conclude the history of insulin research with the presentation in a nutshell of the preparation of the hormone through the processes of biotechnology.

## 7.5 Corticotropin

In the first part of this chapter we have dealt with hormones from the posterior lobe of the pituitary gland. Extraction of the much larger anterior lobe yields several additional hormones, among them growth hormone, a protein with 188 residues, the much smaller melanocyte stimulating hormones $\alpha$- and $\beta$-MSH and also a 39-peptide that prompts the adrenal gland to release regulatory steroids from its cortex. This peptide, adrenocorticotropin or adrenocorticotrophic hormone or corticotropin or ACTH, is applied in medicine for the treatment of hypophyseal insufficiencies and inflammatory processes and accordingly attracted the attention of peptide chemists as a challenging objective for synthesis. The complete sequence of the 39 amino acid residues in the single chain of ACTH was published in 1959 [110a] and slightly revised two decades later (Fig. 5).

No peptide with similar size had been prepared thus far and, therefore, some investigators, (e.g. K. Hofmann et al. [110b]) reduced the task to the synthesis of segments, such as the N-terminal 23-peptide which is about as potent as the full 39-membered chain. The first total synthesis of the porcine hormone, with the sequence regarded as correct at that time, was achieved by Schwyzer and Sieber in 1963 [111]. They prepared the C-terminal 14-peptide portion of ACTH by stepwise chain-lengthening with active esters as acylating agents, and linked to it several shorter segments until the entire chain of the porcine hormone was assembled. The systematic application of blocking groups based on the stability

H – Ser – Tyr – Ser – Met – Glu – His – Phe – Arg – Trp – Gly – Lys – Pro – Val –

Gly – Lys – Lys – Arg – Arg – Pro – Val – Lys – Val – Tyr – Pro – Asp – Gly –

Ala – Glu – Asp – Glu – Ser – Ala – Glu – Ala – Phe – Pro – Leu – Glu – Phe – OH

**Fig. 5.** The amino acid sequence of human ACTH

of the *tert*-butyl cation for semipermanent protection broadened the methodology of peptide synthesis in a major way. For temporary protection of ε-amino groups the benzyloxycarbonyl group was used and it was removed by hydrogenation. In the final deprotection step trifluoroacetic acid was applied for cleavage of the *tert*-butyl-derived blocking groups. Thus, the first synthesis of ACTH was rightly hailed as a major accomplishment.

Several syntheses of ACTH were reported in the following years. Of these the efficient construction of the molecule by Bajusz and his associates [112] is mentioned here. Solid phase synthesis of the hormone on a commercial scale has been attempted but it is more practical at this time to manufacture ACTH in solution.

From the numerous studies aiming at the determination of the conformation of ACTH the work of Schiller [113], based on intramolecular energy transfer should be mentioned here. Substitution of the side chain amino groups of lysine residues in ACTH 1–24 with the 5-dimethylamino-1-naphthalenesulfonyl (dansyl) group provided a suitable material for the energy-transfer experiment. The method has often been used ever since for the study of the geometry of biologically active peptides [114]. Corticotropin itself has low helix content and its geometry can not be described in simple terms.

The two *melanocyte stimulating* hormones, α- and β-MSH or melanotropins were discovered, sequenced and synthesized in the later nineteen-fifties. They enlarge the area occupied by melanocytes in the skin of lizards, frogs and toads and cause a visible darkening of their skin. The physiological role of these peptides, which are present in mammals as well, remains to be determined. In this brief account we merely demonstrate the relationships between the structures of melanotropsins and ACTH. The arrow in Fig. 6 indicates the processing signal for cleavage associated with the formation of a peptide amide:

↓

ACTH (human) SYSMEHFRWG KPVGKKRRPV KVYPNGAEDE SAEAFPLEF

α-MSH acetyl-SYSMEHFRWG KPV-$NH_2$

β-MSH AEKKDEG PYRMEHFRWG SPPKD

**Fig. 6.** Relationship between human ACTH and α- and β-melanotropins

In spite of the conspicuous homology displayed in the MEHFRWG sequence β-MSH is less closely related to ACTH than α-MSH. Difficulties encountered in the synthesis of β-MSH prompted Schwyzer to seek the new tactical approach discussed in connection with the synthesis of ACTH.

## 7.6 Gastrointestinal Hormones

### 7.6.1 Gastrin and Cholecystokinin

Discovery of gastrin, the continued doubt about its existence were mentioned in the introductory part of this chapter. These doubts were laid to rest with the isolation of the hormone, determination of its amino acid sequence and synthesis. Gastrin is an acidic 17-peptide with pyroglutamic acid at the N-terminus and with not less than five glutamic acid residues, in a row, in its chain. Two variants were isolated, (porcine) gastrin I

Pyr-Gly-Pro-Trp-Met-Glu-Glu-Glu-Glu-Glu-Ala-Tyr-Gly-Trp-Met-Asp-Phe-$NH_2$ and gastrin II in which the hydroxyl group of the tyrosine residue is esterified with sulfuric acid.

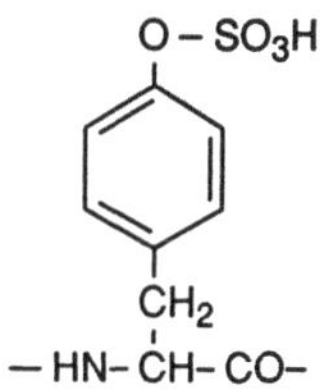


There are minor variations in the amino acid sequence from species to species. For instance in the human hormone leucine rather than methionine is present in position 5. Interestingly, also the second methionine residue (in position 5) can be replaced by leucine without loss of hormonal activity. It is even more noteworthy that not the entire molecule is needed for high potency. Shorter C-terminal segments are quite effective and the C-terminal pentapeptide can replace the 17-peptide in its physiological functions. Therefore the pentapeptide amide Gly-Trp-Met-Asp-Phe-$NH_2$ has lent itself to the study of structure-activity relationships and hundreds of its analogs have been synthesized. During such studies it was accidentally recognized that one of the intermediates, the dipeptide ester Asp-Phe-$OCH_3$, has a sweet taste. Being much sweeter than sucrose it became known as the artificial sweetener **aspartame** and is now manufactured on an industrial scale.

In addition to the 17-peptide also a shorter chain of 14 residues, "mini-gastrin" or "little gastrin" is present in antral extracts and a 34 residue long peptide, obviously the biological precursor of gastrin was obtained from the same source. Its C-terminal 17-peptide portion is identical with respect to sequence to that of gastrin, except that the N-terminal pyroglutamic acid residue of the latter is present as a glutamic acid residue[1].

---

[1] A longer, 34 residue long peptide is clearly the precursor of human gastrins with shorter sequences:

pGlu-Leu-Gly-Pro-Gln-Gly-Pro-Pro-His-Leu-Val-Ala-Asp-Pro-Ser-
Lys-Lys-Gln-Gly-Pro-Trp-Leu-Glu-Glu-Glu-Glu-Glu-Ala-Tyr-Gly-
Trp-Met-Asp-Phe-$NH_2$

The two Lys residues, in positions 16 and 17, are the signal for the specific enzyme that cleaves the chain to produce the 17-peptide gastrin.

As early as in 1856 the observation was made by Claude Bernard that introduction of hydrochloric acid into the duodenum causes the flow of bile. In the nineteen-twenties, Ivy and his coworkers proved the presence of "cholecystokinin" in intestinal extracts, but the gall bladder contracting hormone was isolated, from hog intestines, only several decades later, by the efforts of J.E. Jorpes and V. Mutt (Plate 23) [2] who concluded their study with the elucidation of the structure [115]. The 33-residue sequence of cholecystokinin (CCK, porcine).

Lys-Ala-Pro-Ser-Gly-Arg-Val-Ser-Met-Ile-Lys-
Asn-Leu-Gln-Ser-Leu-Asp-Pro-Ser-His-Arg-Ile-Ser-Asp-Arg-
Asp-Tyr($SO_3H$)-Met-Gly-Trp-Met-Asp-Phe-$NH_2$

shows conspicuous homology with the sequence of gastrin: seven C-terminal residues are identical. Interestingly this C-terminal portion is sufficient for full biological activity. In fact the octapeptide amide or the acetyl-heptapeptide amide are more potent than the 33-membered CCK molecule and the latter [116].

Ac-Tyr($SO_3Na$)-Met-Gly-Trp-Met-Asp-Phe-$NH_2$

closely resembles the parent hormone in its gall bladder contracting action. It must be added here that, unlike in gastrin, the presence of the sulfate ester group in the tyrosine side chain (in position 27) is quite important for full potency. A further noteworthy discovery was the isolation of a 10-peptide from the small Australian frog *Litoria (Hyla) caerules* in the course of the studies led by V. Erspamer on peptides from the skin of amphibians (Cf. Chap. 8). The interesting new peptide, caerulein [117] showed the various activities of CCK and is more potent than the hormone obtained from mammals. Its sequence

Pyr-Gln-Asp-Tyr($SO_3H$)-Thr-Gly-Trp-Met-Asp-Phe-$NH_2$

starts with pyroglutamic acid followed by a glutamine residue, but from here on it is almost identical with the C-terminal octapeptide sequence of CCK the only exception being a methionine residue in CCK which is replaced by threonine in caerulein. Tyrosin is sulfated in both. Simultaneously with the sequence determination [117] also the synthesis of caerulein was reported [118].

In addition to gall bladder contracting activity, intestinal extracts exhibit an enzyme-releasing effect as well. Digestive enzymes, such as trypsin, chymotrypsin and amylase, are set free from the microsomes in pancreatic acinar cells. A specific hormone, pancreozymin (PZ) has been postulated as being responsible for this action. Having pure CCK preparations on hand Jorpes and Mutt were able to show that this 33-peptide has both kind of activities. For a few years the name cholecystokinin-pancreozymin (CCK-PZ) was used but gradually it was simplified to cholecystokinin (CCK), the pancreozymin activity being tacitly understood.

Synthesis of CCK was attempted in several laboratories and while hormonally active highly potent preparations with the C-terminal sequence of the molecule could be secured [119] building of the entire 33-peptide posed serious problems. The simultaneous presence of several residues conducive to side reactions, e.g. tryptophan, methionine, tyrosine, aspartic acid together with the acid sensitive Asp-Pro bond are the source of several difficulties yet these are compounded by the necessary incorporation of the sulfate ester group at the phenolic hydroxyl of a tyrosine residue. Hence the total synthesis of CCK-33 was only accomplished recently [120] in the laboratory of Sakakibara in Osaka, Japan (Plate 37).

The intriguing presence of CCK and related peptides in the brain of animals [121] stimulates numerous investigators to study the effect of the hormone on the central nervous system. Among other activities it leads to the feeling of satiety.

### 7.6.2 The Secretin Family

The discovery of secretin, discussed in the introduction of this chapter, was concluded with the elucidation of the amino acid sequence of the 27 residues constituting the single chain molecule of the hormone:

His-Ser-Asp-Gly-Thr-Phe-Thr-Ser-Glu-Ser-Leu-Arg-Leu-Arg-
1 2 3 4 5 6 7 8 9 10 11 12 13 14
Asp-Ser-Ala-Arg-Leu-Gln-Arg-Leu-Leu-Gln-Gly-Leu-Val-$NH_2$
15 16 17 18 19 20 21 22 23 24 25 26 27

Determination of the structure was accomplished [122] through fragmentation with trypsin and sequencing the individual fragments by the Edman procedure. An additional enzymatic cleavage, selective splitting of the bond between residues 14 and 15 with thrombin provided the still needed information for the positioning of the fragments. While this work was still in progress synthesis of the tryptic fragements was started (based on personal information from Professors Jorpes and Mutt) and concluded with the synthesis of the entire chain of the hormone molecule [123]. The 27-peptide was built stepwise from the C-terminal valine amide and active esters of blocked amino acids were applied exclusively for chain lengthening. In fact, the synthesis of secretin was used by one of the present authors (M.B.) as a test for the applicability of the stepwise approach in the building of chains significantly longer than oxytocin. Since that time several syntheses of the hormone have been reported, mostly through condensation of segments but also by stepwise chain elongation on insoluble polymeric supports and, more recently, by biosynthesis in bacterial cells.

A comparison of the sequence of secretin with that of the blood sugar elevating hormone from the pancreas, glucagon [124], reveals extensive homology between the two peptides. Not less than 14 positions are occupied by identical amino acids (in the porcine sequences):

secretin: His-Ser-Asp-Gly-Thr-Phe-Thr-Ser-Glu-Leu-Ser-Arg-Leu-Arg-
glucagon: His-Ser-Gln-Gly-Thr-Phe-Thr-Ser-Asp-Tyr-Ser-Lys-Tyr-Leu-

secretin: Asp-Ser-Ala-Arg-Leu-Gln-Arg-Leu-Leu-Gln-Gly-Leu-Val-$NH_2$
glucagon: Asp-Ser-Arg-Arg-Ala-Gln-Asp-Phe-Val-Gln-Trp-Leu-Met-Asn-Thr

The first synthesis of glucagon was accomplished by E. Wünsch (Plate 51) and his associates [125] in Munich in 1967. They followed the well-established segment condensation approach and secured the pancreatic hormone in crystalline form.

A few years later Said and Mutt [126] discovered an additional biologically active peptide in extracts from hog gut, the vasoactive intestinal peptide (VIP), that has both secretin like and glucagon like effects and resembles both secretin and glucagon in its sequence. VIP also relaxes smooth muscle preparations from the bronchi and from the stomach but more interestingly it appears to be a neuropeptide: it also occurs in certain well-defined areas of the brain in high concentration. The full biological role of VIP remains clouded in mystery and is the subject of many ongoing studies.

Several more related peptides have been isolated from the intestines of animals. The gastric inhibitory polypeptide or as renamed the glucose-dependent insulinotropic polypeptide (GIP) [127] contains 42 residues in a sequence which shows that it is a member of the secretin family. A 27-peptide with N-terminal histidine and with isoleucine at the C-terminus (accordingly abbreviated as PHI-27 or PHI) was isolated [128] in a novel manner [129]. Instead of biological activity, the presence of the C-terminal residue in amide form was the guiding principle in the fractionation of tissue extracts. The biological role of PHI is being explored. In human tissues the analogous peptide with methionine as the C-terminal residue, PHM-27 was identified [130].

| | | | |
|---|---|---|---|
| Secretin | HSDGTFTSELSRLR | DSARLQRLLQGLV | |
| VIP | HSDAVFTDNYTRFR | LQMAVKKTLNSILN | |
| Glucagon | HSQGTFTSDYSKYL | DSRRAQDFVQWLMNT | |
| GIP | YAEGTFISDYSIAM | DKIRQQDFVNWLLA | QKGKKSDWKHNITQ |
| PHI | HADGVFTSDFSRLL | GQLSAKKYLESLI | |
| PHM | HADGVFTSDFSRLL | GQLSAKKYLESLM | |

Both PHI and PHM inhibit secretion of acidic juice in the stomach and have a pronounced effect in lowering systemic blood pressure. The sequence of GIP, and PHM have been confirmed by synthesis.

### 7.6.3 Motilin

In 1967 J.C. Brown in Vancouver, B.C. (Canada) recognized that the duodenal extract, the source of secretin, cholecystokinin, VIP, etc., also contains a factor which enhances gastric motor activity. The active principle, motilin, was isolated [131] in pure form and the amino acid sequence of the 22-peptide [132]

H-Phe-Val-Pro-Ile-Phe-Thr-Tyr-Gly-Glu-Leu-Gln-
Arg-Met-Gln-Glu-Lys-Glu-Arg-Asn-Lys-Gly-Gln-OH

confirmed in several syntheses.

### 7.6.4 Galanin

The novel method of isolation of peptide amides [129] yielded an additional gastrointestinal peptide with biological activities. Galanine, H-Gly-Trp-Thr-Leu-Asn-Ser-Ala-Gly-Tyr-Leu-Leu-Gly-Pro-His-Ala-Ile-Asp-Asn-His-Arg-Ser-Phe-His-Asp-Lys-Tyr-Gly-Leu-Ala-$NH_2$ raises the blood sugar level and contracts the isolated rat ileum. Synthesis of the (porcine) peptide followed [133] its isolation and sequence determination within a short time.

The remarkable progress in the chemistry and physiology of the gastrointestinal hormones in the last decades is due to the unrelenting efforts of Professors J. Erik Jorpes and Viktor Mutt in the Karolinska Institute, Stockholm, Sweden.

## 7.7 Parathyroid Hormone and Calcitonin

The vital regulation of the level of calcium ions in the blood is under the control of two peptide hormones. The parathyroid hormone raises the concentration of blood calcium by mobilizing calcium uptake from the bones, while the thyroid hormone calcitonin directs the flow of calcium ions in the opposite direction. Thus a constant calcium ion level can be maintained in the blood. Both hormones stimulated extensive research but here we can discuss only some later developments that brought clarity into the picture.

### 7.7.1 Parathyroid Hormone (PTH)

The complete sequence of the human parathyroid hormone (PTHh) [134] was elucidated in 1978 and was soon confirmed by the determination of the nucleotide sequence of its preurcursor, the preparathyroid hormone [135], and also by total synthesis (Sakakibara et al.) [136]. The chain of 84 amino acid residues

Ser-Val-Ser-Glu-Ile-Gln-Leu-Met-His-Asn-Leu-Gly-Lys-His-
Leu-Asn-Ser-Met-Glu-Arg-Val-Glu-Trp-Leu-Arg-Lys-Lys-Leu-
Gln-Asp-Val-His-Asn-Phe-Val-Ala-Leu-Gly-Ala-Pro-Leu-Ala-
Pro-Arg-Asp-Ala-Gly-Ser-Gln-Arg-Pro-Arg-Lys-Lys-Glu-Asp-
Asn-Val-Leu-Val-Gln-Ser-His-Glu-Lys-Ser-Leu-Gly-Glu-Ala-
Asp-Lys-Ala-Asp-Val-Asp-Val-Leu-Thr-Lys-Ala-Lys-Ser-Gln

was constructed from 16 segments, each blocked at its carboxyl terminus by the phenacyl group. The α-amino groups were protected by the *tert*-butyloxycarbonyl group, the side chain functions in the form of benzyl esters and benzyl ethers, the side chain amino group of lysine residues by the 2-chlorobenzyloxycarbonyl group. The guanidine function in arginine was masked by the tosyl group as was the imidazole in the histidine side chain. The latter was deblocked with 1-hydroxybenzotriazole. For the condensation of segments a water soluble carbodiimide was used and racemization was kept at a minimum by the addition of 1-hydroxybenzotriazole. Chiral homogeneity was monitored by analytical high pressure liquid chormatography that was shown to detect the presence of undesired diastereoisomers. In spite of the careful execution of all operations of the process it was necessary to purify the crude product by gel permeation chromatography followed by (repeated) high pressure liquid chromatography and from 2.4 g peptidic material only 10 mg of the pure product could be secured. This is one of the most impressive applications of the segment condensation strategy and of the tactic of complete protection but it also indicates the limits of the approach. The similarly heroic effort leading to synthetic ribonuclease A is mentioned here merely for comparison because the target of this synthesis by Yajima and Fujii [137] is, by our definition, a protein (124 residues) not a peptide.

### 7.7.2 Calcitonin

Since its discovery in 1962 by Copp [138] this hormone became a major objective of investigations not least because of its potential in medicine. It was expected to be of value in the treatment of osteoporosis and other diseases associated with loss of calcium from the bones. Originally isolated from parathyroid glands, calcitonin turned out to be produced in the thyroid and was named thereafter thyrocalcitonin. Yet, the earlier introduced name seems to have reestablished itself once again in the literature. Of the calcitonin from various species, the hormone from salmon shows exceptionally high activity and is used for therapy. The human hormone, however, even if less potent, appears preferable for the treatment of certain conditions because no concern has to be felt in it use about the formation of antibodies. It is interesting to note that the development of peptide chemistry by the late 1960s reached a stage where isolation [139], sequence determination [140] and synthesis [141] could be published within the same year. In the concluding step of the synthesis of human calcitonin, in the closing of the ring through the formation of a disulfide a novel and elegant method [142] was applied: removal of the trityl group from the sulfhydryl functions and simultaneous oxidation to the disulfide by iodine in methanol:

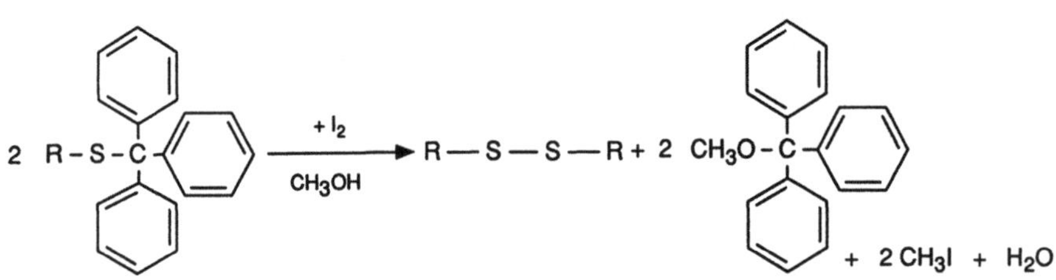


**Fig. 7.** Disulfide formation from tritylthiols

## 7.8 Somatostatin

The well-known, but fortunately, rare, pituitary dwarfism is due to the lack of growth hormone (somatotropin). Normal growth can be restored by the administration of the hormone during adolescence. In spite of its obvious importance, we can not discuss here the history of growth hormone: it is a

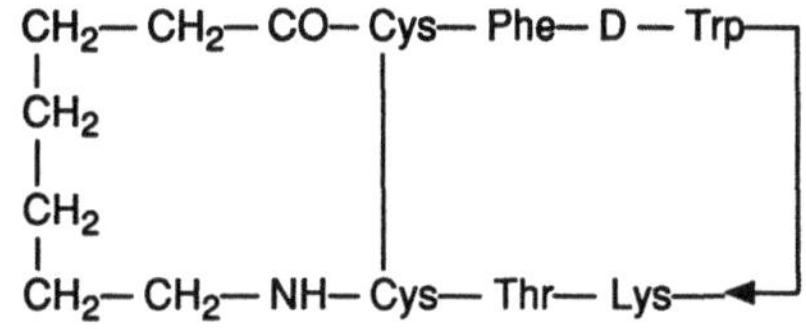

**Fig. 8.** Bicyclic analog of in somatostatin

molecule containing 188 amino acid residues and thus clearly belongs to proteins rather than to peptides. A smaller compound was detected in the **hypothalamus** of several animals, a peptide that inhibits the secretion of somatotropin from the pituitary and that was accordingly named somatostatin. Isolation, structure determination and synthesis of somatostatin were all published [143–149] in the same year, i.e. 1973. Synthesis of the cyclic 14-peptide

H-Ala-Gly-Cys-Lys-Asn-Phe-Phe-Trp-Lys-Thr-Phe-Thr-Ser-Cys-OH

was carried out mostly by the solid phase method but also in solution.

Inhibition of growth hormone release is not the only effect of somatostatin; secretion of several other hormones such as insulin and glucagon is similarly reduced by its action. A lowered level of glucagon is beneficial in certain types of *diabetes mellitus* and therefore somatostatin became an important target in medical research. The molecule can be altered in a major way without loss of biological activity. Even the dimers formed during ring closure are active. Several analogs were designed in which an additional bridge creates conformational restrictions in the expectation that such compounds might provide a better fit to the receptors and be more resistant to the hydrolytic action of proteolytic enzymes. A potent bicyclic analog is shown in Fig. 8 but compounds with high efficacy were also constructed without the disulfide bridge. In addition to such carba-analogs, cyclic peptides, in which the stabilizing effect of to the disulfide bridge or a bridging aliphatic chain is replaced by non-covalent interaction for instance between two aromatic side chains, also show enhanced potency [150–152].

Peptide hormones are often found in more than one form. Thus the 28-peptide, presomatostatin

H-Ser-Ala-Asn-Ser-Asn-Pro-Ala-Met-Ala-Pro-Arg-Glu-Arg-Lys-
Ala-Gly-Cys-Lys-Asn-Phe-Phe-Trp-Lys-Thr-Phe-Thr-Ser-Cys-OH

was isolated [153] from 630000 pig hypothalami. Its amino acid sequence was confirmed by synthesis [154].

## 7.9 Releasing Hormones (Releasing Factors)

The plethora of hormones of which just a selection was discussed in this chapter raises the question: qui custodiat custodes? or what prompts the release of these regulators when they are needed? Hormone release can be induced by stimulation through the nervous system, for instance secretion of gastric juice producing factors by the sight or smell of food. Oxytocin can be released and milk ejected from the mammary glands of the cow by showing her a calf. Between stimuli through the sensory system and the actual physiological event, however, there can be several steps mediated by humoral factors. These releasing factors or releasing hormones are probably quite numerous, although only a limited number of them are known so far. Also, some hormones perform as releasing factors as well: several of the gastrointestinal hormones have an insulin-releasing effect and vasopressin is a potent secretagog of ACTH. A more specific releasing factor is the thyrotropin releasing hormone (TRH or TRF or thyroliberin), that

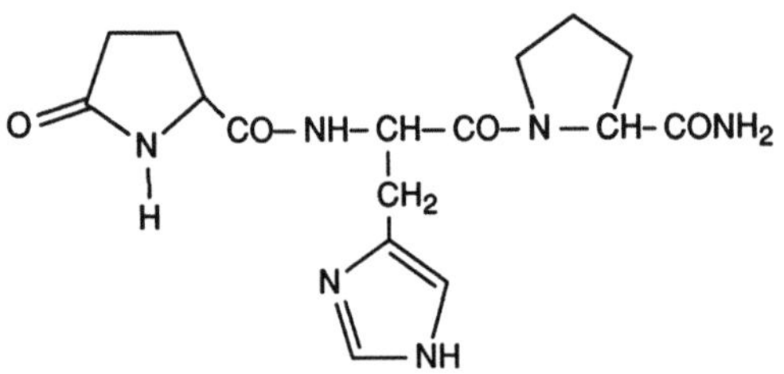


was first isolated from sheep hypothalami [155]. From 25 kg starting material only one mg of pure material was obtained. Elucidation of the structure of such a small molecule might appear simple, but it was hampered by the absence of a free amino group at the N-terminus and a free carboxyl group at the C-terminus. Thus neither exopeptidases nor the usual methods of N-terminal tagging could be applied, nor the best approach, Edman degradation. The sequence was determined by mass spectrometry [156] and also by an ingenious application of peptide synthesis [157]. Nine tripeptides containing the 3 amino acid constituents in all possible order were prepared and tested for biological activity: only one for them, pyroglutamyl-histidyl-proline, was active and its amide had full potency. Synthesis of thyroliberin was achieved in several laboratories [158, 159].

A second, at least equally important example of releasing factors is the decapeptide LH-RH, luteinizing hormone releasing hormone [160–162]

$$\text{Pyr-His-Trp-Ser-Tyr-Gly-Leu-Arg-Pro-Gly-NH}_2$$

isolated simultaneously (1972) in the laboratories of R. Guillemin and of A.V. Schally. In 1977, both researchers were awarded the Nobel Prize in recognition of their work.

The medical importance of releasing factors can be illustrated with the more recently isolated [163] growth hormone-releasing hormone, a 44-peptide that might replace, at least in certain instances, the much larger molecule of human growth hormone.

## References

1. W.M. Bayliss, E.H. Starling, On the causation of the so-called "peripheral reflex secretion" of the pancreas. Proc. Roy. Soc. 60: 352–353 (1902)

2. J.E. Jorpes, V. Mutt, Secretin and cholecystokinin. In Handbook of Exptl. Pharmacology, vol. 34 (J.E. Jorpes and V. Mutt eds., Springer Verlag, Berlin 1973) pp. 1–179
3. J.S. Edkins, On the chemical mechanism of gastric secretion. Proc. Roy. Soc. 76: 376–379 (1905)
4. R.A. Gregory, H.J. Tracy, The preparation and properties of gastrin. J. Physiol. 156: 523–543 (1961); R.A. Gregory, H.J. Tracy, The contribution and properties of two gastrins extracted from hog antral mucosa. Gut 5: 103–117 (1964)
5. H. Gregory, P.M. Hardy, D.S. Jones, G.W. Kenner, R.C. Sheppard, The antral hormone gastrin. Structure of gastrin. Nature 204: 931–933 (1964)
6. J.C. Anderson, M.A. Barton, R.A. Gregory, P.M. Hardy, G.W. Kenner, J.K. McLeod, J. Preston, R.C. Sheppard, J.S. Morley, The antral hormone gastrin. II. Synthesis of gastrin. Nature (Lond.) 204: 933–934 (1964)
7. O. Kamm, T.B. Aldrich, I.W. Grote, L.W. Rowe, E.P. Bugbee, The active principles of the posterior lobe of the pituitary gland. I. The demonstration of the presence of two active principles. II. The separation of the two principles and their concentration in the form of potent solid preparations. J. Amer. Chem. Soc. 50: 573–601 (1928)
8. A.H. Livermore, V. du Vigneuad, Preparation of high potency oxytocic material by the use of countercurrent distribution. J. Biol. Chem. 180: 365–373 (1949)
9. J.G. Pierce, S. Gordon, V. du Vigneaud, Further distribution studies on the oxytocic hormone of the posterior lobe of the pituitary gland and the preparation of an active crystalline flavianate. J. Biol. Chem. 199: 929–940 (1952)
10. F. Sanger, The free amino group in gramicidin S. Biochem. J. 40: 261–262 (1946)
11. P. Edman, Determination of the amino acid sequence in peptides. Arch. Biochem. 22: 475–476 (1949); also: Method for the determination of the amino acid sequence in peptides. Acta Chem. Scand. 4: 283–293 (1950); Stepwise degradation of peptides via phenylthiohydantoins. Acta Chem. Scand. 7: 700–701 (1953)
12. V. du Vigneaud, C. Ressler, S. Trippet, The sequence of amino acids in oxytocin with a proposal for the structure of oxytocin. J. Biol. Chem. 205: 949–957 (1953)
13. F. Sanger, The chemistry of insulin. Science 129: 1340–1344 (1959)
14. H. Tuppy, Amino acid sequence in oxytocin. Biochim. Biophys. Acta 11: 449–450 (1953); also H. Tuppy, H. Michl, Über die chemische Struktur des Oxytocins. Monatshefte der Chemie 84: 1011–1020 (1953)
15. S. Gordon, V. du Vigneaud, Preparation of S,S′-dibenzyloxtocin and its reconversion to oxytocin. Proc. Soc. Exptl. Med. 84: 723–725 (1953)
16. V. du Vigneaud, L.F. Audrieth, H.S. Loring, The reduction of cystine in liquid ammonia by metallic sodium. J. Amer. Chem. Soc. 52: 4500–4504 (1930)
17. R.H. Sifferd, V. du Vigneaud, A new Synthesis of carnosine with some observations on the splitting of the benzyl group from carbobenzoxy derivatives and benzyl thioethers. J. Biol. Chem. 108: 753–761 (1935)
18. V. du Vigneaud, A trail of Research. Cornell University Press, Ithaca, New York (1952)
19. C.R. Harington, R.V. Pitt-Rivers, The synthesis of cysteine-(cystine-) tyrosine peptides and the action thereon of crystalline pepsin. Biochem. J. 38: 417–428 (1944)
20. J.R. Vaughan, Jr., R.L. Osato, Preparation of peptides using mixed carboxylic acid anhydrides. J. Amer. Chem. Soc 73: 5553–5555 (1951)
21. G.W. Anderson, J. Blodinger, A.D. Welcher, Tetraethyl pyrophosphite as reagent in peptide synthesis. J. Amer. Chem. Soc. 74: 5309–5312 (1952)
22. V. du Vigneaud, C. Ressler, J.M. Swan, C.W. Roberts, P.G. Katsoyannis, S. Gordon, The synthesis of an octapeptide with the hormonal activity of oxytocin. J. Amer. chem. Soc. 75: 4879–4880 (1953)
23. C. Ressler, V. du Vigneaud, The synthesis of the tetrapeptide amide S-benzyl-L-cysteinyl-L-prolyl-L-leucyl amide. J. Amer. Chem. Soc. 76: 3107–3109 (1954); J.M. Swan, V. du Vigneaud, The synthesis of L-glutaminyl-L-asparagine and L-isoglutamine from p-toluenesulfonyl-L-glutamic acid. ibid. 76: 3109–3113 (1954); P.G. Katsoyannis, V. du Vigneaud, The synthesis of p-toluenesulfonyl-L-isoleucyl-L-glutaminyl-L-asparagine and related peptides. ibid. 76: 3113–3115 (1954); V. du Vigneaud, C. Ressler, J.M. Swan, C.W. Roberts, P.G. Katsoyannis, The synthesis of oxytocin. ibid. 76: 3115–3121 (1954)
24. V. du Vigneaud, A trail of sulfur research: from insulin to oxytocin. Les Prix Nobel en 1955. Stockholm 1956; cf. also Science 123: 967–974 (1956)
25. R.A. Boissonnas, S. Guttmann, P-A. Jaquenoud, J.P. Waller, Une nouvelle synthèse de l'oxytocine. Helv. Chim. Acta 38: 1491–1501 (1955)

26. J. Rudinger, J. Honzl, M. Zaoral, Synthetic studies in the oxytocin field III. An alternative synthesis of oxytocin. Coll. Czechoslov. Chem. Commun. 21: 202–210 (1956)
27. Th. Curtius, Synthetische Versuche mit Hippurazid. Ber. dtsch. Chem. Ges. 35: 3226–3228 (1902)
28. M. Bodanszky, V. du Vigneuad, An improved synthesis of oxytocin. J. Amer. Chem. Soc. 81: 2504–2507 (1959)
29. M. Bodanszky, V. du Vigneaud, Synthesis of oxytocin by the nitrophenyl ester method. Nature (Lond.) 183: 1324–1325 (1959); also A method of synthesis of long peptide chains using a synthesis of oxytocin as an example. J. Amer. Chem. Soc. 81: 5688–5691 (1959)
30. M. Bodanszky, Stepwise synthesis of peptides by the nitro-phenyl ester method. Ann. N.Y. Acad. Sci. 88: 655–664 (1960)
31. S. Sakakibara, Y. Shimonishi, A synthesis of oxytocin. Bull. Chem. Soc. Jpn. 38: 120–123 (1965)
32. M. Manning, Synthesis by the Merrifield method of a protected nonapeptide amide with the amino acid sequence of oxytocin. J. Amer. Chem. Soc. 90: 1348–1349 (1968)
33. M. Bodanszky, V. Du Vigneaud, Synthesis of a biologically active analog of oxytocin with phenylalanine replacing tyrosine. J. Amer. Chem. Soc. 81: 1258–1259; 6072–6075 (1959)
34. P.A. Jaquenoud, R.A. Boissonnas, Synthèse de la Phe$^2$-oxytocine. Helv. Chim. Acta 42: 788–793 (1959)
35. V. du Vigneaud, G. Winestock, V.V.S. Murti, D.B. Hope, R.D. Kimbrough Jr., Synthesis of $\beta$-mercaptopropionic acid oxytocin (desaminooxytocin), a highly potent analogue of oxytocin. J. Biol. Chem. 235: P.C. 64–66 (1960); also D.B. Hope, V.V.S. Murti, V. du Vigneaud, A highly potent analogue of oxytocin, desaminooxytocin. ibid. 237: 1563–1566 (1962)
36. D. Jarvis, V. du Vigneaud, Crystalline deamino-oxytocin. Science 143: 545–548 (1964)
37. W.B. Lutz, C. Ressler, D.E. Nettleton Jr., V. du Vigneaud, Isoasparagine oxytocin: the isoasparagine isomer of oxytocin. J. Amer. Chem. Soc. 81: 167–173 (1959)
38. C. Ressler, V. du Vigneaud, The isoglutamine isomer of oxytocin: its synthesis and comparison with oxytocin. J. Amer. Chem. Soc. 79: 4511–4515 (1957)
39. D. Jarvis, M. Bodanszky, V. du Vigneaud, The synthesis of 1-(hemihomocystine)-oxytocin and a study of some of its pharmacological properties. J. Amer. Chem. Soc. 83: 4780–4784 (1961)
40. C. Ressler, The cyclic disulfide ring of oxytocin. Proc. Soc. Exptl. Biol. Med. 92: 725–730 (1956)
41. R.R. Sealock, V. du Vigneaud, Studies on the reduction of pitressin and pitocin with cysteine. J. Pharmacol. Exptl. Therapeutics 54: 433–447 (1935)
42. I.L. Schwartz, H. Rasmussen, L.M. Livingston, J. Marc-Aurele, Neurohypophyseal hormone-receptor interaction in Oxytocin, vasopressin and their structural analogs. Proc. 2nd intern. pharmacol. meeting. Prague, 1963 (Pergamon 1964) 10: 125–133; H. Rasmussen, I.L. Schwartz, The interaction between neurohypophyseal hormones and the amphibian urinary bladder. ibid. 41–45
43. J. Rudinger, K. Jost, A biologically active analogue of oxytocin not containing a disulfide group. Experientia 20: 570–571 (1964); O. Keller, J. Rudinger, Synthesis of [1,6-$\alpha,\alpha'$-diaminosuberic acid] oxytocin ('dicarbaoxytocin'). Helv. Chim. Acta 57: 1253–1259 (1974)
44. F.H. White Jr., Reduction and reoxidation of disulfide bonds Methods of Enzymology 11 (C.H.W. Hirs Ed.) Academic, New York 1967, pp. 481–484
45. D.W. Urry, M. Danishi, R. Walter, Secondary structure of the cyclic moiety of the peptide hormone oxytocin and its deamino analog. Proc. Natl. Acad. Sci. US 66: 111–116 (1970); also D.W. Urry, R. Walter, Proposed conformation of oxytocin in solution. ibid. 68: 956–958 (1971)
46. S.P. Wood, I.J. Tickle, A.M. Treharne, J.E. Pitts, Y. Mascarenhas, J.Y. Li, J. Husain, S. Cooper, T.L. Blundell, V.J. Hruby, A. Buku, A.J. Fischman, H.R. Wyssbrod, Crystal structure of deamino-oxytocin. Conformational flexibility and receptor binding. Science 232: 633–636 (1986)
47. V. du Vigneaud, G.S. Denning Jr., S. Drabarek, W.Y. Chan, The effect of replacement of the carboxamide group by hydrogen in glutamine or asparagine residues of oxytocin on its biological activity. J. Biol. Chem. 238: PC 1560–1561 (1963)
48. M. Wälti, D. B. Hope, Synthesis of [1-(L-2-hydroxy-3-mercapto-propionic acid)] oxytocin. A highly potent analog of oxytocin. J.C.S. Perkin I: 1946–1950 (1972)
49. V. du Vigneaud, P.S. Fitt, M. Bodanszky, M. O'Connell, Synthesis and some pharmacological properties of a peptide derivative of oxytocin:glycyl-oxytocin. Soc. Exptl. Biol. Med. 104: 653–656 (1960)

50. H.D. Law, V. du Vigneaud, Synthesis of 2-**p**-methoxyphenylalanine oxytocin (**O**-methyl-oxytocin) and some observations on its pharmacological behavior. J. Amer. Chem. Soc. 82: 4579–4581 (1960)
51. H. Schulz and V. du Vigneaud, Synthesis of 1-L-penicillamine-oxytocin, 1-D-penicillamine-oxytocin and 1-deaminopenicillamine-oxytocin, potent inhibitors of the oxytocic response of oxytocin. J. Med. Chem. 9: 647–650 (1966)
52. J.J. Nestor Jr., M.F. Ferger, V. du Vigneaud, [1-$\beta$-mercapto-$\beta$-$\beta$-pentamethylenepropionic acid] oxytocin, a potent inhibitor of oxytocin. J. Med. Chem. 18: 284–287 (1975)
53. J. Lowbridge, M. Manning, J. Haldar, W.H. Sawyer, Synthesis and some pharmacological properties of [4-threonine-7-glycine] oxytocin, [1-(L-2-hydroxy-3-mercaptopropionic acid)-4-threonine-7-glycine] oxytocin and [7-glycine] oxytocin, peptides with high oxytocic-antidiuretic selectivity. J. Med. Chem. 20: 120–123 (1977)
54. H.B. van Dyke, B.F. Chow, R.O. Greep, A. Rothen, Isolation of a protein from the pars neuralis of the ox pituitary with constant oxytocic, pressor and diuresis inhibiting activities. J. Pharmacol. Exper. Therap. 74: 190–209 (1942)
55. J.E. Stouffer, D.B. Hope, V. du Vigneaud, Neurophysin, oxytocin and desaminooxytocin. In perspectives in Biology (C.F. Cori, V.G. Foglia, L.F. Leloir, S. Ochoa eds.) Elsevier, Amsterdam 1962, pp. 75–80
56. R. Acher, J. Chauvet, M.T. Chauvet, D. Crepy, Phylogeny of peptides of the neurohypophysis. Isolation of mesotocin (8-iso-leucine oxytocin) of the frog, intermediate between 4-serine-8-isoleucine oxytocin of the bony fish and mammalian oxytocin. Biochim. Biophys. Acta 90: 613–615 (1964)
57. R. Acher, J. Chauvet, M.T. Chauvet, D. Crepy, Isolation of a new neurohypophyseal hormone, isotocin, present in vertebrate fish. Biochim. Biophys. Acta 58: 624–625 (1962)
58. R. Acher, J. Chauvet, M.T. Chauvet, D. Crepy, Phylogeny of neurohypophyseal peptides; isolation of a new hormone, glumitocin, (4-serine-8-glutamine oxytocin), present in cartilaginous fish, the ray (Raja clavata). Biochim. Biophys. Acta 107: 393–396 (1965)
59. A.E. Wilhelmi, G.E. Pickford, W.H. Sayer, Initiation of the spawning reflex response in Fundulus by the administration of fish and mammalian neurohypophyseal preparations and synthetic oxytocin. Endocrinology 57: 243–252 (1955)
60. J. Brodeurs, A. Beaulnes, Effets de l'oxytocine sur les arythmies digitaliques chez le chien. Rev. Can. Biol. 22: 275–279 (1963)
61. M. Bodanszky, S.L. Engel, Oxytocin and the life-span of male rats. Nature (Lond.) 210: 751 (1966)
62. R.L. Katz, Antiarrhythmic action of synthetic oxytocin in man. Experientia 19: 160–161 (1963)
63. V. du Vigneaud, H.C. Lawler, E.A. Popenoe, Enzymatic cleavage of glycinamid from vasopressin and a proposed structure for this pressor-antidiuretic hormone of the pituitary. J. Amer. Chem. Soc. 75: 4880–4881 (1953)
64. R. Acher, J. Chauvet, Structure of vasopressin. Biochim. Biophys. Acta 12: 487–488 (1953)
65. V. du Vigneaud, D.T. Gish, P.G. Katsoyannis, A synthetic preparation possessing biological activities associated with arginine vasopressin. J. Amer. Chem. Soc. 76: 4751–4752 (1954); also V. du Vigneaud, D.T. Gish, P.G. Katsoyannis, G.P. Hess, Synthesis of the pressor-antidiuretic hormone arginine vasopressin. ibid. 80: 3355–3358 (1958)
66. V. du Vigneaud, M.F. Bartlett, A. Jöhl, The synthesis of lysine vasopressin. J. Amer. Chem. Soc. 79: 5572–5575 (1957)
67. J. Meienhofer, V. du Vigneaud, Preparation of lysine-vasopressin through a crystalline protected nanapeptide intermediate and purification of the hormone by chromatography. J. Amer. Chem. Soc. 82: 2279–2282 (1960)
68. M. Bodanszky, J. Meienhofer, V. du Vigneaud, Synthesis of lysine-vasopressin by the nitrophenyl ester method. J. Amer. Chem. Soc. 82: 3195–3198 (1960)
69. R.J. Huguenin, R.A. Boissonnas, Synthèses de la Phe$^2$-arginine-vasopressine et de la Phe$^2$-arginine vasotocine et nouvelles synthèses de l'arginine-vasopressine et de l'arginine-vasotocine. Helv. Chim. Acta 45: 1629–1643 (1962); also M. Bodanszky, M.A. Ondetti, C.A. Birkhimer, P.L. Thomas, Synthesis of arginine containing peptides through their ornithine containing analogs. Synthesis of arginine vasopressin, arginin vasotocin and L-histidyl-L-phenylalanyl-L-arginyl-L-tryptophyl-glycine. J. Amer. Chem. Soc. 86: 4452–4449 (1964)
70. J. Meienhofer, Y. Sano, A solid phase synthesis of [lysine]-vasopressin through crystalline protected intermediate. J. Amer. Chem. Soc. 90: 2996–2997 (1968)

71. J. Meienhofer, A. Trzeciak, R.T. Havran, R. Walter, A solid phase synthesis of [8-arginine]-vasopressin through a crystalline nonapeptide intermediate and biological properties of the hormone. J. Amer. Chem. Soc. 92: 7199–7202 (1970)
72. D.A. Jones Jr., R.A. Mikulec, R.H. Mazur, A simple, high yield synthesis of arginine vasopressin. J. Org. Chem. 38: 2865–2869 (1973)
73. M.L. Fink, M. Bodanszky, Synthesis and hormonal activities of 8-L-homonorleucine vasopressin. J. Med. Chem. 16: 1324–1326 (1973)
74. P. Dreyfuss, Synthesis and some pharmacological properties of 8-L-hydroxynorleucine vasopressin. J. Med. Chem. 17: 252–255 (1974)
75. M. Bodanszky, C.A. Birkhimer, 8-L-citrulline vasopressin and 8-L-citrulline oxytocin. J. Amer. Chem. Soc. 84: 4943–4948 (1962)
76. R.L. Huguenin, R.A. Boissonnas. Synthèse de l'Orn$^8$-vasopressine et de l'Orn$^8$-oxytocine. Helv. Chim. Acta 46: 1669–1676 (1963)
77. M. Bodanszky, G. Lindeberg, Synthesis and hormonal activities of 8-L-homolysine vasopressin. J. Med. Chem. 14: 1197–1199 (1971)
78. M. Manning, W.H. Sawyer, Development of selective agonists and antagonists of vasopressin and oxytocin. In Vasopressins (R.W. Schrier ed.) Raven, New York, 1985 pp. 131–144
79. J.P.H. Burbach, G.L. Kovacs, D. de Wied, J.W. van Nispen, H.M. Greven, A major metabolite of arginine vasopressin in the brain is a highly potent neuropeptide. Science 221: 1310–1312 (1983)
80. P.G. Katsoyannis, V. du Vigneuad, Arginine vasotocin, a synthetic analogue of the posterior pituitary hormones containing the ring of oxytocin and the side chain of vasopressin. J. Biol. Chem. 233: 1352–1354 (1958); also Active principles of the neurophysis in cold-blooded vertebrates. Nature (Lond.) 184: 1465 (1959); also R.D. Kimbrough and V. du Vigneaud, Lysine vasotocin, a synthetic analog of the posterior pituitary hormones, containing the ring of oxytocin and the side chain of lysine-vasopressin. J. Biol. Chem. 236: 778–780 (1961)
81. B.T. Pickering, H. Heller, Chromatographic and biological characteristics of fish and frog neurophysial extracts. Nature (Lond.) 184: 1463–1464 (1959)
82. W.H. Sawyer, R.A. Munsick, H.B. van Dyke, Pharmacological evidence for the presence of arginine vasotocin and oxytocin in neurohypophysial extracts from cold-blooded vertebrates. Nature (Lond.) 184: 1464–1465 (1959)
83. S. Pavel, I. Dimitru, I. Klepsh, M. Dorescu, Gonadotropin inhibiting principle in the pineal gland of human fetuses. Evidence for its identity with arginine vasotocin. Neuroendocrinology 1973 41–46
84. L.C. Craig, D. Craig, Extraction and distribution; in Technique of Organic Chemistry vol. 3 (A. Weisberger ed.) Wiley, New York, 1950 p. 171–311
85. R. Robinson, The interpretation of the reactions of pencillin and remarks on the constitution of penicillin in The Chemistry of Penicillin (H.T. Clarke, J.R. Johnson, Sir R. Robinson eds.) Princeton) University Press, Princeton, N.J. 1949 pp. 450–454
86. V. du Vigneaud, F.H. Carpenter, R.W. Holley, A.H. Livermore, J.R. Rachele, Synthetic penicillin. Science 104: 431–433 (1946)
87. D. Yamashiro, Partition chromatography of oxytocin on 'Sephadex' Nature (Lond.) 201: 76–77 (1964); also D. Yamashiro, D. Gillessen, V. du Vigneaud, Oxytocein and deamino-oxytocein. Biochemistry 5: 3711–3720 (1966)
88. E. Brand, M. Sandberg, The lability of the sulfur in cystine derivatives and its possible bearing on the constitution of insulin. J. Biol. Chem. 70: 381–395 (1926)
89. H. Jensen, O. Wintersteiner, V. du Vigneaud, Studies on crystalline Insulin IV. The isolation of arginine, histidine and leucine. J. Pharmacol. exp. Ther. 32: 387–396 (1928)
90. V. du Vigneaud, H. Jensen, O. Wintersteiner, Studies on crystalline insulin III. Further observations on the crystallization of insulin and on the nature of the sulfur linkage. The isolation of cystine and tyrosine from hydrolyzed crystalline insulin. J. Pharmacol. Exp. Ther. 32: 367–385 (1928)
91. K. Freudenberg, W. Dirscher, H. Eyer, The chemistry of insulin. Hoppe–Seylers Z. physiol. Chem. 187: 89–117 (1930); also K. Freudenberg, T. Wegman, The sulfur of insulin. ibid. 233: 159–171 (1935)
92. V. du Vigneaud, The role which insulin has played in one concept of protein hormones and a consideration of certain phases of the chemistry of insulin. Cold Spring Harbor Symposia on Quantitative Biology VI. 275–285 (1938)
93. F. Sanger, E.O.P. Thompson, Amino acid sequence of the glycyl chain of insulin. I. Identification of lower peptides from partial hydrolysates. Biochem. J. 53: 353–366 (1953)

94. F. Sanger, H. Tuppy, Amino acid sequence in the phenylalanine chain of insulin. I. Identification of lower peptides from partial acid hydrolysates. Biochem. J. 49: 463–481 (1951)
95. A.P. Ryle, F. Sanger, L.F. Smith, R. Kitai, Disulfide bonds in insulin. Biochem. J. 60: 541–556 (1955)
96. M.V. Adams, T.L. Blundell, E.J. Dodson, G.G. Dodson, M. Vijayan, E.N. Baker, M.M. Harding, D.C. Hodgkin, B. Rimmer, S. Sheaf, Structure of rhombohedral 2 zinc insulin cyrstals. Nature (Lond.) 224: 491–495 (1969)
97. W. Kauzmann, Relative probabilities of isomers in cystine-containing randomly coiled polypeptides. In Sulfur in Proteins, Proc. of Symposium, Falmouth, Mass. 1958, p. 93–108 (1959)
98. G.H. Dixon, A.C. Wardlaw, Regeneration of insulin activity from the separated and inactive A and B chains. Nature (Lond.) 188: 721–724 (1960)
99. Y.C. Du, R.Q. Jiang, C.L. Tsou, Conditions for successful resynthesis of insulin from its glycyl and phenylalanine chains. Sci. Sin. (Peking) 14: 229–236 (1965)
100. J. Meienhofer, E. Schnabel, H. Bremer, O. Brinkhoff, R. Zabel, W. Sroka, H. Klostermeyer, D. Brandenburg, T. Okuda, H. Zahn, Synthese der Insulinketten und ihre Kombination zu insulinaktiven Präparaten. Z. Naturforschung 18b: 1120–1121 (1963)
101. P.G. Katsoyannis, A. Tometsko, K. Fukuda, Insulin peptides IX. The synthesis of the A-chain of insulin and its combination with natural B-chain to generate insulin activity. J. Amer. Chem. Soc. 85: 2863–2865 (1963); also P.G. Katsoyannis, K. Fukuda, A. Tometsko, K. Suzuki, M. Tilak, Insulin peptides X. The synthesis of B-chain of insulin and its combination with natural or synthetic A-chain to generate insulin activity. ibid. 86: 930–932 (1964)
102. Y-t. Kung, Y-c. Du, W-t. Huang, C-c. Chen, L-t. Ke, S-c. Hu, R-q. Jiang, S-q. Chu, C-i. Niu, J-z. Hsu, W-c. Chang, L-l. Cheng, H-s. Li, Y. Wang, T-p. Loh, A-h. Chi, C-h. Li, P-t. Shi, Y-h. Yich, K-l. Tang, C-y. Hsing, Total synthesis of crystalline bovine insulin, Sci. Sin. (Peking) 14: 1710–1716 (1965)
103. A. Marglin, R.B. Merrifield, The synthesis of bovine insulin by the solid phase method. J. Amer. Chem. Soc. 88: 5051–5052 (1966)
104. P. Sieber, B. Kamber, A. Hartmann, A. Jöhl, B. Riniker, W. Rittel, Totalsynthese von Humaninsulin unter gezielter Bildung der Disulfidbindungen. Helv. Chim. Acta 57: 2617–2621 (1974)
105. G. Weitzel, U. Weber, J. Martin, K. Eisele, Structure and activity of insulin X. Participation of arginine B 22 in the action of insulin. Hoppe Seylers Z. physiol. Chem. 352: 1005–1013 (1971)
106. M. Bodanszky, J. Fried (to E.R. Squibb) Process for preparing human insulin. U.S. Patent 3276961 (1966)
107. K. Morihara, T. Oka, H. Tzuzuki, Semi-synthesis of human insulin by trypsin-catalyzed replacement of Ala-B 30 by Thr in porcine insulin. Nature (Lond.) 280: 412–413 (1979)
108. D.F. Steiner, P.E. Oyer, The biosynthesis of insulin and a probable precursor of insulin by a human islet cell carcinoma. Proc. Nat. Acad. Sci. USA 57: 473–480 (1967)
109. B.H. Frank, R.E. Chance, Two routes for producing human insulin utilizing recombinant DNA technology. München Med. Wschr. 125(Suppl. 1): 14–20 (1983)
110a. P.H. Bell, Purification and structure of $\beta$-corticotropin. J. Amer. Chem. Soc. 76: 5565–5567 (1954); K.S. Howard, R.G. Shepperd, E.A. Eigner, D.S. Davies, P.H. Bell, Structure of $\beta$-corticotropin. Final sequence studies. ibid. 77: 3419–3420 (1955); B. Riniker, P. Sieber, W. Rittel, H. Zuber, Revised amino acid sequence for porcine and human adrenocorticotropic hormone. Nature (Lond.) New Biol. 235: 114–115 (1972)
110b. K. Hofmann, H. Yajima, N. Yanaihara, T.Y. Liu, S. Lande, The synthesis of a tricosapeptide possessing essentially the full biological activity of natural adrenocorticotropin (ACTH). J. Amer. Chem. Soc. 83: 487–489 (1967)
111. R. Schwyzer, P. Sieber, Total synthesis of adrenocorticotrophic hormone, Nature (Lond.) 199: 172–174 (1963); P. Sieber, W. Rittel, B. Riniker, Die Synthese von menschlichem adrenocorticotropem Hormon ($\alpha_h$-ACTH) mit revidierter Aminosauresequenz. Helv. Chim. Acta 55: 1243–1248 (1972)
112. S. Bajusz, Z. Paulay, Z. Lang, K. Medzihradszky, L. Kisfaludy, M. Löw, Synthesis and biological properties of human corticotropin and its fragments. Acta Chim. Acad. Sci. Hung. 52: 335–348 (1967)
113. P.W. Schiller, Study of adrenocorticotropic hormone conformation by evaluation of intramolecular resonance energy transfer in N-dansyllysine$^{21}$-ACTH (1-24) tetrakosipeptide. Proc. Nat. Acad. Sci. USA 69: 975–979 (1972)

114. P.W. Schiller, The use of steady-state fluorescence techniques in the conformational analysis of polypeptides; in Perspectives in Peptide chemistry (A. Eberle, R. Geiger, Th. Wieland eds.) S. Karger, Basel 1981, pp. 236–248
115. V. Mutt, J.E. Jorpes, Hormonal polypeptides of the upper intestine. Biochem. J. 125: 57P–58P (1971)
116. M. Bodanszky, J.C. Tolle, J.D. Gardner, M.D. Walker, V. Mutt, Cholecystokinin (Pancreozymin) 6. Synthesis and properties of the N-acetyl derivative of cholecystokinin 27–33. Int. J. Pept. Prot. Res. 16: 402–411 (1980)
117. A. Anastasi, V. Erspamer, R. Endean, Isolation and structure of caerulein, an active decapeptide from the skin of Hyla Caerulea. Experientia 23: 699–700 (1967)
118. L. Bernardi, G. Bosisio, R. de Castiglione, O. Goffredo, Synthesis of Caerulein, Experientia 23: 700–702 (1967)
119. M.A. Ondetti, J. Pluscec, E.F. Sabo, J.T. Sheehan, N. Williams, Synthesis of cholecystokinin-pancreozymin I. The C-terminal dodecapeptide. J. Amer. Chem. Soc. 92: 195–199 (1970)
120. Y. Kurano, T. Kimura, S. Sakakibara, Total synthesis of porcine cholecystokinin-33 (CCK-33). J. Chem. Soc. Chem. Comm. 1987, 323–325
121. G.J. Dockray, Immunological evidence of cholecystokinin-like peptides in the brain. Nature (Lond.) 264: 568–570 (1976)
122. V. Mutt, J.E. Jorpes, S. Magnusson, Structure of porcine secretin. The amino acid sequence. Eur. J. Biochem. 15: 513–519 (1970)
123. M. Bodanszky, M.A. Ondetti, S. Levine, V.L. Narayanan, M. von Saltza, J.T. Sheehan, N.J. Williams, E.F. Sabo, Synthesis of a heptacosapeptide amide with the hormonal activity of secretin. Chemistry & Industry 42: 1757–1758 (1966): M. Bodanszky, N.J. Williams, Synthesis of secretin. I. The protected tetradecapeptide corresponding to sequence 14–27. J. Amer. Chem. Soc. 89: 685–689 (1967); M. Bodanszky, M.A. Ondetti, S.D. Levine, N.J. Williams, Synthesis of secretin. II. The stepwise approach. ibid. 89: 6753–6757 (1967); M.A. Ondetti, V.L. Narayanan, M. von Salza, J.T. Sheehan, E.F. Sabo, M. Bodanszky, The synthesis of secretin III. The fragment condensation approach. ibid. 90: 4711–4716 (1968)
124. W.W. Bromer, L.G. Sinn, O.K. Behrens, The amino acid sequence of glucagon. V. Location of the amide groups, degradation studies and summary of sequential evidence. J. Amer. Chem. Soc. 79: 2807–2810 (1957)
125. E. Wünsch, Die totale Synthese des Pankreas-Hormons Glukagon. Z. Naturforschung (B) 22: 1269–1276 (1967)
126. S.I. Said, V. Mutt, Polypeptide with broad biological activity: isolation from the small intestine. Science 169: 1217–1218 (1970)
127. J.C. Brown, J.R. Dryburgh, A gastric inhibitory polypeptide II. The complete amino acid sequence. Can. J. Biochem. 49: 867–872 (1971); H. Jornvall, M. Carlquist, S. Kwank, S.C. Otte, C.H.S. McIntosh, J.C. Brown, V. Mutt, Amino acid sequence and heterogeneity of gastric inhibitory polypeptide (GIP). FEBS Lett. 123: 205–210 (1981)
128. K. Tatemoto, V. Mutt, Isolation and characterization of the intestinal porcine PHI (PHI-27), a new member of the glucagon-secretin family. Proc. Natl. Acad. Sci. USA 78: 6603–6607 (1981)
129. K. Tatemoto, V. Mutt, Chemical determination of polypeptide hormones. Proc. Natl. Acad. Sci. USA 75, 4115–4119 (1978)
130. N. Itoh, K. Obata, N. Yanaihara, H. Okamoto, Human pre-provasoactive polypeptide contains novel PHI-27-like peptide, PHM-27. Nature 304: 547–549 (1983)
131. J.C. Brown, M.A. Cook, J.R. Dryburgh, Motilin, a gastric motor-activity stimulating polypeptide: the complete amino acid sequence. Can. J. Biochem. 51: 533–537 (1973)
132. J.C. Brown, M.A. Cook, J.R. Dryburgh, Motilin, a gastric motor-activity stimulating polypeptide: final purification, amino acid composition and C-terminal residues. Gastroenterology 62: 401–404 (1972)
133. K. Tatemoto, A. Rökaens, H. Jörnvall, T.J. Mc Donald, V. Mutt, Galanin, a novel biologically active peptide from porcine intestines. FEBS Letters 164: 124–128 (1983); H. Yajima, S. Futaki, N. Fujii, K. Akaji, S. Funakoshi, M. Sakurai, S. Katakura, K. Inoue, R. Hosotani, T. Tobe, T. Segawa, A. Inoue, K. Tatemoto, V. Mutt, Synthesis of galanine, a new gastrointestinal Polypeptide. J. Chem. Soc. Chem. Commun. 1985, 877–878.
134. H.J. Keutmann, M.M. Sauer, G.N. Hendy, J.L.H. O'Riordan, J. Potts, Complete amino acid sequence of human parathyroid hormone. Biochemistry 17: 5723–5729 (1978)
135. G.N. Hendy, H.M. Kronenberg, J.T. Potts, A. Rich, Nucleotide sequence of cloned cDNA-s encoding human preparatyroid hormone. Proc. Natl. Acad. Sci. USA 78: 7365–7369 (1981)

136. T. Kimura, T. Morikawa, M. Takai, S. Sakakibara, Total synthesis of human parathyroid hormone (1–84). J. Chem. Soc. Chem. Commun. 1982, 340–341
137. H. Yajima, N. Fujii, Chemical synthesis of ribonuclease A with full enzymatic activity; in Chemical Synthesis and Sequencing of Peptides and Proteins (t-Y. Liu, A.N. Schechter, R.L. Heinrikson, P.G. Condliffe eds.) Elseview/North Holland, New York, 1981 p. 21–39
138. D.H. Copp, E.C. Cameron, B. Cheney, G.F. Davidson, K.G. Henze, Evidence for calcitonin, a new hormone from the parathyroid that lowers blood calcium. Endocrinology 70, 638–649 (1962)
139. B. Riniker, R. Neher, R. Maier, F.W. Kahnt, P.G.H. Byfield, L. Galante, I. MacIntyre, T.V. Gudmundsson, Menschliches Calcitonin I. Isolierung und Charakterisierung. Helv. Chim. Acta 51, 1738–1742 (1968)
140. R. Neher, B. Riniker, W. Rittel, H. Zuber, Menschliches Calcitonin II. Studien von Calcitonin M und D. Helv. Chim. Acta 51: 1900–1905 (1968)
141. P. Sieber, M. Brugger, B. Kamber, B. Riniker, W. Rittel, Menschliches Calcitonin IV. Die Synthese von Calcitonin M. Helv. Chim. Acta 51: 2057–2061 (1968)
142. B. Kamber, W. Rittel, Eine neue, einfache Methode zur Synthese von Cystinpeptiden. Helv. Chim. Acta 51: 2061–2064 (1968)
143. P. Brazeau, W. Vale, R. Burgus, N. Ling, M. Butcher, J. Rivier, R. Guillemin, Hypothalamic polypeptide that inhibits the secretion of immunoreactive pituitary growth hormone. Science 179: 77–79 (1973)
144. R. Burgus, N. Ling, M. Butcher, R. Guillemin, Primary structure of somatostatin, a hypothalamic peptide that inhibits the secretion of pituitary growth hormone. Proc. Natl. Acad. Sci. USA 70: 684–688 (1973)
145. A.V. Schally, W.Y. Huang, R.C.C Chang, A. Akimura, T.W. Redding, R.P. Millar, M.W. Hunkapiller, L.E. Hood, Isolation and structure of pro-somatostatin: putative precursor from pig hypothalamus. Proc. Natl. Acad. Sci. USA 77: 4489–4493 (1980)
146. D. Yamashiro, C.H. Li, Synthesis of a peptide with full somatostatin activity. Biochem. Biophys. Res. Comm. 1973, 882–888
147. D.H. Coy, E. Coy, A. Arimura, A.V. Schally, Solid phase synthesis of growth hormone-release inhibitory factor. Biochem. Biophys. Res. Comm. 1973, 1267–1273
148. D. Sarantakis, W.A. Mc Kinley, Total synthesis of hypothalamic somatostatin. Biochim. Biophys. Res. Comm. 1973, 234–238
149. J.E.F. Rivier, Somatostatin. Total solid phase synthesis. J. Amer. Chem. Soc. 96: 2986–2992 (1974)
150. D.F. Veber, Conformational considerations in the design of somatostatin analogs showing increased metabolic stability. Proc. of the Sixth Amer. Peptide Symp. (Pierce Chem. Co., Rockford, Ill) 1979, pp. 409–419
151. D.F. Veber, R.M. Freidinger, D.S. Perlow, W.J. Paleveda Jr., F.W. Holly, R.F. Strachan, R.F. Nutt, G.H. Arison, C. Homnick, W.C. Randall, M.S. Glitzer, R. Saperstein, R. Hirschmann, A potent cyclic hexapeptide analogue of somatostatin. Nature 292: 55–58 (1981)
152. W. Bauer, U. Briner, W. Doepfner, H. Haller, R. Huguenin, P. Marbach, T.J. Petcher, J. Pless, A very potent and selective analogue of somatostatin with prolonged action. Life Sci. 31: 1133–1140 (1982)
153. L. Pradayrol, J.A. Chayvialle, M. Carlquist, V. Mutt, Isolation of a porcine intestinal peptide with C-terminal somatostatin. Biochem. Biophys. Res. Comm. 85: 701–708 (1978)
154. L. Moroder, M. Gemeiner, W. Göring, E. Jaeger, J. Musiol, R. Scharf, H. Stocker, E. Wünsch, L. Pradayrol, N. Vaisse, A. Ribet, Totalsynthese von somatostatin-28. Hoppe Seylers Z. Physiol. Chem. 362: 697–716 (1981)
155. R. Burgus, Isolation and structural elucidation of ovine hypothalamic thyrotropin (TSH) releasing factor (TRF). Proc. of the Second Amer. Peptide Symp. (Gordon and Breach, New York) 1972, pp. 287–294
156. D.M. Desiderio, The elucidation of primary structure of oligopeptides of biological importance via mass spectrometry. Proc. of the Second Amer. Peptide Symp. (Gordon and Breach, New York) 1972, pp. 159–168
157. D. Gillessen, A.M. Felix, W. Lergier, R.D. Studer, Synthese des "tyrotropin-releasing" Hormons (TRH) (Schaf) und verwandter Peptide. Helv. Chim. Acta 53: 63–72 (1970)
158. S. Bajusz, I. Fauszt, Improved method for the synthesis of the thyrotropin releasing hormone TRH. Acta Chim. Acad. Sci. Hung 75: 419–422 (1973)
159. J. Rivier, Total synthesis of the hypothalamic thyrotropin releasing factor. Methods Enzymol. 37 (Pt.B): 408–415 (1975)

160. R. Burgus, M. Butcher, M. Amoss, N. Ling, M. Monahan, J. River, P. Fellows, R. Blackwell, W. Vale, R. Guillemin, Primary structure of the ovine hypothalamic luteinizing hormone releasing factor (LRF) Proc. Natl. Acad. Sci. USA 69: 278–282 (1972)
161. A.V. Schally, Aspects of hypothalamic regulation of the pituitary gland (Nobel Lecture) Science 202: 18–28 (1978)
162. R. Guillemin, Peptides in the brain: the new endocrinology of the neuron. (Nobel Lecture) Science 202: 390–402 (1978)
163. R. Guillemin, P. Brazeau, P. Böhlen, F. Esch, N. Ling, W.B. Wehrenberg, Growth hormone-releasing factor from a human pancreatic tumor that causes acromegaly. Science 218: 585–587 (1982)

# 8 Biologically Active Fragments of Proteins

In the preceding chapter the definition of hormones, compounds which transmit instructions from one organ to another, was not strictly followed, but there are numerous biologically active peptides that do not fit even into a loose hormone-definition. They exhibit great variation in their activities and it is difficult to find a simple way to classify them. Hence we intend to discuss on the following pages compounds containing only proteinogen amino acid constituents, more or less according to the chronology of their discovery. Peptides with unusual building components, generally obtained from microorganisms, will be the subject of a separate discussion (Chap. 9).

Two comments need to be added at this point. The title of this chapter should not be understood as a suggestion that all biologically active peptides which are formed through selective fragmentation of proteins will be treated below. The hormones described in the preceding chapter are formed in the same manner, through the peptide-bond splitting action of specific enzymes on prohormones and preprohormones and finally on proteins. Also, the number of newly isolated biologically active peptides increases so rapidly that major changes, probably even important additions, can be expected during the preparation of this manuscript and further growth while the book is in production. Thus, it would be a futile undertaking to attempt an exhaustive presentation of the field and our aim is merely to illustrate an important principle of Nature, the generation of a large variety of functional compounds through the combination of a limited number of amino acids.

## 8.1 Angiotensin

Since 1934 it has been well known (from the observations of H. Goldblatt) that extracts from the cortex of kidneys cause the release of a blood-pressure elevating substance in the blood plasma [1]. The active principle was designated as hypertensin or angiotonin then renamed [2] angiotensin. Two enzymes are involved in the process: renin, so called because of its renal origin, and the angiotensin converting enzyme (ACE) that splits the primary product, the weakly active angiotensin I, a decapeptide, into two fragments, a dipeptide and the potent octapeptide angiotensin II:

plasma protein $\xrightarrow{\text{renin}}$

H-Asp-Arg-Val-Tyr-Ile-His-Pro-Phe-His-Leu-OH
angiotensin I

angiotensin I $\xrightarrow{\text{ACE}}$

H-Asp-Arg-Val-Tyr-Ile-His-Pro-Phe-OH + H-His-Leu-OH
angiotensin II (or angiotensin)

The sequences shown are those of porcine and human angiotensins; in the bovine peptide position 5 is occupied by valine rather than isoleucine. Treatment of the intact plasma protein, which is the source of the angiotensins, with trypsin yields a 14-peptide, angiotensinogen, itself a good substrate for renin [2]:

precursor protein $\xrightarrow{\text{trypsin}}$

(renin ↓ between Leu and Leu)

⟶ H-Asp-Arg-Val-Tyr-Ile-His-Pro-Phe-His-Leu-Leu-Val-Tyr-Ser-OH
angiotensinogen

Determination of the amino acid sequence of angiotensin [3] in 1956 was followed quickly, in 1957, by the synthesis of the octapeptide [4] and of an analog [5] in which the N-terminal aspartyl residue was replaced by asparagine, without loss of biological activity. In both instances a segment condensation pattern was followed: the octapeptides were obtained from two tetrapeptide segments constructed from dipeptide units. This pioneering work stimulated intesive research worldwide, not only because the molecule offered a challenge to peptide chemists, but also because of the obvious importance of angiotensin in medicine.

In a major part of the patients suffering from hypertension, angiotensin was found to participate in the etiology of this dangerous condition. Therefore the synthesis of analogs which, while inactive per se, have affinity to the angiotensin receptors and thus might block the binding of the agonist, became a widely accepted approach. Cleveland, Ohio, was, and still is, one of the major centers of this line of research. At the Cleveland Clinic Research Foundation M. Bumpus, M.C. Khosla, I.H. Page, H. Schwarz, R.R. Smeby and their associates made important contributions through the synthesis of a large number of angiotensin antagonists while at the Veterans Administration Hospital in Cleveland L.T. Skeggs, Jr., K.E. Lentz and their coworkers investigated an alternative approach based on the inhibition of renin by competitive substrates. Similar efforts in other laboratories produced hundreds of antihypertension-drug candidates. For instance an octapeptide with the sequence of angiotensin but having sarcosin rather than aspartic acid at the N-terminus and alanine rather than phenylalanine as the C-terminal residue,

H-Sar-Arg-Val-Tyr-Val-His-Pro-Ala-OH,

turned out to be an effective competitive inhibitor [6]. Nevertheless, none of the countless angiotensin analogs is used in the management of high blood pressure because selective inhibitors of the active site of the angiotensin converting enzyme (ACE) proved themselves the drugs of choice for this purpose.

Several peptides isolated from the venom of the South American snake *Bothrops jararaca* are potent ACE inhibitors and were briefly used for the treatment of hypertension, but were soon superseded by surprisingly simple molecules with high inhibitory effect. At the Squibb Institute for Medical Research in Princeton, New Jersey, Miguel A. Ondetti (Plate 32) recognized that ACE is similar in its substrate specificity to the well studied protease carboxypeptidase A. He designed, therefore, molecules that should fit into the active site of ACE (presumably similar to the active site of carboxypeptidase A) and form complexes with the enzyme. The "dipeptide", 2-D-methyl-3-mercapto-propionyl-L-proline, *captopril* [7] strongly associated with the enzyme and

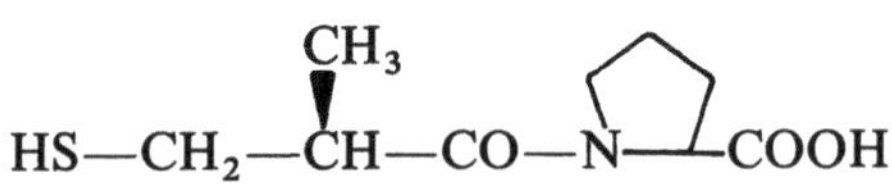


fulfilled all expectations both in vitro and in vivo experiments and in the treatment of hypertension and of certain heart conditions as well.

The almost unique success of captopril stimulated extensive research in the area of enzyme inhibitors, since enzymes are implicated in many pathological conditions. Inhibition of the release of angiotensin still attracts considerable interest and many peptide-like pseudosubstrates are being constructed to serve as renin inhibitors.

## 8.2 Bradykinin and Kallidin

Incubation of a protein fraction from blood plasma with trypsin gives rise to peptides with conspicuous biological effects. Pain, dilation of peripheral blood vessels, increased coronary flow and enhanced capillary permeability were observed on administration of these protein fragments [8]. In the early sixties the nonapeptide bradykinin and its precursors, kallidin and methionyl-kallidin were isolated in pure form and their amino acid sequences determined soon after.

| | |
|---|---|
| bradykinin | H-Arg-Pro-Pro-Gly-Phe-Ser-Pro-Phe-Arg-OH |
| kallidin | H-Lys-Arg-Pro-Pro-Gly-Phe-Ser-Pro-Phe-Arg-OH |
| methionyl-kallidin | H-Met-Lys-Arg-Pro-Pro-Gly-Phe-Ser-Pro-Phe-Arg-OH |

In the sequencing of bradykinin one of the proline residues was overlooked. The error was corrected both by a reexamination of the degradation study [9] and through experiments with peptides synthesized for this purpose [10]. From the various syntheses of bradykinin the scheme of Nicolaides and de Wald, stepwise chain-lengthening with active esters [11] is mentioned here. Also, bradykinin was the first biologically active peptide synthesized by the solid phase method [12] (Chap. 5).

From the plethora of bradykinin analogs prepared for and applied in studies of structure-activity relationships it seems worthwhile to point out one in which the serine residue in position 6 was replaced by glycine (6-glycine bradykinin). The high potency of this analog in several tests shows that the serine residue is not essential for some of the biological effects of bradykinin. An even larger number of analogs were prepared in the hope that they will inhibit, through competition for the receptor, the effects of bradykinin. While the physiological role of bradykinin and kallidin is still unknown, their known effects are rather noxious, resembling symptoms, such as inflammation and pain, seen in rheumatoid arthritis. Therefore bradykinin antagonists held considerable promise. The long search for useful antagonists was vindicated in recent years with the design and synthesis [13] of a bradykinin (or perhaps rather kallidin) analog

H-D-Arg-Arg-Pro-Hyp-Gly-Phe-Ser-D-Phe-Phe-Arg-OH

(where Hyp = hydroxyproline). In this peptide, which is a pain-killer, replacement of the proline residue in position 7 of bradykinin by D-phenylalanin or by the isosteric $\beta$-(2-thienyl)-alanine is the key modification that converts agonists into antagonists.

An important line of research was initiated by the study of the enzymatic inactivation of bradykinin [14]. Inhibitors of carboxypeptidases involved in the process appeared as "potentiators" of bradykinin and were then shown to be inhibitors of the angiotensin converting enzyme (ACE, cf. p. 182) as well.

## 8.3 Substance P

In 1931 U.S. von Euler and J.H. Gaddum discovered, in alcoholic extracts of horse brain and intestinal tissues, a blood-pressure lowering material that also contracted smooth muscle and caused pain [15]. They named it "substance P". The active substance, that resembles bradykinin in several biological tests, occurs in low concentration in the tissues and therefore it could be isolated in pure form only decades later [16]. Determination of the structure and synthesis of the 11-peptide

H-Arg-Pro-Lys-Pro-Gln-Gln-Phe-Phe-Gly-Leu-Met-$NH_2$

were reported [17] forty years after its discovery as an active principle. Substance P with its relatively simple structure stimulates considerable research which has already produced several antagonists with potential usefulness in medicine [18].

## 8.4 Biologically Active Peptides from Mollusks, Frogs, Toads, Snakes and Insects

### 8.4.1 Eledoisin

In 1952 vasodilator substance was discovered in octopods and in a snail by Vittorio Erspamer, at that time professor of pharmacology at the University of Bari, later in Parma and then at the University of Rome (Plate 17) [19]. The active material, eledoisin (after the sources *Eledon moschata* and *Eledon aldrovandi*), was isolated in pure form and ten years after the first observations Erspamer and Anastasi reported the sequence of the 11-peptide [20]:

pyroGlu-Pro-Ser-Lys-Asp-Ala-Phe-Ile-Gly-Leu-Met-$NH_2$.

Two characteristic features of the structure of eledoisin, the N-terminal pyroglutamyl residue and the presence of the C-terminal amino acid as the amide can be recognized in numerous other biologically active peptides as well and the impression is created that Nature designed these posttranslational changes to protect the active molecules against the action of certain proteolytic enzymes (exopeptidases). Also, a noteworthy, albeit not readily interpretable, analogy (homology) exists between eledoisin and substance P: the two 11-peptides:

| | | | | | | | | | | | |
|---|---|---|---|---|---|---|---|---|---|---|---|
| eledoisin | pGlu- | Pro- | Ser- | Lys- | Asp- | Ala- | Phe- | Ile- | Gly- | Leu- | Met-$NH_2$ |
| substance P | H-Arg- | Pro- | Lys- | Pro- | Gln- | Gln- | Phe- | Phe- | Gly- | Leu- | Met-$NH_2$ |
| | 1 | 2 | 3 | 4 | 5 | 6 | 7 | 8 | 9 | 10 | 11 |

have identical residues in positions 2, 7, 9, 10 and 11. Eledoisin indeed resembles substance P, and also bradykinin in their physiological effects.

The first synthesis of eledoisin by Sandrin and Boissonnas in 1962 [21] was followed by the preparation of a large number of analogs which were then used in the exploration of the relationships between structure and biological activities.

### 8.4.2 Physalaemin and Other Peptides from the Skin of Amphibians

A potent hypotensive agent, *physalaemin*, was obtained by Erspamer and his associates in 1964 from the skin of the South American frog *Physalaemus fuscumaculatus* [22]. Practically at the same time the amino acid sequence of the 11-peptide was also elucidated and proven by synthesis [23]. Gradually a whole series of closely related peptides with similar biological activities were isolated, in the same laboratory, from the skins of amphibians, for instance *phyllomedusin* from the skin of Amazonian frog *phyllomedusa bicolor* and several other frogs of the *Phyllomedusa* species, *kassinin* from the African frog *Kassina senegaliensis*, *uperolein* from the frog *Uperoleia rugosa*, etc. [24]. A comparison of the amino acid sequences of these peptides reveals considerable homology; the C-terminal 5 residues are Phe-X-Gly-Leu-Met-$NH_2$ in all of them, where X is a variable amino acid:

| | |
|---|---|
| eledoisin | pGlu-Pro -Ser -Lys-Asp-Ala -Phe-Ile -Gly-Leu-Met-$NH_2$ |
| physalaemin | pGlu-Ala -Asp-Pro-Asn-Lys -Phe-Tyr-Gly-Leu-Met-$NH_2$ |
| phyllomedusin | pGlu-Asn-Pro-Asn-Arg-Phe-Ile -Gly-Leu-Met-$NH_2$ |
| kassinin | H-Asp-Val-Pro -Lys-Ser -Asp-Glu-Phe-Val-Gly-Leu-Met-$NH_2$ |
| uperolein | pGlu-Pro -Asp-Pro-Asn-Ala -Phe-Tyr-Gly-Leu-Met-$NH_2$ |

Some members of this group of peptides (*tachikinins*) show less similarity. Thus *litorin* [25] from the Australian frog *Litoria aurea* is a nonapeptide, pGlu-Gln-Trp-Ala-Val-Gly-His-Phe-Met-$NH_2$ and a heptapeptide, *tryptophillin* [26] H-Val-Pro-Pro-Leu-Gly-Trp-Met-$NH_2$ was isolated from the South American frog *Phyllomedusa rhodei*, while *sauvagine* from *Phyllomedusa sauvagei* [27] contains 40 amino acid residues. The same frog and also *Phyllomedusa rhodei* are the source of a heptapeptide [28] *dermorphin* that has long lasting opiate activity (cf. p. 187). In an important aspect dermorphin is unique: so far it is the only peptide of protein origin that contains a D-amino acid residue,

H-Tyr-D-Ala-Phe-Gly-Tyr-Pro-Ser-$NH_2$

clearly the result of a post-translational process. A natural analog of dermorphin in which proline is replaced by hydroxyprolie had also been detected and both peptides secured by synthesis [29]. A study of analogs revealed that the D-configuration of the second residue is crucial for biological activity.

So far the misleading impression might have been created that the source of tachikinins must be frogs from Australia, Africa or the New World but this is not really so. A 14-peptide, *bombesin*, and the related *alytesin* are present in the skin of the European frog *Biombina bombina*, and the similar *ranatensin* in the common *Rana pipiens*. Bombesin [30] which occurs also in *B. variegata* and *B. orientalis*, exhibits a certain homology with eledoisin, physalaemin and other tachikinins

pGlu-Gln-Arg-Leu-Gly-Asn-Gln-Trp-Ala-Val-Gly-His-Leu-Met-$NH_2$

but has pharmacological effects different from them. It stimulates gastric secretion and the release of cholecystokinin and other gastro-intestinal hormones (cf. p. 165) from the duodenum [31].

### 8.4.3 Venom Peptides

The hemolytic principle of bee venom, **mellitin**, was isolated in 1967 by E. Habermann and J. Jentsch, who also determined the amino acid sequence of the 26 peptide amide as

H-Gly-Ile-Gly-Ala-Val-Leu-Lys-Val-Leu-Thr-Thr-Gly-Leu-
Pro-Ala-Leu-Ile-Ser-Trp-Ile-Lys-Arg-Lys-Arg-Gln-Gln-$NH_2$

and attributed the permeability changes caused by mellitin to the unusual distribution of hydrophobic and hydrophilic amino acid residues along the chain [32].

The bradykinin potentiating effect of a series of peptides present in the venom of the South American poisonous snake *Bothrops jararaca* has already been mentioned in connection with the inhibition of the angiotensin converting enzyme (cf. p. 183) [33]. These proline rich peptides

pGlu-Trp-Pro-Arg-Pro-Gln-Ile-Pro-Pro-OH
pGlu-Asn-Trp-Pro-Arg-Pro-Gln-Ile-Pro-Pro-OH

pGlu-Ser-Trp-Pro-Gly-Pro-Asn-Ile-Pro-Pro-OH
pGlu-Trp-Pro-Arg-Pro-Thr-Pro-Gln-Ile-Pro-Pro-OH
pGlu-Gly-Gly-Trp-Pro-Arg-Pro-Gly-Pro-Glu-Ile-Pro-Pro-OH

are not the toxic principles in the venom. It is interesting to note that they have polyproline-like helical conformation in solution [34] (cf. pp. 40, 122).

Several biologically active peptides were isolated from the venom of the American lizard *Gila monster* [35]. The remarkable homology of these venom peptides with the intestinal peptides of the secretin family is demonstrated here through the comparison of *helospectin* and *helodermin* with the vasoactive intestinal peptide (VIP, p. 168)

| | |
|---|---|
| helospectin | HSDATFTAEYSKLLAKLALQKYLESILGSSTSPRPPSS |
| helodermin | HSDAIFTEEYSKLLAKLKYALQLASILGSRTSPPP* |
| VIP | HSDAVFTDNYTRLRKQMKYAVKLNSILN* |

(where * designates a C-terminal amide group). The close relationship with VIP is also reflected in the similar pharmacological effects.

Recent discoveries of biologically active peptides in venoms, such as *mastoparan* from the Japanese hornet or *vespulakinin* from wasps, suggest that many more such interesting compounds should be found in various insects in the not too distant future.

## 8.5 Opioid Peptides

In 1975 a discovery of major importance was reported by J. Hughes, H.W. Kosterlitz and associates [36]. They found, in the brain, two closely related pentapeptides with potent opiate activity. The *enkephalins*,

| | | |
|---|---|---|
| Leu-enkephalin | H-Tyr-Gly-Gly-Phe-Leu-OH | and |
| Met-enkephalin | H-Tyr-Gly-Gly-Phe-Met-OH | |

displace naloxon, a compound with well established affinity for morphine receptors, from brain tissues. Not much later a whole series of opioid peptides were isolated, $\alpha,\beta$- and $\gamma$-endorphins [37] and the very potent dynorphin [38]. In the sequence of (human) $\beta$-endorphin

H-Tyr-Gly-Gly-Phe-Met-Thr-Ser-Glu-Lys-Ser-Gln-Thr-Pro-Leu-Val-
Thr-Leu-Phe-Lys-Asn-Ala-Ile-Lys-Asn-Ala-Tyr-Lys-Lys-Gly-Gln-OH

the first 16 residues represent the total sequence of $\alpha$-endorphin and the first 17 residues correspond to the entire molecule of $\gamma$-endorphin. The sequence of (porcine) dynorphin,

H-Tyr-Gly-Gly-Phe-Leu-Arg-Arg-Ile-Arg-
Pro-Lys-Leu-Lys-Trp-Asp-Asn-Gln-OH,

is not so closely related to that of the endorphins. The first 13 residues of the 17-peptide are responsible for its high potency.

The opioid peptides stem from a large precursor molecule in which several copies of the enkephalins are present, the ratio between Met-enkephalin and Leu-enkephalin being 6:1 [39]. The same precursor, pro-opiocortin, also contains a modified form of the enkephalin sequence in which the N-terminal tyrosine is present as the sulfate ester [40]. This kind of post-translational change has been discussed in connection with gastrin, cholecystokinin and caerulein (p. 165). The last mentioned peptide prompts us to recall dermorphin (p. 186), an opioid peptide found in the skin of an amphibian.

Opioid peptides soon became the subject of extensive studies, perhaps the broadest experienced in peptide chemistry so far. Innumerable enkephalin analogs were synthesized not least in the hope of finding a harmless substitute for morphine. Peptides were designed and prepared that could be applied instead of the initially used intraventricular route by intravenous injection, then others that could be administered intramuscularly or subcutaneously and finally orally active analogs were obtained as well. Conformationally restricted analogs, e.g. cyclic derivatives of the enkephalins were prepared for the study of the role of geometry in opioid activity and some of these were selective enough to allow differentiation between distinct, specific morphine receptors. Yet, in spite of such intensive research no useful morphine substitute emerged. It is also quite possible that morphine substitutes in general have the potential of becoming addiction drugs. Nevertheless, the discovery of the opioid peptides has already had a major consequence: it became obvious that the body itself provides materials to control pain and that morphine and other pain killers are merely active because they resemble the endogenous agonists in molecular geometry and in the location of interacting sites. Thus it was logical to assume that a similar relationship exists between other pharmacologically active agents, for instance tranquilizers, and yet to be discovered endogeneous peptides. A wide search for such biologically active peptides is in progress.

## 8.6 An Era of Discoveries

The unprecedented acceleration in the discovery of new biologically active peptides in the last two decades makes it impractical to dedicate a separate section to each novel compound. Just to indicate the multitude of peptides with interesting and widely different physiological or pharmacological activities we mention the 49 residue long human *epidermal growth factor* [41] obtained from human urine and the related *urogastrone* [42], the *delta-sleep inducing peptide* [43], H-Trp-Ala-Gly-Gly-Asp-Ala-Ser-Gly-Glu-OH or the *cerebellum specific peptide* [44] H-Ser-Gly-Ser-Ala-Lys-Val-Ala-Phe-Ser-Ala-Ile-Arg-Ser-Thr-Asn-His-OH.

During the early period of structure elucidation and synthesis of biologically active peptides in the 1950s and 1960s, those who were involved in this area of research were convinced that oxytocin, vasopressin, angiotensin or ACTH will be followed by numerous other potent compounds. Nevertheless, the actual growth of the field could not be fully anticipated. Inspection of a single catalog of a research supply company shows that in 1988 it offered more than 800, mostly synthetic, peptides for the investigator. Among the peptides listed one finds some with interesting names, such as inhibin, cerebellin, magainins, morphin modulating peptide, bursin, delicious peptide and so on. Entire groups appear, for instance the leucokinins, growth hormone releasing factors of various length, immunopeptides such as tuftsin and rigin, neuromedins and so on. Of course the biologically active fragments of proteins discussed in the preceding sections of this chapter together with their many analogs form a major part of the list as do the hormones presented in Chap. 7, the corresponding prohormones, preprohormones and the antagonists of the hormonally active peptides. While many of these materials will remain scientific curiosities, some should gain considerable significance in medicine. Recent developments in the preparation and study of synthetic vaccines leave little doubt in this respect. Treatment of the role of peptides in immunochemistry would require an article to itself. At this point we wish to stress only one aspect of peptide research: the finding of new sources of biologically active materials.

The classical sources of hormones were the endocrine glands, such as the adrenals, the thyroid or the pituitary gland. Organs previously not regarded "endocrine", for instance the gut, turned out to be abundantly rich sources of peptides with various physiological effects. Furthermore the gut peptides were shown to have influence on functions controled by the central nervous system and then were discovered to be present in the brain as well. The designation cerebrogastrointestinal hormonal polypeptides [45] is indeed justified. In a similar way the lung, thought to be merely an organ of breathing, is the site of production of several peptides, one with smooth muscle relaxing and another, related to the vasoactive intestinal peptide VIP, with muscle contracting effect, compounds being investigated by V.I. Said. Perhaps the most surprising discovery in this area are the atrial or natriuretic peptides because they prove that even the heart, regarded by some surgeons a mere pump, should be considered an endocrine organ. Extracts of auricular tissues were shown to have diuretic, natriuretic and smooth muscle relaxing effect [46]. The active materials were first obtained from the atria of rat hearts and subsequently, of course, from human tissues as well. Sequencing of a 28-peptide (of the rat) [47] was soon followed by the determination of the human sequence [48]. The 27-peptide

H-Ser-Leu-Arg-Arg-Ser-Ser-Cys-Phe-Gly-Gly-Art-Met-Asp-Arg-Ile-
Gly-Ala-Gln-Ser-Gly-Leu-Gly-Cys-Asn-Ser-Phe-Arg-OH

is only one example of the natriuretic factors of different length. The promise of possible application in medicine prompted numerous laboratories to dedicate considerable effort to the study of peptides from the heart and accordingly synthesis of atrial peptides, cloning of the corresponding DNA-s and determination of the preproatrial natriuretic factor were reported at rapid intervals [49], within the same year, 1984.

While predictions are generally not to be trusted, it would still be surprising if the atrial peptides turned to be not just the first but also the last compounds with regulatory activity in the heart. Also, it became clear that no organ should a priori be excluded from the search for biologically active peptides.

## References


1. H. Goldblatt, The renal origin of hypertension. Physiol. Rev. 27: 120–165 (1947)
2. L.T. Skeggs, F.E. Dorer, M. Levine, K.E. Lentz, J.R. Kahn, The biochemistry of the angiotensin system. In advances in Exptl. Med. and Biol. Vol. 130 (J.A. Johnson, R.R. Anderson eds., Plenum, New York 1980); L.T. Skeggs, Jr., J.R. Kahn, K. Lentz, N.P. Shumway, Preparation, purification and amino acid sequence of a polypeptide renin substrate. J. Exptl. Med. 106: 439–453 (1957); L.T. Skeggs Jr., K. Lentz, J.R. Kahn, N.P. Shumway, Synthesis of a tetradecapeptide renin substrate. J. Exptl. Med. 108: 283–297 (1958)
3. D.F. Elliott, W.S. Peart, Amino acid sequence in a hypertensin. Nature (Lond.) 177: 527–528 (1956); L.T. Skeggs, Jr., K.E. Lentz, J.R. Kahn, N.P. Shumway, K.R. Woods. The amino acid sequence of hypertensin II. J. Exptl. Med. 104: 193–197 (1956)
4. H. Schwarz, M. Bumpus, I.H. Page, Synthesis of a biologically active octapeptide similar to natural isoleucine angiotensin octapeptide. J. Amer. Chem. Soc. 79: 5697–5703 (1957)
5. W. Rittel, B. Iselin, H. Kappeler, B. Riniker, R. Schwyzer, Synthese eines hochwirksamen Hypertensin-II-amids. Helv. Chim. Acta 40: 614–624 (1957)
6. D.T. Pals, F.D. Masucci, G.S. Denning Jr., F. Sipos, D.C. Fessler, Role of the pressor action of angiotensin II in experimental hypertension. Circulation Research 29: 673–687 (1971)
7. M.A. Ondetti, B. Rubin, D. Cushman, Design of specific inhibitors of angiotensin converting enzyme:New class of orally active antihypertensive agents. Science 196: 441–444 (1977)
8. M. Rocha e Silva, W.T. Beraldo, G. Rosenfeld, Bradykinin a hypotensive and smooth muscle stimulating factor released from plasma globulin by snake venoms and by trypsin. Am. J. Physiol. 156: 261–273 (1949)
9. D.F. Elliott, G.P. Lewis, E.W. Horton, The structure of bradykinin—a plasma kinin from ox blood. Biochem. Biophys. Res. Commun. 3: 87–91 (1960)
10. R.A. Boissonnas, St. Guttmann, P.A. Jaquenoud, Synthèse de la L-arginyl-L-prolyl-L-prolyl-glycyl-L-phenylalanyl-L-seryl-L-prolyl-L-phenylalanyl-L-arginine, un nonapeptide presentant les proprietees de la bradykinine. Helv. Chim. Acta 43: 1349–1358 (1960)
11. E.D. Nicolaides, H.A. de Wald, Studies on the synthesis of polypeptides. J. Org. Chem. 26: 3872–3876 (1961)
12. R.B. Merrifield, Solid phase peptide synthesis II. Synthesis of bradykinin. J. Amer. Chem. Soc. 86: 304–305 (1964)
13. R.J. Vavrek, J.M. Stewart, Competitive antagonists of bradykinin Peptides 6: 161–164 (1985)
14. E.G. Erdös, E.M. Sloane, A.G. Renfrew, J.R. Wohler, Enzymatic studies on bradykinin and similar peptides, Ann. N.Y. Acad. Sci. 104: 222–235 (1963)
15. U.S. von Euler, J.H. Gaddum, An unidentified depressor substance in certain tissue extracts. J. Physiol. 72: 74–87 (1931)
16. J. Franz, R.A. Boissonnas, E. Stürmer, Isolierung von Substanz P aus Pferdedarm und ihre biologische und chemische Abgrenzung gegenüber Bradykinin. Helv. Chim. Acta 44: 881–883 (1961); H. Zuber, R. Jaques, The isolation of substance P from bovine brain. Angew. Chem. Int. Ed. 1: 160. (1962); K. Vogler, W. Haefely, A. Hürliman, R.O. Studer, W. Lergier, R. Strässle, K.H. Berneis, A new purification procedure and biological properties of substance P. Ann. N.Y. Acad. Sci. 104: 378–390 (1963)
17. G.W. Tregear, H.D. Niall, J.T. Potts, S.E. Leeman, M.M. Chang, Synthesis of substance P. Nature (Lond.) New Biol. 232: 86–89 (1971); R.O. Studer, A. Trzeciak, W. Lergier, Isolierung und Aminosäuresquenz von Substanz P aus Pferdedarm. Helv. Chim. Acta 56: 860–866 (1973)
18. A. Fournier, R. Couture, D. Regoli, M. Gendrau, S. St. Pierre, Synthesis of peptides by the solid phase method 7. Substance P and analogues. J. Med. Chem. 25: 64–68 (1982)
19. V. Erspamer, Active constituents of the posterior salivary glands of octopods and of the hypobranchial gland of the purple snail. Arzneimittelforschung 2: 253–258 (1952)
20. V. Erspamer, A. Anastasi, Eledoisin. Experientia 18: 58–59 (1962)
21. E. Sandrin, R.A. Boissonnas, Synthesis of Eledoisin. Experientia 18: 59–60 (1962)
22. V. Erspamer, A. Anastasi, G. Bertaccini, J.M. Cei, Structure and pharmacological action of physalaemin, the main active polypeptide in the skin of *Physalaemus fuscumaculatus*. Experientia 20: 489–490 (1964)

23. L. Bernardi, G. Bosisio, O. Goffredo, R. de Castiglione, Synthesis of physalaemin, Experientia 490–492 (1964)
24. A. Anastasia, V. Erspamer, R. Andean, Structure of uperolein a physalaemin-like undecapeptide occuring in the skin of *Uperoleia rugosa* and *Uperoleia marmorata*. Experientia 31: 394–395 (1974)
25. A. Anastasi, V. Erspamer, R. Andean, Amino acid composition and sequence of litorin, a bombesin-like nonapeptide from the skin of the Australian leptodactylid frog *Litora aurea*. Experientia 31: 510–511 (1975)
26. R. de Castiglione, M. Gigli, L. Gozzini, P.C. Montecucchi, G. Perseo, V. Erspamer, Structure, synthesis and preliminary biological results of tryptophyllin-7 and analogues. Peptides 1984, Proceedings of the 18th European Peptide Symposium (Almquist and Wiksell Internat. Stockholm, Sweden) 1984, pp. 533–536
27. P.C. Montecucchi, A. Anastasi, R. de Castiglione, V. Erspamer, Isolation and amino acid composition of sauvagine. Internatl. J. of Pept. and Prot. Res. 16: 191–199 (1980)
28. P.C. Montecucchi, R. de Castiglione, S. Piani, L. Gozzini, V. Erspamer, Amino acid composition and sequence of dermorphin, a novel opiate-like peptides from the skin of *Phyllomedusa sauvagei*. Int. J. Pept. Prot. Res. 17: 275–283 (1981)
29. P.C. Montecucchi, R. de Castiglione, V. Erspamer, Identification of dermorphin and $Hyp^6$-dermorphin in skin extracts of the Brazilian frog *Phyllomedusa rhodeii*. Int. J. Pept. Protein Res. 17: 316–321 (1981); R. de Castiglione, F. Faoro, G. Perseo, S. Piani, Synthesis of dermorphins, a new class of opiate like peptides. Int. J. Pept. Protein Res. 17: 263—272 (1981)
30. A. Anastasi, V. Erspamer, M. Bucci, Isolation and amino acid sequences of alytesin and bombesin, two analogous active tetradecapeptides from the skin of the European dicoglossid frog. Arch. Biochem. Biophys. 148: 443–446 (1972)
31. V. Erspamer, Peptides of amphibian skin active on the gut. II. Bombesin like peptides: isolation, structure and basic functions. In Gastrointestinal Hormones (G.B. Jerzy Glass ed.) Raven, New York, 1980, pp. 344–361
32. E. Habermann, J. Jentsch, Sequential analysis of melittin from tryptic and peptic fragments. Hoppe Seyler's Z. Physiol. Chem 348: 37–50 (1967)
33. M.A. Ondetti, N.J. Williams, E.F. Sabo, J. Pluscec, E.R. Weaver, O. Kocy, Angiotensin-converting enzyme inhibitor from the venom of *Bothrops jararaca*. Isolation, elucidation of structure and synthesis. Biochemistry 10: 4033–4039 (1971)
34. A. Bodanszky, M.A. Ondetti, C.A. Ralofsky, M. Bodanszky, Optical rotatory dispersion of prolin-rich peptides from the venom of *Bothrops jararaca*. Experientia 27: 1269–1270 (1971)
35. D.S. Parker, J.P. Raufman, T.L. O'Donohue, M. Bledsoe, H. Yoshida, J.J. Pisano, Amino acid sequence of helospectin, new member of the glucagon superfamily found in Gila monster venom. J. Biol. Chem. 259: 11751–11755 (1984)
36. J. Hughes, T.W. Smith, H.W. Kosterlitz, L.A. Fotherhill, B.A. Morgan, H.R. Morris, Identification of two related peptides from the brain with potent opiate agonist activity. Nature (Lond.) 258: 577–579 (1975)
37. C.H. Li, D. Chung, Isolation and structure of an untriakontapeptide with opiate activity from camel pituitary glands. Proc. Natl. Acad. Sci. U.S. 73: 1145–1148 (1976); N. Ling, R. Burgus, R. Guillemin, Isolation, primary structure and synthesis of $\alpha$-endorphin and $\gamma$-endorphin, two peptides of hypothalamic-hypophysial origin with morphinomimetic activity. Proc. Natl. Acad. Sci. U.S. 73: 3942–3946 (1976); M. Robintsen, S. Stein, S. Udenfriend, Isolation, characterization of the opioid peptide from rat pituitary: $\beta$-endorphin. Proc. Natl. Acad. Sci. U.S. 74: 4969–4972 (1977)
38. A. Goldstein, S. Tachibana, L.I. Lowney, M. Hunkapiller, L. Hood, Dynorphin (1–13) an extraordinarily potent opioid peptide. Proc. Natl. Acad. Sci. U.S. 76: 6666–6670; A. Goldstein, W. Fischli, L.I. Lowney, M. Hunkapiller, L. Hood, Porcine pituitary dynorphin: complete amino acid sequence of the biologically active heptadecapeptide. ibid. 78: 7219–7223 (1981)
39. S. Kimura, R.V. Lewis, L.D. Gerber, L. Brink, M. Rubinstein, S. Stein, S. Udenfriend, Purification to homogeneity of camel pituitary pro-opiocortin, the common precursor of opioid peptides and corticotropin. Proc. Natl. Acad. Sci. U.S. 76: 1756–1759 (1979)
40. C.D. Unsworth, J. Hughes, J.S. Morley, O-sulfated Leu-enkephalin in brain. Nature (Lond.) 295: 519–522 (1982)
41. S. Cohen, G. Carpenter, Human epidermal growth factor. Isolation and chemical and biological properties. Proc. Natl. Acad. Sci. U.S. 72: 1317–1321 (1975)
42. H. Gregory, Isolation and structure of urogastrone and its relationship to epidermal growth factor. Nature 257: 325–327 (1975)

43. G.A. Schoenenberger, M. Monnier, Characterization of a deltaelectroencephalogram (-sleep)-inducing peptide. Proc. Natl. Acad. Sci. U.S. 74: 1282–1286 (1977)
44. J.R. Slemmon, R. Blacher, W. Danho, J.L. Hemstead, J.I. Morgan, Isolation and sequence of two cerebellum-specific peptides. Proc. Natl. Acad. Sci. U.S. 81: 6866–6870 (1984)
45. V. Mutt, Questions answered and raised by work on the chemistry of gastrointestinal and cerebrogastrointestinal hormonal polypeptides. Chemical Scripta 26B: 191–207 (1986)
46. A.J. de Bold, H.B. Borenstein, A.T. Veress, H.A. Sonnenberg, A rapid and potent natriuretic response to intravenous injection of atrial miocardial extracts in rats. Life Sci. 28: 89–94 (1981); A.J. de Bold, Atrial natriuretic factor of the rat heart. Studies on isolation and properties. Proc. Soc. Exptl. Biol. Med. 1982: 133–138
47. T.G. Flynn, M.L. de Bold, A.J. de Bold, The amino acid sequence of an atrial peptide with potent diuretic and natriuretic properties. Biochem. Biophys. Res. Commun. 117: 859–865 (1983); M.G. Currie, D.M. Geller, B.R. Cole, N.R. Siegel, K.F. Fok, S.P. Adams, S.R. Eubanks, C.R. Galluppi, P. Needleman, Purification and sequence analysis of bioactive atrial peptides (atriopeptines). Science 223: 67–69 (1984)
48. K. Nakayama, H. Ohkubo, T. Hirose, S. Inayama, S. Nakanishi, mRNA sequence for human cardiodilatin-atrial natriuretic factor and regulation of precursor mRNA in rat atria. Nature (Lond.) 310: 699–701 (1984); S. Oikawa, M. Imai, A. Ueno, S. Tanaka, T. Noguchi, H. Nakazato, K. Kangawa, A. Fukuda, H. Matsuo, Cloning and sequence analysis of cDNA encoding a precursor for human atrial natriuretic polypeptides. ibid. 309: 724–726 (1984)
49. S.A. Atlas, H.D. Kleinert, M.J. Camargo, A. Januszewicz, J.E. Sealy, J.H. Laragh, J.W. Schilling, J.A. Lewicki, L.K. Johnson, T. Maack, Purification, sequencing and synthesis of a natriuretic and vasoactive rat atrial peptide. Nature (Lond.) 309: 717–719 (1984); M. Yamanaka, B. Greenberg, T. Friedmann, J. Miller, S. Atlas, L. Laragh, J. Lewicki, J. Fiddes, Cloning and sequence analysis of the c DNA for the rat atrial natriuretic factor precursor. ibid. 309: 719–722 (1984); M. Maki, R. Takayanagi, K.S. Misono, K.N. Pandey, C. Tibbets, T. Inagami, Structure of rat atrial factor precursor deduced from c DNA sequence. ibid. 309: 722–724 (1984); R.A. Zivin, J.H. Condra, R.A.F. Dixon, N.G. Seidah, M. Chretien, M. Nemer, Molecular cloning and characterization of DNA sequences encoding rat and human atrial natriuretic factors. Proc. Natl. Acad. Sci. U.S. 81: 6325–6329 (1984); N.G. Seidah, C. Lazure, M. Chretien, G. Thibault, R. Garcia, M. Cautin, J. Genest, R.F. Nutt, S.T. Brady, T.A. Lyle, W.J. Paleveda, C.D. Colton, T.M. Cicerone, D.F. Veber, Amino acid sequence of homologous rat atrial peptides: natriuretic activity of native and synthetic forms. ibid. 81: 2640–2644 (1984); C.E. Seidman, A.D. Duby, E. Choi, R.M. Graham, E. Haber, C. Honcy, J.G. Seidman, The structure of rat preproatrial natriuretic factor as defined by a complementary DNA sequence. Science 225: 324–326 (1984)

# 9 Biologically Active Peptides from Microorganisms and Fungi

In Chap. 1 (p. 14) a few naturally occurring peptides were presented as substances which quite early aroused the attention of chemists at a time when peptide chemistry still led a hidden existence. With the recognition that several hormones consist of amino acids (see Chap. 7), and with the discovery of microbial products which exhibit antibiotic or toxic properties and were also built from amino acids, interest in such materials greatly increased. In the following years amino acids were identified as building blocks of biologically active substances which had been recognized and isolated many years before pharmacologists and physicians knew about their peptide-like nature.

## 9.1 Ergot Alkaloids

Ergot is a growth on the grains of rye and other cereal plants caused by the attack of a fungus (*Claviceps purpurea*). When the rye with this disease is used for food, it sometimes causes a kind of gangrene, or severe convulsions. On chronic ingestion of ergot deadly, even epidemic intoxications have been observed in previous times when contaminated cereals were ground without inspection and removal of the ergot from the corn. *Ergotismus gangraenosus* begins with dizziness, spasms, diarrhea, nausea and ends with burning pains (ignis sacer, St. Anthony's Fire) and gangrene of the limbs. Because of its uterus-contracting activity, ergot has been used since early times in gynecology and prompted several studies to reveal its components. It was Arthur Stoll, who shortly after joining the Swiss Sandoz AG in 1920 started a systematic investigation of the ergot drug. An early isolation and characterization of "ergotinine crystallisée" and of "ergotinine amorphe" by Ch. Tanret in 1875–1879, as well as the preparation of "Hydro-ergotinin" by F. Kraft and of the identical "Ergotoxin" by George Barger and Carr (1906) had not led to a clear picture of the chemistry of the active substances. The famous pharmacologist H.H. Dale had, besides the typical uterus-contracting activity, extensively studied the sympatholytic effect of the drug, the clinical results, however, were disappointing.

Arthur Stoll after entering this much researched field, soon made important progress. By particularly mild extraction methods he succeeded in isolating crystals of a homogenous alkaloid, ergotamine, the component mainly responsible for the therapeutically useful activity. Besides ergotamine, in the following years a great number of additional substances with various effects were

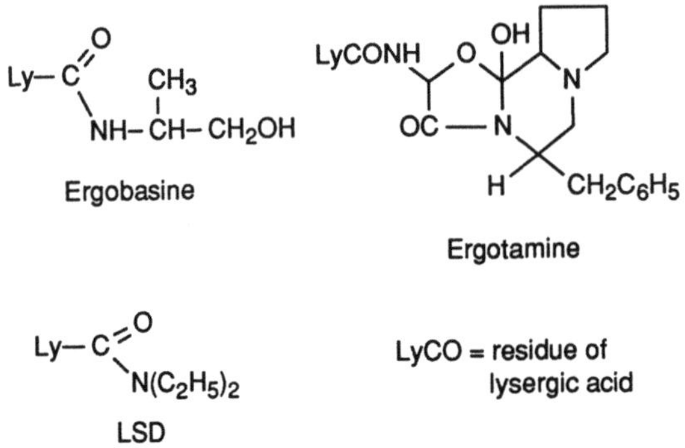


Fig. 1.

obtained, all of them derivatives of lysergic acid. Lysergic acid (Ly-$CO_2H$), a complicated four-ring heterocycle, not formulated here, biosynthetically derived from the amino acid tryptophan, forms two series of ergot alkaloids: 1. relatively simple amides for instance with L-alaninol (1-hydroxy-2-aminopropane): ergobasine (the synthetic diethylamide is the extremely strong hallucinogen LSD), and 2. The peptide series of which ergotamine is the most important representative (Fig. 1).

As was shown by Jacobs and Craig at the Rockefeller Institute and by Stoll and coworkers at Sandoz, the peptide moiety consists of the diketopiperazine from the amino acids L-proline and L-phenylalanine, combined with an α-hydroxy-L-alanine residue with its oxygen bound to the 6-membered ring thus forming a cyclol structure.

### 9.1.1 Cyclols

As mentioned on p. 3, cyclols are potentially formed by addition of an NH of a peptide bond to an opposite carbonyl (C O); in the present case a hydroxyl group is added to a peptide bond, hence, generating an "oxa-cyclol".

In the synthesis of ergotamine by A. Hofmann et al. (1961) [1] the cyclol moiety is formed by addition of the OH group, set free by hydrogenation, in the α-carboethoxy-α-benzyloxypropionyl residue bound to L-phenylalanine-L-proline anhydride (diketopiperazine) (Fig. 2). The precursor of the oxacyclol part in the biosynthesis by *Claviceps* is an alanyl moiety into which aerial oxygen is introduced by enzymatic hydroxylation.

The unique structure of the ergot peptides prompted several laboratories to investigate closer the chemistry of cyclols. The Swiss chemists by following the course of oxacyclol formation with simple model components obtained cystalline products from α-hydroxyacyl lactames as indicated in Fig. 3; in a supposed equilibrium reaction the cyclols here seem to be energetically favored.

β-Hydroxyacyl lactames under the same conditions yielded macrocylic lactones, an insertion reaction of the acyl residue bound to the amide nitrogen into the newly formed ring [2].

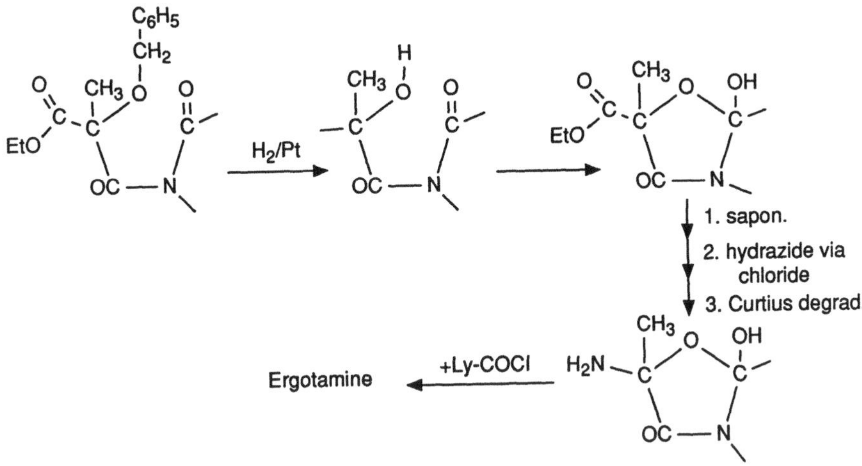


**Fig. 2.** Final steps in synthesis of ergotamine [1]

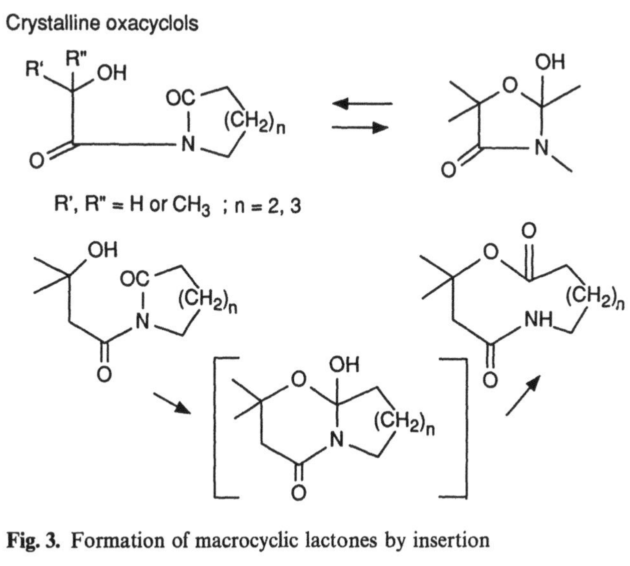


**Fig. 3.** Formation of macrocyclic lactones by insertion

In a parallel study Soviet chemists engaged in cyclodepsipeptide research (p. 20), studied hydroxyacyl incorporation in more detail [3] and arrived at a 14-membered cyclic depsipeptide, serratamolide, by double insertion of *β*-hydroxyacyl residues into a diketopiperazine (Fig. 4). Here, we would like to remind the reader of the early experiments of Max Bergmann in 1929 with *N*-acetyl-diketopiperazine, through which the "activation" of acyl residues on amide nitrogen first became apparent (p. 46).

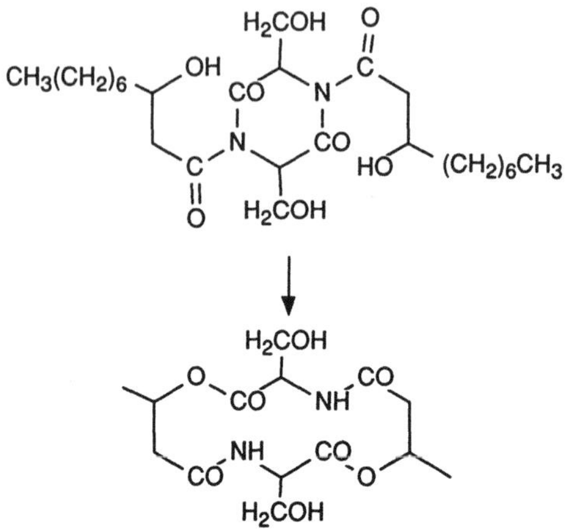


**Fig. 4.** Serratamolide by double-insertion

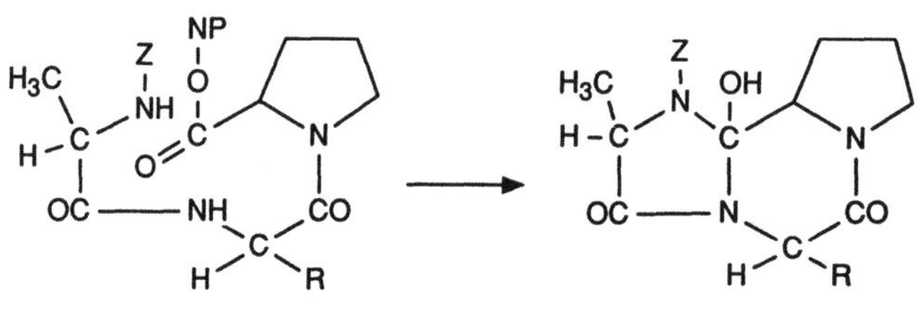


**Fig. 5.** Azacyclol from activated tripeptide

Genuine cyclols, azacyclols, were obtained in 1971 by Lucente and Romeo [4] from N-protected (Z-group) tripeptide *p*-nitrophenylesters, e.g. Z-L-Ala-L-Phe-L-Pro-ONP (Fig. 5). Presumably a cyclic tripeptide intermediate, due to the close proximity of the reactive groups, spontaneously forms the azacyclol.

Early attempts to achieve analogous syntheses with unprotected tripeptides failed. As M. Rothe et al. [5] showed, e.g. with *N*-glycylglycine anhydride (*N*-glycyl-diketopiperazine), azacyclols having hydrogen-atoms adjacent to the cyclol group will eliminate $H_2O$ to yield imidazolones or anhydrocyclols. Quite recently, however, Rothe and Roser reported the successful isolation of four tautomers, two cyclols and two cyclotripeptides from tripeptides of the type pro-X-X [6]. Thus, *N*-prolyl-cyclo-dialanine forms the "primary" cyclol which, via transannular ring opening (cyclotripeptides) isomerizes to yield the more stable "secondary" cyclol (Fig. 6).

Aminoacyl insertion as a peptide forming reaction was studied as early as 1956 when it was shown [7] that *N*-bisaminoacylamides undergo extremely facile

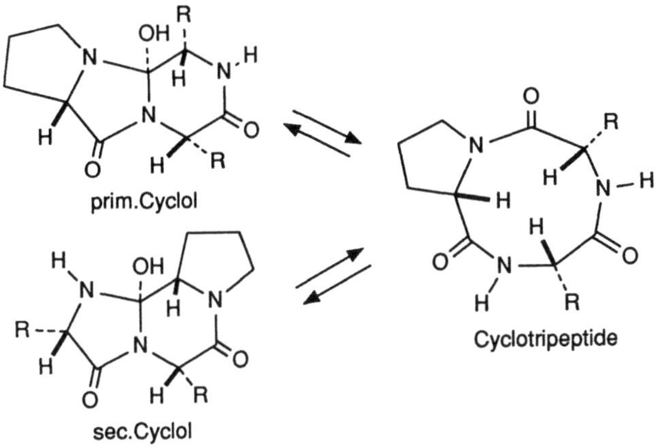


**Fig. 6.** Azacyclols

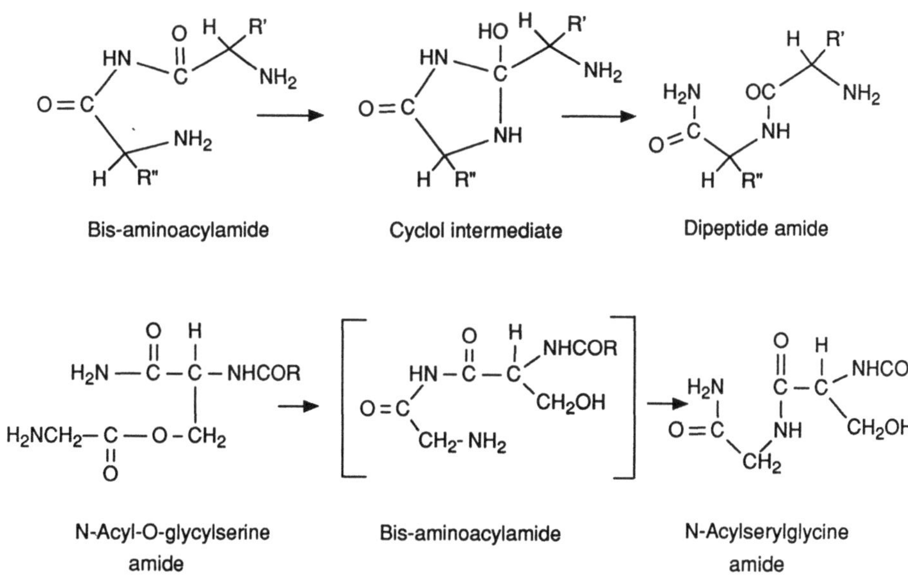


**Fig. 7.** Possible mechanisms of aminoacyl insertion

rearrangement to peptide amides. M. Brenner's group engaged in amino acid insertion reactions, first in 1958 with peptides of salicylic acid, later reported similar rearrangements e.g. of *N*-acyl-*O*-glycylserine amide in the presence of *tert*-butoxide yielding *N*-acylseryl-glycine amide (Fig. 7) [8].

Thiacyclols containing a sulfur atom instead of the oxygen or nitrogen bridge are the compounds which complete the class of cyclols. The tripeptide corresponding to the peptide part of ergotamine was prepared by G. Zanotti et al.

from D,L-α-mercapto-propionyl-L-phenyl-alanyl-L-proline *p*-nitrophenylester. The diastereomers were separated by chromatography; they crystallized and thus could be subjected to X-ray structure analysis [9]. Thioergotamine has not been synthesized so far.

## 9.2 Peptide Antibiotics

### 9.2.1 Penicillins and Cephalosporins

The observation that two different microorganisms in the same nutrient medium can influence their mutual growth was made by L. Pasteur and Joubert as far back as 1877. More interesting than symbiosis was the phenomenon of "antibiosis", a consequence of excretion by one organism of substances that are highly toxic to another, a pathogenic microbe. Clinical application of such substances, however, was delayed until extraction and purification techniques had reached an appropriate state. The most famous instance is the discovery of penicillin. In 1929 Alexander Fleming described the antibiotic effect of a fungus, defined by him as *Penicillium rubrum* (later identified as *Penicillium notatum, Westling*) on numerous bacteria such as Staphylococci, Streptococci, Gonococci; he named the antimicrobial agent "penicillin". Attempts at isolation of a well defined substance from the growth medium, however, failed because of the instability of the penicillins, and thus work in this area was not continued. With the outbreak of the war in 1939 an intensified search for antiinfectious drugs was started. The pathologist H.H. Florey in Oxford suggested resuming the investigation of penicillin with the strain still kept in culture by Fleming. Together with E. Chain and other chemists, the "Oxford group" in 1940 succeeded in isolating a product which, although containing only about 3% active substance, showed high antibacterial effect in infected animals. This was the signal for several laboratories in the USA to join forces with the scientists in Oxford. This led in the following years to a parallel effort of hundreds of researchers in numerous laboratories and to the large scale production of pure penicillins in USA and in Great Britain. The rather difficult elucidation of the structure of the American Penicillin G after several erroneous formulas, finally established by X-ray diffraction analysis, revealed a hitherto unknown thiazolidin-β-lactam system coupled to a phenylacetic acid residue (Fig. 8). In natural analogs as well as in derivatives obtained by adding artificial components to the growing microorganism in the fermentors, and also in analogs obtained by chemical coupling (semisynthesis), the penicillin nucleus (6-aminopenicillanic acid) appears combined with various acyl components like phenoxyacetic acid (Penicillin V), D-phenylglycine (Ampicillin) or L-α-aminoadipic acid (Isopenicillin N). The almost countless penicillins differ greatly in their stability against acids and enzymatic inactivation by "penicillinase" and also in the spectrum of their antibiotic effectivity against gram-positive and gram-negative bacteria. The early history of penicillin has been presented in a large collection of papers. The Chemistry of Penicillin [10].

Penicillins, Cephalosporin

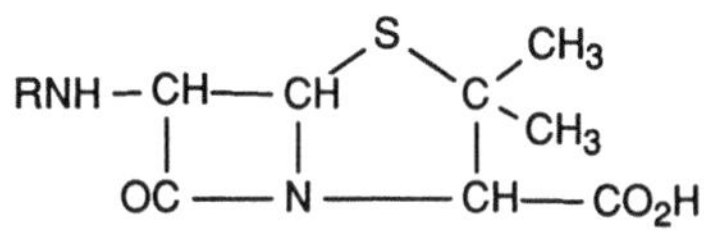

| R | Name |
| --- | --- |
| H | 6-Aminopenicillanic acid |
| $C_6H_5CH_2$ CO | Penicillin G |
| $C_6H_5$- O - $CH_2CO$ | Penicillin V |
| $C_6H_5CH(NH_2)CO$ | Ampicillin |
| $HO_2C$ $CH(NH_2)$- $(CH_2)_3CO$ | Penicillin N |

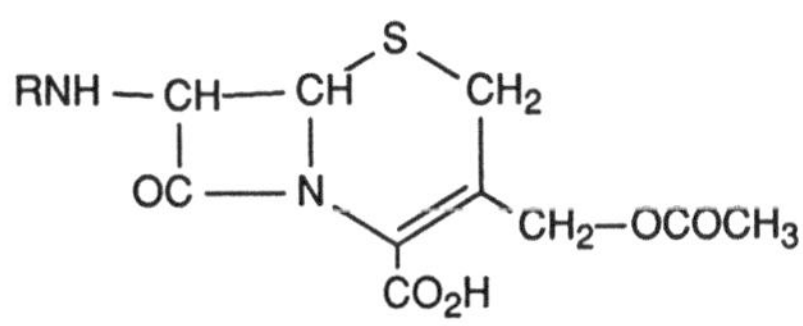

**Fig. 8.** R like in Penicillin N: Cephalosporin C. R = H; 7-Aminocephalosporanic acid

The peptidic nature of penicillin was not recognized immediately after its isolation. In its $\beta$-lactam bicyclic ring structure the tripeptide $\delta$-(L-$\alpha$-aminoadipoyl)-L-cysteinyl-D-valine is hidden but in 1960 it was discovered in extracts of Penicillium by Arnstein et al. [11]. Experiments on biosynthesis of penicillins in the intact mycelium of the producing fungi did not lead to unequivocal results, due to the poor permeability of the cell wall. Protoplasts, "naked" cells obtained after enzymatic removal of the outer wall, however, were able to serve in studies with radioactively labeled potential precursors, and in cell-free systems from *Cephalosporium acremonium*, the route to isopenicillin N was finally revealed [12]. The tripeptide L-$\alpha$-aminoadipoyl-L-cysteinyl-D-valine first forms the $\beta$-lactam moiety and a second, oxidative step closes the thiazolidine ring between the $\beta$-carbon of valine and the thiol of cysteine (Fig. 9).

The chemical synthesis of penicillin was attempted at a time when no correct structural formula was yet established. V. du Vigneaud who was involved in the venture from the beginning (see p. 154) tried to synthesize a putative azlactone compound and eventually, by empirical experimentation came to a small amount of penicillin G. J.C. Sheehan's group at the Massachusetts Institute of Technology (MIT) in Cambridge, USA, in the second half of the 1950s worked out several preparatively rewarding syntheses, making use of protecting groups of amino- and carboxyl functions just invented at that time. One important step in their syntheses was the closure of the $\beta$-lactam ring between the intermediate carboxylic acid and the thiazolidine nitrogen atom which succeeded surprisingly

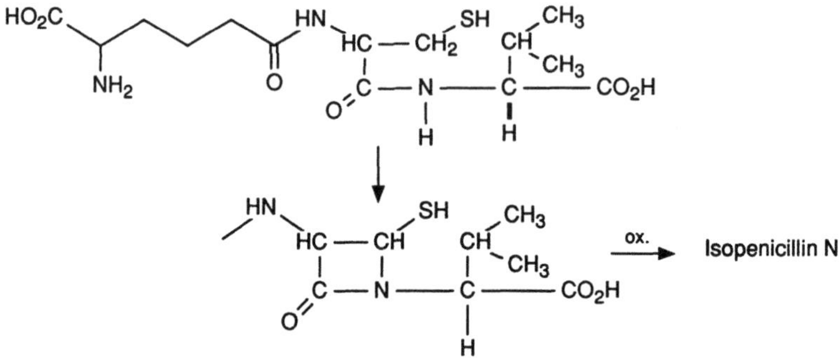


**Fig. 9.** Biosynthesis of penicillin

well with Sheehan's carbodiimide reagents (see p. 88). Since in all syntheses several separations of isomeric intermediate products are required (penicillins contain three chiral centers and the "wrong" diastereomers are inactive), none of the processes can compete with production by fermentation. The *N*-acyl residues of the different natural penicillins can be removed without destruction of the ring system; the 6-amino-penicillanic acid so obtained can be coupled with any desired acid. Most of the commercially available penicillins are manufactured by this semisynthetic route.

During a chemical study of the structure of penicillin N, produced by a species of *Cephalosporium* from Sardinia, an acid-stable accompanying substance was discovered by E.P. Abraham and G.G.F. Newton in Oxford in the early fifties. It exhibited a certain antibacterial activity against penicillin-resistent strains and closely resembled the penicillins in some of its chemical and biological properties, but differed from them strikingly in others, e.g. by its resistance against the $\beta$-lactam-splitting penicillinase. Nevertheless, Cephalosporin C, as it was named [13], turned out to be a $\beta$-lactam. It contains a $\beta$-lactam-dihydrothiazine ring system as shown in Fig. 8. In the structure the same tripeptide as in the penicillins, D-$\alpha$-aminoadipoyl-L-cysteinyl-D-valine is hidden. The cephalosporins, in position 3 of the dihydrothiazine ring, contain an acetoxymethyl group which has been formed by oxygenation of the original methyl group after the establishment of the six-membered ring. Experiments with cell-free extracts of *Cephalosporium acremonium* and several mutant strains indicate that the thiazolidine ring of penicillin N is converted to desacetoxy cephalosporin C (3-methyl group) which is then oxidized and acetylated to yield the final product. Thus it is very probable that the cephalosporins in the cell are formed via penicillin. Chemically, a great deal of experimentation has also been carried out with cephalosporins. As with penicillins it was possible to remove the side chain acyl residue from the amino group to produce the antibiotically inactive 7-aminocephalosporanic acid. This compound has been converted by acylation with a variety of different acids to clinically useful antibiotic drugs.

Cephalosporin C was prepared through total synthesis by the famous organic chemist R.B. Woodward (Plate 50) and his associates in a sequence of ingenious reactions steps. To mention only the β-lactam-dihydro-thiazine ring system, this has been constructed not, as in the penicillins by closure of the β-lactam ring in the last step, but by closing the six membered thiazine ring as attached to the preformed β-lactam moiety [14].

Penicillins and cephalosporins specifically inhibit late stages of the enzymatic construction of the bacterial peptidoglycan cell wall component, a network of peptides and polysaccharides. Since mammalian cells do not possess such a cell wall, the β-lactam antibiotics are very specific and virtually non-toxic. This ideal property is not shared by other bactericidal substances, which impair or prevent growth of microorganisms by other, less specific, reaction mechanisms. This is true, e.g. for compounds affecting ion transport in lipid membranes.

### 9.2.2 Valinomycin

In 1955, H. Brockmann and Schmidt-Kastner [15] isolated an antibiotic substance from extracts of *Streptomyces fulvissimus*. They named it valinomycin after valine having been found as the only amino acid in the acid-hydrolyzate. Since no amino group nor carboxyl group could be detected in the substance which was almost insoluble in water, a cyclic structure had to be assumed. Valinomycin has a macrocyclic molecular structure consisting of three identical tetradepsipeptide fragments with alternating peptide and ester bonds between D-α-hydroxyisovaleric acid, D-valine, L-lactic acid, L-valyl residues (Fig. 10).

The antibiotic has been found to be active against a number of bacteria, yeasts, and fungi, and later to uncouple oxidative phosphorylation of mitochondria.

In 1964, Moore and Pressman [16] discovered that valinomycin induces $K^+$-uptake in mitochondria. By various methods it was then demonstrated that in alcoholic solution the depsipeptide forms very stable complexes with $K^+$, $Rb^+$ and $Cs^+$-ions. Since then the investigation of mechanisms by which certain substances facilitate ion transport in lipid membranes has developed into a major field in biophysics. Besides the "carrier" transport mentioned here, there also exists a "channel" mechanism.

The stability constant, $K_{stab}$, of the potassium-valinomycin complex in ethanol is rather high (ca. $10^6$/M vs $10^1$/M with $Na^+$), hence the conformation of the $K^+$ complex in solution is rigid, therefore stable and well defined. The geometry of the complex has been determined by the combined use of spectroscopic methods in several laboratories of which here only the results of

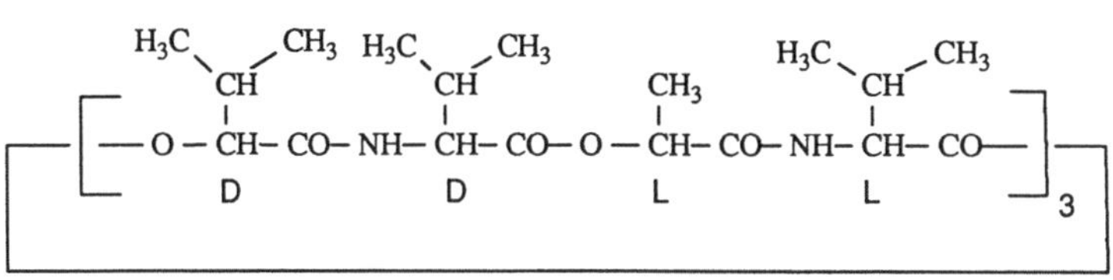

**Fig. 10.** Valinomycin

Yu.A. Ovchinnikov's group in the Shemyakin Institute of the Soviet Academy of Sciences in Moscow (Plate 33) will be mentioned [17]. Figure 11 shows the structures proposed from physico-chemical data (A) compared with the structure obtained by X-ray diffraction analysis (B) [18].

In both structures, which agree rather well, the $K^+$-ion is totally encapsulated, coordinated with six carbonyl oxygen atoms of ester groups in a nearly regular octahedron.

Non-complexed valinomycin shows considerable conformational variability, at least three principal forms whose predominances in an equilibrium depend on the polarity of the solvent. The antibiotic activity of valinomycin is certainly due to impairment of alkali ($K^+$) ion transport in bacterial membranes. This follows, for instance from the fact that none of the numerous synthetic non-complexing analogs has antibiotic activity.

Valinomycin was first synthesized by Shemyakin (Plate 42) and coworkers in 1963 [19]. In their first attempt they tried to confirm the structure as originally published [15], corresponding to an incorrect molecular weight pointing to four instead of six valines per molecule. The discrepancy between the properties of the synthetic and the natural product, finally led to the correct formula of the depsipeptide.

The synthesis of depsipeptides follows the principles of the synthesis of homodetic peptides. The amino acid components are protected in the same way, here in the first synthesis by the phthaloyl group. The hydroxy acids have to be transiently protected only at the carboxyl group, by esterification, in this case as benzyl esters. For coupling of amino acids to hydroxyl groups of the α-hydroxy acids the usual peptide coupling methods are applied, the coupling of the carboxyl end of a hydroxy acid unit to an amino group of an amino acid occurs as

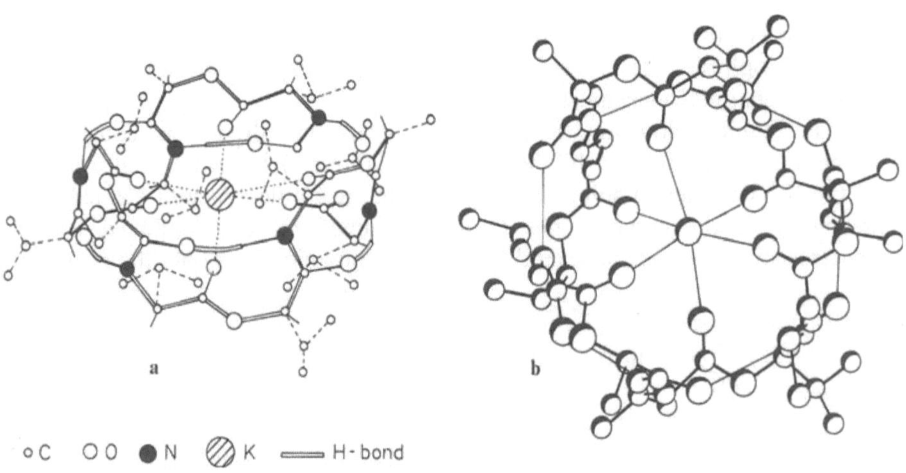


**Fig. 11 a, b.** Space formula of valinomycin-$K^+$; **a** from Ref. 17, **b** from Ref. 18

in peptide coupling. Activation in the form of acid chlorides was the method used for cyclization preferentially by the USSR chemists.—The all-peptide analog of valinomycin, cyclo(L-Val-Gly-Gly-L-Pro)$_3$ was synthesized in E.R. Blout's group. In acetonitril solution the stability of the peptide-$K^+$ complex was comparable to that of the valinomycin-$K^+$ complex [20].

Ionophoric ligands other than depsipeptides and peptides have been found in nature or have been prepared by synthesis. As nitrogen-free macrocycles the *depsides* may be mentioned, compounds of similar geometry in which instead of amide bonds hydroxy-acid building blocks are joined by ester linkages. A most prominent instance is the series of "nactins" (nonactin, monactin, dinactin etc.), produced by *Actinomyces* species which are, according to H. Gerlach and V. Prelog cyclic ester-tetramers of the unit

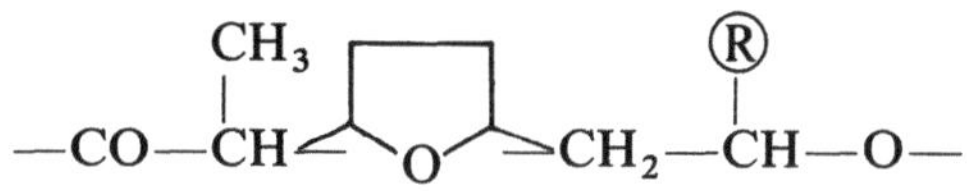


($R = CH_3$ or $C_2H_5$) [21]

Artificial ionophores have been known since 1967 when C.J. Pedersen published his "crown ethers", macrocyclic polyethers of any desired number of members, built according to the general principle $[\text{—O—}\overset{|}{\text{C}}\text{H—}\overset{|}{\text{C}}\text{H—}]_n$. These, depending of ring size, form complexes with a large number of mono-, di-, and trivalent metal ions [22].

The interesting complexone research, in the following years, spread to may laboratories, by constructions of macrobicyclic and -tricyclic (3-dimensional) cage-like ether- or amino "cryptands" mainly by J.M. Lehn and coworkers, efforts eventually leading to the Nobel Prize in 1987 [22].

### 9.2.3 Gramicidins

The eldest and most thoroughly studied peptide antibiotic, gramicidin S, does not belong to the ionophoric compounds but exerts its bactericidal effect by membrane interaction the exact mechanism of which is yet not exactly known. Historically, the term "tyrothricin" has to be mentioned before "gramicidin". It was as early as in 1939 that René Dubos of the Rockefeller Institute, New York, published his discovery in soil samples of an antibiotic producer, *Bacillus brevis*, that secreted a peptide mixture with antibiotic properties against pathogens. Pneumococci infected mice, after injection indeed recovered but later died from the hemolytic action of the drug. From the impure extract, homogenous substances, tyrocidines and gramicidins were isolated in the following years by Craig's counter-current distribution (p. 55). At about the same time from another strain of *Bacillus brevis* a crystalline antibiotic substance was obtained in the Soviet Union. In the Moscow Central Institute of Malaria and Parasitology, Georgyi Gause and his wife Mariya Brazhnikova isolated a substance with high effectivity against grampositive cocci, that was named gramicidin S (Soviet), although later it was shown to be of the tyrocidin type.

In the 1940s, gramicidin S became a popular model in peptide chemistry, for testing separation methods, degradation procedures, characterization, theories

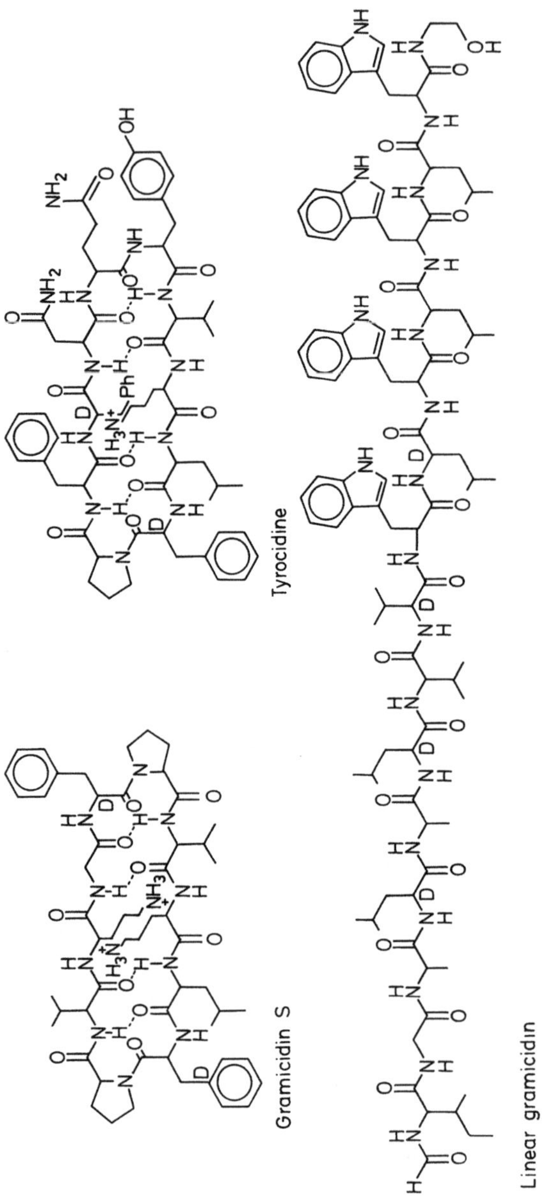


**Fig. 12.** Antibiotics from *Bacillus brevis*

of structure, etc. The analysis of its amino acid components after hydrolysis was one of the first applications of the methods of Synge, and, particularly, of the paper chromatography methods of Consden et al. (see p. 53). The arrangement of the five constituting amino acids—valine, ornithine, leucine, D-phenylalanine, proline—was determined in a classical study by Consden, Gordon, Martin and Synge, [24], who partially hydrolyzed the peptide with HCl-acetic acid, separated, by paper chromatography, the di- and tripeptides formed, determined their composition, before and after deamination by nitrous fumes, and combined the information so obtained, mainly by overlapping of the fragments, to conclude with the above sequence. The D-configuration of phenylalanine was recognized by its optical rotation. The occurrence of L-ornithine, an amino acid not present in proteins was a little surprising at a time when uncommon amino acids in natural peptides were not yet trivial. Since the antibiotic lacked a carboxyl- or amino terminal group, it was clearly a cyclic peptide, but whether the ring contained the sequence only once or twice had to be decided by molecular weight determination. Gramicidin S turned out to be a cyclic decapeptide with the repeated sequence shown in Fig. 12.

The synthesis of gramicidin S was accomplished in 1957 by Schwyzer and Sieber [25]. At that time the synthesis of the linear decapeptide that finally had to be cyclized, was by no means a trivial task. American Scientists (Erlanger, Sachs, Brand of Columbia University in New York) had already developed a synthesis of HVal-(Tos)Orn-Leu-D-Phe-Pro-Val-(Tos)Orn-Leu-D-Phe-Pro-OME in which the free $\delta$-amino group of ornithin had been protected by the tosyl residue. The main task now was to generate the cyclic structure of the antibiotic. Only a few cyclic peptides had been obtained prior to this attempt. The Swiss scientists approached this goal by preparing for cyclization the above decapeptide through the introduction of the triphenylmethyl (Trt) building block, saponification of the methylester, activation of the carboxyl end as *p*-nitrophenylester (by reaction with di-*p*-nitrophenylsulfite), and liberation of the Trt-amino end by weak acid.

$$\text{HVal.....ProOCH}_3 \rightarrow \text{TrtVal.....ProOCH}_3 \rightarrow$$
$$\text{TrtVal...ProOH} \rightarrow \text{TrtVal...Pro-ONP} \rightarrow \text{HVal.....Pro-ONP} \rightarrow \text{cycle}$$

## Cyclization of Linear Peptides

As an intramolecular peptide coupling reaction cyclization can be achieved by the general methods of peptide synthesis, i.e. by any kind of activation of the carboxyl terminal group. Principally it is essential to work at high dilution to suppress bimolecular reactions leading to polycondensation. For practical purposes about $10^{-3}$ M concentrations are applied, yet the yields of cyclic peptides are far below 50% in most cases. The high dilution can conveniently be achieved by adding the reactive compound dropwise over several hours to a large volume of the solvent. Activation of the carboxyl group of the linear peptide to be cyclized must be carried out with the terminal amino group being protected in such a way that deprotection, which permits internal peptide synthesis, can take place without affecting the activated carboxyl group (Fig. 13). This is possible, for

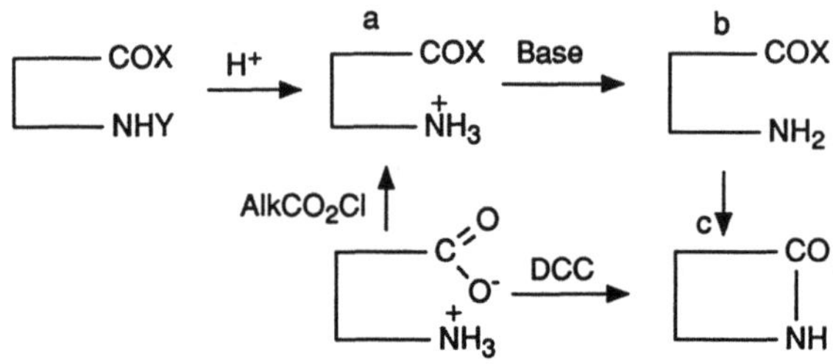


**Fig. 13.** Cyclization of linear peptides

instance, with the combination of active esters (—CO X), which are relatively stable against acids (Chap. 4) and acidolytically removable N-protecting groups (—NHY) (Chap. 3). In this case the amino group simultaneously will be reprotected by protonation (a). Cyclization then is conducted by adding the substance to a large volume of base (pyridin, triethylamine) in which the free amino end (b) can react with the activated carboxyl end to generate the cyclic peptide (c). In the synthesis of gramicidin S, the solution of the linear decapeptide (a) was dropped into pyridin at 55 °C for several hours.

The intermediate (a) can also be arrived at directly from the unprotected linear peptide by applying the mixed anhydride method. After addition of one equivalent of acid, e.g. trifluoroacetic acid, the amino group will be protonated and the (still deprotonated) carboxyl anion will react with alkyl chloroformate to form (a), X = CO—O—Alk. Very conveniently, carbodiimides can be reacted with linear peptides unprotected at both ends to form cyclic peptides in satisfactory yields [26]. Since the carbodiimide method, particularly in the presence of *N*-hydroxybenzotriazole, is causing little racemization (see p. 93) this system is preferred in most laboratories.

An interesting observation was the so called "doubling reaction". This effect was noted after the first synthesis of a cyclic peptide. Sheehan and Richardson in 1954 claimed to have obtained cyclotriglycine from triglycine azide, but found a little later that their product, in fact, was cyclohexaglycine [27]. Schwyzer, who studied the phenomenon in detail reproduced this result with various active esters of the tripeptide HGlyGlyPheOH all of which yielded the identical cyclic hexapeptide. This is easy to understand by considering the conformations of the reactants. A tripeptide, as a consequence of the prevailing *trans*-configuration at the peptide bonds, most probably will exist in solution in a relatively stretched shape so that a mutual reaction between the two ends is quite unlikely. Instead, even at high dilution, the reaction between two molecules will prevail to produce a hexapeptide intermediate, that easily forms a cyclic peptide containing six *trans*-peptide bonds. Maintaining three *trans*-peptide bonds a ring can not be formed, with *cis*-peptide bonds, however, a tricyclic peptide is very stable; for tertiary amide bonds, $—CO—N\begin{smallmatrix}\diagup R\\ \diagdown R'\end{smallmatrix}$, e.g. of proline or of sarcosine (*N*-methylglycine), the *cis*- and *trans*-configurations are of comparable stability, the energy barrier to a reversible conversion is low. Thus H-Pro-Pro-Pro-*p*-nitrophenylester yielded cyclotriproline in 90% yield and the same held for cyclotrisarcosine.

For the existence of a cyclic *tripeptide* one proline together with two α-amino acids can be sufficient; M. Rothe succeeded in the preparation of cyclo(Pro-Val-Val), yet not by cyclization of the corresponding linear tripeptide but by the insertion reaction of L-proline into the diketo-piperazine of L-valine [6]. For cyclotripeptide-cyclol equilibrium see page 197.

In linear *tetrapeptides* for the ends to meet, only one of the four peptide bonds needs to have the *cis*-configuration. Thus the exclusive presence of imino acids is not essential, cyclic tetrapeptides were found in nature even with exclusively α-amide bonds. R.O. Studer in 1969 synthesized the natural *fungisporin*, cyclo-(D-Val-L-Val-D-Phe-L-Phe) [28]. When discussing cyclization of *pentapeptides*, we shall come back to the doubling reaction and to gramicidin S.

The antibiotic gramicidin S consists of two identical *pentapeptide* moieties. Schwyzer and Sieber, who carried out the first synthesis of the cyclic peptide by cyclization of the linear decapeptide (see above) in their attempts to obtain a cyclic pentapeptide, half gramicidin, observed cyclodimerization of two identically activated pentapeptides, H-Val-(Z)Orn-Leu-D-Phe-Pro-*p*-nitrophenylester, even in rather dilute solution, in a yield even better than with the linear decapeptide [29]. In cyclodimerization experiments with numerous analogous pentapeptide *p*-nitrophenylesters were carried out mainly by the group of Izumiya in Japan at the end of the 1960s and it was found that the ratio of cyclic pentapeptide to decapeptide is very sensitive to the nature of the terminal amino acids. The bulkier the N-terminal amino acid side chain the more dimeric product was found. It is clear that steric factors, among them conformations that allow more or less close association of the monomers, also through hydrogen bridges, play an essential role. Interestingly, this doubling reaction has not been observed with linear tetra- or hexa-peptides, but it occurs readily with tetra- and hexa-depsipeptides as the Moscow peptide chemists found during their studies of syntheses of ionophoric cyclic depsipeptides (see p. 202).

Preparation of cyclic peptides from linear *hexapeptides* and from longer ones does not offer special problems. Such compounds occur in great numbers in Nature. Some of them exhibit conspicuous biological activities; these became the subject of investigation of the molecular mechanism of their interaction with receptors and of their conformations. The mechanisms of their biosynthesis are also extremely interesting.

Elucidation of the biosynthesis of gramicidin S and of tyrocidin (Fig. 12), an additional component in Dubos' tyrothricin, represents an important period in the development of our understanding of biochemistry. It took place at the time when the mechanism of protein-biosynthesis had just been recognized. In the 1950s, P. Zamecnik discovered that the synthesis of proteins in the cell occurs in organelles, later designated as ribosomes. Then M. Hoagland found a type of RNA, the transfer RNAs (t-RNAs) to which, in the presence of ATP, amino acids are attached in activated form. H.G. Zachau, G. Acs and F. Lipmann identified the mode of activation as esters of an adenosin residue in the t-RNA [30].

Building of the peptide chain takes place through aminolysis of the ester group by the amino group of the next actively bound amino acid. Thus the chain grows "from below": the first amino acid (formylmethionine) remains N-terminal throughout the process, the last t-RNA-bound amino acid forms the C-terminal, unlike in peptide synthesis in the laboratory, either in solution or on a solid support, where the chain is lengthened from the protected (anchored) carboxyl end toward the amino end of the peptide.

There is ample information available in any textbook of biochemistry about the template mechanism that starts with the gene (a DNA) and, through a specific t-RNA and an appropriate messenger RNA, provides for the strictly determined sequence of the building components in various proteins.

The cyclic microbial peptides are produced in *non-ribosomal* systems. The classical example is the biosynthesis of gramicidin S and of the tyrocidins, elucidated in three laboratories. The synthesis of gramicidin S in cell-free solution was first accomplished in Oslo (S.G. Laland and his associates). Kurahashi and his coworkers, in Osaka (1966/67) determined the consumption of one mole ATP per mole of peptide bond formation and separated the enzyme system, gramicidin S-synthetase, into two components. Two years later at a meeting in Cold Spring Harbor with Fritz Lipmann and his group, who had been very successful in their study of ribosomal protein-synthesis, it was established that the activation of amino acids in the synthesis of microbial peptides is different from their activation in protein-biosynthesis. It might be interesting to note that one of the present authors, impressed by the "activated acetic acid" of Lynen (1950), as early as 1952 used thiol esters of amino acids as activated peptide components in experimental work (see p. 78) and gave consideration to their possibly being the activated intermediates in the then still unclarified biosynthesis of proteins in the cell [31].

A key contribution to the understanding of gramicidin biosynthesis was the characterization of the intermediate aminoacyl- and peptidyl-residues as thiolesters by W. Gevers and H. Kleinkauf in Lipmanns laboratory [32]. These compounds were found to fulfill the stability criteria for activated *S*-acyl intermediates of fatty acid biosynthesis as summarized by Feodor Lynen: resistance against acids, lability against alkali, formation of hydroxamic acids with hydroxylamine (yielding a cherry-red color with $Fe^{3+}$) and cleavage by mercury ions. The authors incubated the purified synthesizing enzymes with the amino acid components of the antibiotic in the presence of ATP and $Mg^{++}$, isolated the loaded protein and detached the covalently found amino acids and peptides from the protein. Analysis by paper chromatography and by paper electrophoresis revealed single amino acids and a series of peptides with amino acid sequences of the nascent chain as D-Phe-Pro, D-Phe-Pro-Val, D-Phe-Pro-Val-Orn and D-Phe-Pro-Val-Orn-Leu. This pointed to a "thiotemplate" mechanism in which the respective amino acids were actively bound to the enzyme in the required sequence. In order to react one amino acid with the next distant one a carrier is required which was soon discovered to be phosphopantetheine covalently linked to the larger of the two enzymes functioning as a rotating

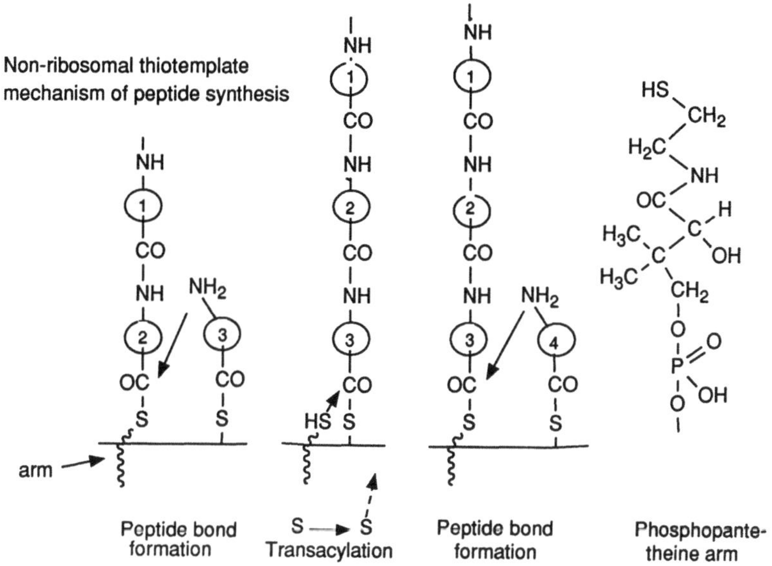


**Fig. 14.** Non-ribosomal thiotemplate mechanism of peptide synthesis

arm that accepts and transports the amino acids and growing peptides (Fig. 14).

This mechanism also holds for the biosynthesis of the tyrocidins and many additional cyclopeptides, depsipeptides and for the gramicidins A, B and C, as well.

The antibiotic mixture produced by *Bacillus brevis* contains a group of related peptides, the gramididins A, B and C. Due to the lack of a free amino and carboxyl end, they too were first assumed to have cyclic structures but R. Sarges and B. Witkop (Plate 49) showed that they are linear pentadecapeptides, with a formyl group blocking the N-terminal and ethanolamide at the carboxyl end [33]. One structure is shown in Fig. 12. Note the great number of six D-amino acids in the chain.

The gramicidins possess antibiotic activity against many Gram-positive microorganisms. At biological and artificial membranes they produce large and stepwise changes of ion conductance. A direct investigation of the ion-binding properties has, however, shown that the peptides bind cations very weakly. Therefore they are believed to act by forming ion conducting channels in the membrane. Interesting hypotheses have been developed on the conformational states of the linear gramicidins involved in transmembrane channel formation. The ion conducting channel may be formed by two gramicidin molecules in helical conformation connected head to head by hydrogen bonds (D.W. Urry, 1971). The tube is long enough to span across a lipid bilayer membrane and has an internal cavity that can accomodate metal ions. According to another model (E.R. Blout et al., 1974) two gramicidin molecules form an antiparallel $\beta$-sheet wound into a helical channel whose length and central hole are compatible with the requirements of ion conductance.

### 9.2.4 Alamethicin

Following the first recognition of transmembrane ion-migration without metal complexing carriers, this fascinating field, with its close connection to nerve physiology, has been widely explored with the help of newly discovered ion conducting drugs. Alamethicin, a product of the fungus *Trichoderma viride* [34] is a weak antibiotic which on natural and artificial membrane bilayers exerts voltage-dependent ion conductance phenomena. Alkali metal ions are transported through pores which can adopt discrete states depending on the pore size. Formation and variation of pores in the membranes is assumed to be based entirely on (voltage-dependent) dipole-dipole interactions within aggregates of α-helices formed by the polypeptides [35].

Alamethicin originally considered to be a homogenous peptide turned out to be a mixture of closely related nonadecapeptides (19 amino acids). Since both a free amino end group and a carboxyl terminus were lacking, a cyclic structure was first assumed until an acetyl group at the N-terminus and phenylalaninol (Phol) at the C-terminus were detected. Alamethicin A, the most thoroughly investigated member of the group has the structure shown in Fig. 15.

Ac— Aib— Pro— Aib— Ala— Aib— Ala— Glu— Aib— Val— Aib—

Gly— Leu— Aib— Pro— Val— Aib— Aib— Glu— Glu— Phol

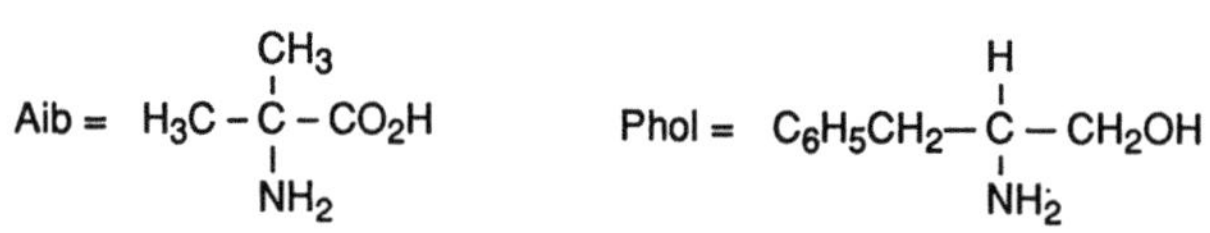


**Fig. 15.** Formula of alamethicine A

Remarkably, nearly half of the amino acid components are L-α-amino-isobutyric acid residues (Aib), an amino acid not present in ribosome-manufactured proteins (p. 8). Peptides containing Aib-residues have been recognized to form α-helices particularly readily, an explanation of the pore forming properties of alamethicins and of similar natural and artificial polypeptides. In the natural analogs alanine or valine is found to be replaced by Aib; closely related natural compounds are among others, suzukacillin, trichotoxin (mould products), and less similar, a component of bee venom, the 25-peptide mellitin. In the synthesis of peptides containing α-amino-isobutyric acid certain difficulties are encountered due to the poor steric accessibility of the amino—as well as of the carboxyl group.

### 9.2.5 Cyclosporin A

During the last few years a rather exciting development in chemistry and pharmacology started with the isolation of an antifungal compound from a mould *Trichoderma polysporum* [36] later from *Tolypocladium inflatum* in the

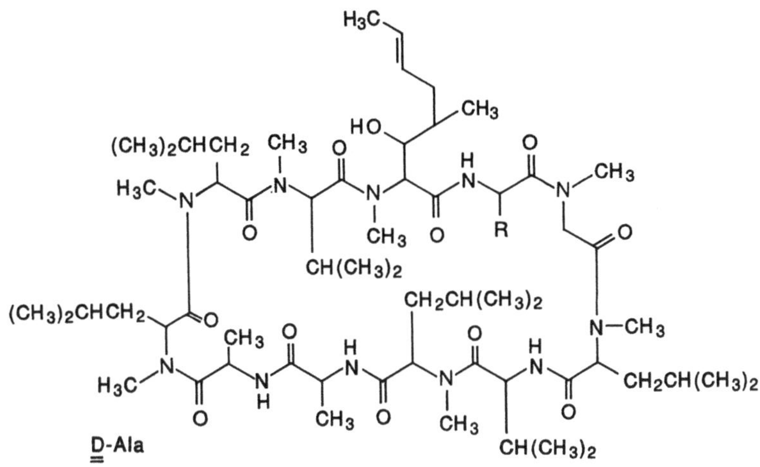


**Fig. 16.** Cyclosporin A

laboratories of Sandoz Ltd. in Basle [37]. The peptide called cyclosporin A turned out to have not only an inhibitory effect on fungal and yeast growth, but was also antiphlogistic, and—most importantly—an inhibitor of humoral and cellular immunity. Today wide use is made of this immunosuppressive effect in organ transplant surgery, and in the treatment of autoimmuno diseases. As with other fungal peptides, numerous naturally occurring cyclosporins have been described which are derived by amino acid substitutions in different positions of the cyclic molecule. Cyclosporin A, the main component, has the structure shown in Fig. 16 ($R = C_2H_5$).

The cyclic 11-peptide contains mainly "normal" amino acids, one D-alanine and one L-α-aminobutyric acid, but, strikingly, not less than seven *N*-methyl groups and a quite uncommon amino acid, a $C_9$-compound: (4*R*)-4[(*E*)-2-butenyl]-4-methyl-*N*-methyl-L-threonine.

The total synthesis of cyclosporin A has been described by R.M. Wenger [38]; biosynthetically it is produced by adding appropriate amino acids to the fermentation with the fungus [39] or to cell free preparations therefrom [40]. Various cyclosporins with different activities have been obtained. The activation and connection of the amino acids occurs, as recognized in the synthesis of gramicidin S (p. 208), by the thiotemplate mechanism; the methyl groups are introduced into the already formed peptide bonds by *S*-adenosyl-methionine.

## 9.3 The Peptides of *Amanita* Mushrooms [41]

The investigation of the deadly poison of Amanita mushrooms was initiated at the beginning of this century in the USA. At Johns Hopkins University in Baltimore, MD, the bacteriologist William W. Ford attempted the isolation of the toxin, "amanitatoxin", and succeeded in obtaining a preparation containing about 5% of the fatal agent without recognizing its peptide nature. The efforts were resumed in the thirties in the laboratory of Heinrich Wieland in Munich,

Bavaria, where in 1937 Ulrich Wieland and Feodor Lynen were able to crystallize for the first time a substance which was called "phalloidin" that rapidly killed mice after intraperitoneal application. Three years later Heinrich Wieland and Rudolf Hallermayer crystallized from *Amanita phalloides* extracts a second toxin, "amanitin", which with smaller doses killed the experimental animals only after several days like a deadly dish of *A. phalloides* kills humans. Bernhard Witkop (Plate 49) 1940 in H. Wieland's laboratory recognized that phalloidin is a peptide and isolated a new imino acid, allo-hydroxyproline. Then the Second World War interrupted Amanita research which was resumed in Heidelberg by one of the present authors only ten years later and continued in Mainz, Frankfurt and again in Heidelberg through the following decades.

In the search for minor phallotoxic components in the lipophilic part of extracts of *A. phalloides*, a fraction was obtained which, although containing phallotoxins (color reaction on paper chromatograms), proved non-toxic in animal tests. An antitoxic principle, therefore, was suspected of being present in those fractions and was eventually purified and crystallized in 1968. It was called antamanide (anti-amanitatoxin, AA) and characterized as a cyclic decapeptide [42]. On account of its relatively simple composition, its cation-binding properties, and interesting antitoxic activity, antamanide has been the subject of extensive investigations in several laboratories. It will be the first peptide from the Amanita species to be discussed.

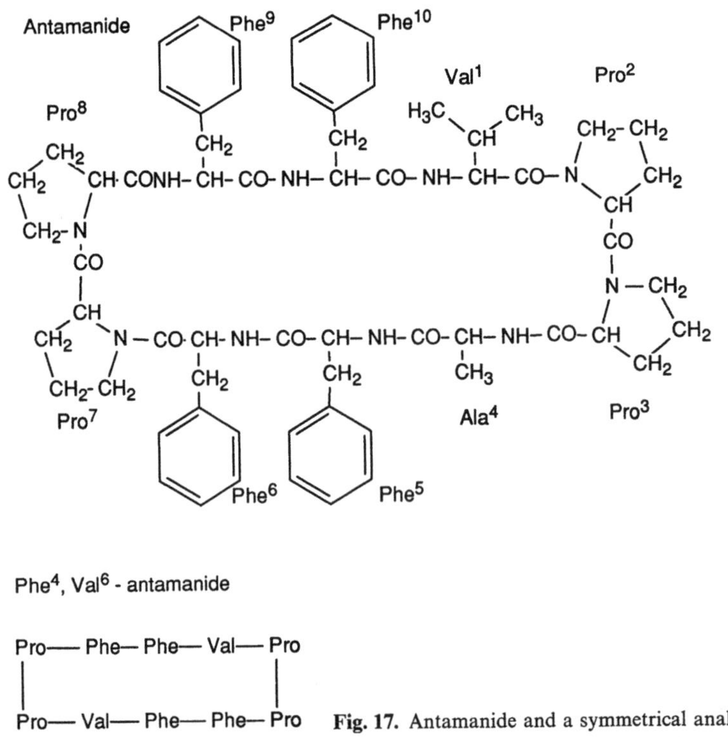


**Fig. 17.** Antamanide and a symmetrical analog

### 9.3.1 Antamanide

Antamanide is a lipophilic neutral compound that forms colorless crystals which are easily soluble in alcohols but almost insoluble in water; it has a molecular mass of 1147 (mass spectroscopy). On hydrolysis, antamanide is split to yield the L-amino acids alanine, valine, phenylalanine, proline in a molar ratio of 1:1:4:4. Its structure, determined by gas chromatographic separation of the fragments obtained by partial methanolysis and their analysis by mass spectroscopy is displayed in Fig. 17.

The appearance of an ion corresponding to AA-$Na^+$ in the mass spectrum was the first indication of ion-complexes of the cyclopeptide. An increase of carbonyl adsorption in the presence of $Na^+$ in the infrared spectra, showed that the peptide interacts with cations through its amide carbonyl groups. The stability of AA-complexes with various cations was then determined, by various physical methods with the result that it exhibits the order of preference.

$$Li^+ < Na^+ > K^+ > NH_4{}^+ > Rb^+ > Cs^+$$

and that $Ca^{++}$ is bound about as strongly as $Na^+$.

Antamanide exerts its protective effect against phalloidin (PHD) (see p. 217) in mice when given one hour before or at the latest simultaneously with the drug. Five milligram PHD per kg body weight will kill a white mouse within 2–4 hours with certainty but the animal is fully protected from death by 0.6 mg AA per kg.

Shortly after its structure had been determined syntheses of AA were described. In the laboratory of one of the authors, two pentapeptide segments consisting of amino acids Nos. 6–10 and 1–5 were synthesized mainly by the mixed anhydride method (p. 79), connected to form a linear decapeptide which, after deprotection of its ends, was cyclized (Fig. 18).

6 10 1 5
BocPheProProPhePheOMe BocValProProAlaPheOMe
↓ deprotection
Boc(6–10)OH + H(1–5)OMe
↓ combination
Boc(6,7,8,9,10,1,2,3,4,5)OMe
↓ deprotection
H(6————————5)OH
↓ cyclization
antamanide

Synthesis of antamanide

**Fig. 18.**

The solid phase technique of Merrifield (Chap. 5) proved very useful for obtaining a series of analogs of AA which were compared with respect to their antitoxic activities, ion binding capacities and molecular structures.

**Metal-complexing properties.** The stability constants $K_{stab}$ of AA complexes with different metals strongly depend on the nature of the solvent. They are greater in lipophilic non-polar solvents than in more hydrophilic ones and are practically zero in the presence of water. It may be sufficient to show the $K_{stab}$-values for AA in acetonitrile with $Na^+$: $3 \times 10^4$, $K^+$: $1.9 \times 10^2$, Li: $1 \times 10^3$ and in methanol with $Na^+$: $5 \times 10^2$, $K^+$: $1 \times 10^2$, Li: $1 \times 10^1$. The affinity for $Na^+$ is not far from the range of complexones mentioned before (p. 203). Nevertheless AA is not an ionophore, it does not transport $Na^+$-ions through bilayer membranes. The reason is that the rate of formation of the AA-$Na^+$ complex (in methanol) is 100-times slower, and the rate of its dissociation even $10^4$-times slower than the corresponding rates for other macrocyclic complexones like valinomycin-$K^+$ (p. 202). With AA, apparently, slow rates of the conformational transitions, required for forming the fitting cavity or releasing the metal ion, exclude it from functioning as an ionophore.

Analogs of AA exhibit different stability constants of their Na-complexes, although the differences do not exceed a factor of about 20. All biologically, highly effective analogs exhibited (in ethanol) comparable $K_{stab}$ values around $10^3$ $M^{-1}$ whereas those with protecting doses 10–20 times higher (e.g. $Ala^1$, $Gly^4$-AA) form about tenfold less stable complexes with sodium ions. Therefore, it was tempting to suggest a direct correlation between the protective mechanism and the complexing ability of the cyclic peptides. Metal ions, e.g. $Ca^{++}$ could be responsible for the protective binding of AA to the membrane of liver cells, the target organ of phalloidin intoxication, in a sandwich-like structure as V.T. Ivanov from the Shemyakin Institute in Moscow proposed in 1975.

Today we know that AA exerts its protective activity by competitive binding to a rather multifunctional membrane carrier protein of molecular mass of 48000 dalton [43] which also binds phalloidin, bile acids etc. Since the latter substances have no affinity to $Na^+$ or $Ca^{++}$-ions the binding of AA does not seem to require these ions. Nevertheless the striking coincidence of complexing and protecting properties of AA variants is quite unlikely to be accidental. Possibly it reflects the readiness of an AA molecule to adopt the proper binding conformation.

The conformation of metal-free and of complexed AA in solution as well as in the crystalline state aroused interest from the beginning of research [44]. As can easily be seen from changes of the circular dichroism (CD) spectra (p. 122), the shapes of the backbones of cyclic peptides in many cases change depending on the nature of the solvent, or in the absence and the presence of metal ions. Two methods are available for the determination of the conformations of peptides (see p. 125): X-ray structure analysis of suitable crystals and NMR-spectroscopy of a dissolved sample. The former method produces an exact picture of the location of the atoms in the solid state. This conformation may be the same as that of the substance in solution, but free peptides are extremely flexible molecules with rotations possible about each N—C and C—C′ bond in the backbone (see p. 12). Even for cyclic peptides where the number of degrees of freedom is reduced by the constraint of the closed ring many different possibilities remain. Thus it is more probably that the conformations in solution differ from those in the crystal unless a particular force stabilizes the shape of the molecule. NMR and X-ray methods have been applied with AA and an analog in which $Ala^4$ has been replaced by Phe, and $Phe^6$ by Val. $Phe^4$, $Val^6$ AA, (Fig. 17), a c-2 symmetrical cyclopeptide has about 50% of the protective effectivity of AA, binds $Na^+$-ions with the same affinity and displays solvent-dependent changes of CD spectra in the same way as AA.

The conformational studies were started with AA in chloroform in 1970 with D.J. Patel and A.E. Tonelli from Bell Laboratories, Murray Hill, USA., using $^1H$-NMR exclusively, while at the same time the Russian scientists V.T. Ivanov, Yu.A. Ovchinnikov et al. investigated free AA as well as its Na-complex using ORD and infrared (IR) techniques in addition. Combined with minimal energy calculations bracelet-like ring structures with defined transannular hydrogen bonds were proposed and striking differences between free and complexed conformation were stated. A *cis*-peptide bond from $Pro^2$ to $Pro^3$ and from $Pro^7$ to $Pro^8$, in both conformations, has been clearly demonstrated by Isabella Karle, who in 1973 began a series of X-ray diffraction analyses with the crystalline LiBr complex of AA (Fig. 19) and the isostructural NaBr-complex of the symmetrical analog [45]. In the following years the structure of uncomplexed AA crystallized from various solvents has also been solved [46].

In the complexed form, the backbone is folded in a fashion somewhat similar to the seam in a tennis ball, the $Pro^2Pro^3$ and $Pro^7Pro^8$ peptide bonds have the **cis** configuration, the $Li^+$ or $Na^+$ ion is located in a shallow cup formed by the folding of the backbone, and four ligands between the metal ion and carbonyl oxygens are formed. The fifth ligand involves a N or O atom from the solvent that may be $CH_3CN$ or $C_2H_5OH$. The decapeptide is not large enough to completely encapsulate an ion, hence a solvent molecule not only provides an additional ligand for the metal ion in the interior, but also completes the lipophilic exterior

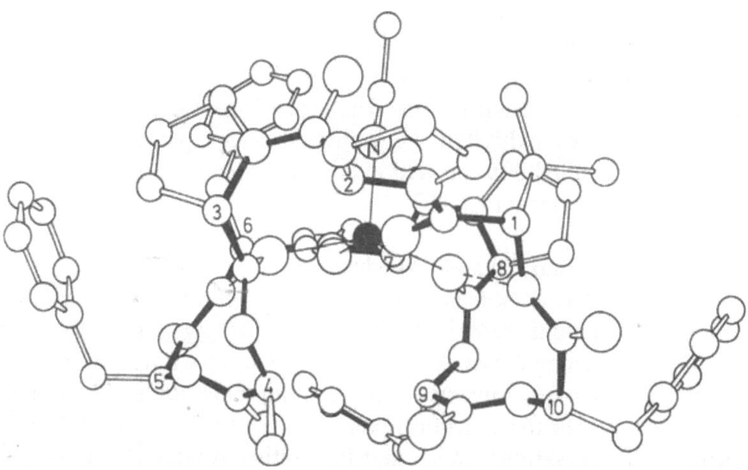


**Fig. 19.** The lithium antamanide $CH_3CN$ complex. The peptide backbone is depicted with *heavy lines*; the *numbers* refer to the C-atoms in the ten residues. Five-coordinated $Li^+$ is represented by the *black dot*. One of the ligands is acetonitrile (solvent)

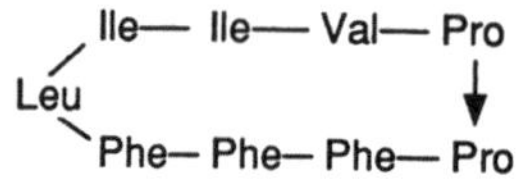

Cyclolinopeptide A

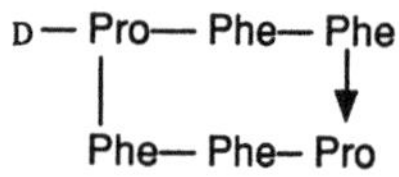

c ( D-PFFPFF)

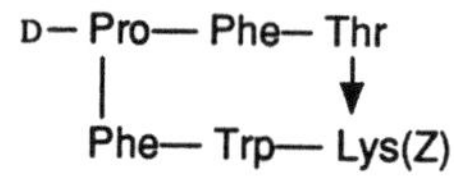

008

**Fig. 20.** Cytoprotective cyclopeptides

of the complex created by the side chains. The uncomplexed form has an elongated planar ring, O-atoms, in the complex participating in the ligation with the metal ion, are here directed outward with respect to the backbone ring.

The conformations of the alkali metal complex of AA and of its c-2 symmetric analog are identical, and this holds also for their solid and dissolved state. Complexation, apparently, renders the structure rigid. For the uncomplexed cyclopeptides, however, it turned out by measurement of CD, ultrasonic absorption [47a] and, mainly, NMR spectroscopy [47b] that there is a fast conformational equilibrium between up to four individual molecules depending on the polarity of the solvents. The stability of one conformer will be directed i.a. by the competing formation of hydrogen bonds intramolecularly vs H-donors or acceptors in the solvents. The important question as to the "bio-active" conformation has not been strictly answered so far.

A cyclo-nonapeptide, isolated from linseed by Kaufmann and Tobschirbel [48] which also prevents the uptake of phalloidin (and other substances) by hepatocytes, cyclolinopeptide A, shows a certain similarity with AA in its amino acid sequence (Fig. 20) but does not form alkali-metal complexes. With the accumulation of aromatic amino acids between proline residues as an active region in mind, Kessler et al. designed some smaller cyclopeptides of which one, c(D-ProPhePheProPhePhe), exhibited somewhat stronger protection and a second less related one, called 008, showed even better protection of liver cells against phalloidin than antamanide [49].

### 9.3.2 Poisonous Peptides from Amanita Mushrooms

In the search for the lethal poison in *A phalloides*, in the absence of specific indicators, animal experiments were used, first with guinea pigs then with white mice. The fractions obtained in the work-up of extracts from the fungus were, as is the practice in pharmacology, injected into the animals under their skin (subcutaneously, s.c.) or into the veins (intravenously, i.v.) or into the abdomen (intraperitoneally, i.p). A sample was regarded active (toxic) if on its administration the mouse died within a few hours. Guided by this assay it became possible to isolate in 1937, as mentioned above, crystalline phalloidin and later several related peptides, the phallotoxins. The smallest dose which kills a 20 g mouse within a few hours is 50 μg, that is 2.5 mg per kg of body weight. In the course of the purification experiments the observation was made that in some instances the experimental animals that survived the test nevertheless died a few days later. They succumbed to a second, slow acting toxin, which, in 1940, was similarly secured in crystalline form and was named amanitin (cf. p. 219).

Fatal poisonings in man occur not only after the ingestion of the green *Amanita phalloides* (death cup, deadly agaric), but also more often in North America, after the consumption of a white species, *A. virosa*, known there under the name "destroying agnel", that is almost certainly identical with the toxic white mushrooms *A. bisporigera* or *A. tenuifolia*. From these H. Faulstich et al. in 1979 isolated, in addition to the already known toxic peptides present in *A. phalloides*, a new class of peptide-toxins, the virotoxins [50]. They exert a fast lethal effect like phalloidin and the accompanying peptides and are similar to them in other respect as well. Therefore we will treat the phallotoxins and virotoxins, one example of each, in one section and conclude with the peptides which cause fatal mushroom-poisoning in man, the amatoxins.

Phallotoxins and Virotoxins

On hydrolysis with acids phalloidin yields L-alanine, 4-*cis*-L-hydroxyproline, D-threonine, L-γ, δ-dihydroxyleucine, L-cysteine and β-oxindolalanine ("oxytryptophan"). The tryptophan derivative is not a genuine building component of the cyclic peptides: it is formed by hydrolysis of a thioether crosslink between position 2′ of tryptophan and the side chain of cysteine (tryptathionine). The structure of the bicyclic heptapeptide is shown in Fig. 21.

The subsequently discovered virotoxins are, in all likelihood, biogenetically related to phallotoxins. They are monocyclic heptapeptides, in which the thioether bridge of the phallotoxins has evidently undergone oxidative cleavage. The sulfur is found in the form of a methylsulfonyl- or methylsulfinyl group ($CH_3SO_2$—, $CH_3SO$—) in position 2′ of the indole ring, the original cysteine residue appears as D-serine and a second hydroxyl group is introduced into the

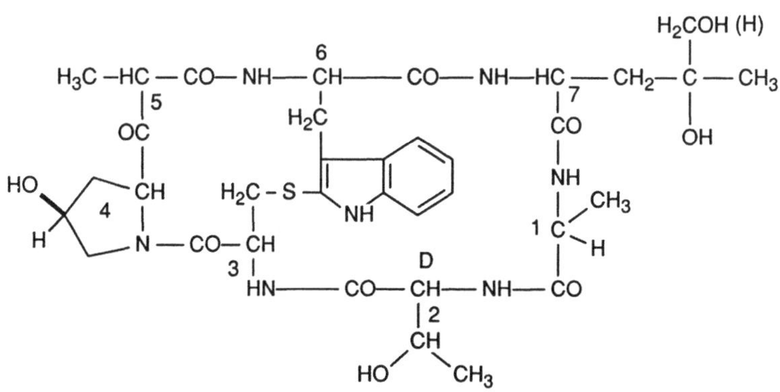

**Fig. 21.** Phalloidin (H instead of OH: phalloin)

proline ring of the phallotoxins to give 2,3-*trans*-3,4-*trans*-3,4-dihydroxy-L-proline.

In mammals, phallotoxins, and certainly also virotoxins, are absorbed from the intestines into the blood only slowly, if at all (in contrast to the amatoxins). After intraperitoneal injection they rapidly reach the liver and exert their similar toxic effect there. They weaken the structure of the hepatocyte membrane, hence blood penetrates the liver cells and the periphery is depleted of blood to such an extent that anemic shock occurs from internal bleeding. A cause of damage to the membranes is supposedly the very strong binding of phallotoxins and virotoxins to F-actin that under the lipid-bilayer stabilizes the membrane. At this point a brief discussion of actin is necessary.

*Actin*, first known from the physiology of muscle contraction, has been established in the last 20 years as a ubiquitous protein in all eukaryotic cells. The molecules of G-actin in various cells, from yeast to man, are not too different from each other. They have a molcular weight of about 43 000 dalton and the ability to polymerize, that is to combine with each other reversibly to long double-wound filaments of F-actin. The equilibrium G-actin$\rightleftharpoons$F-actin, a function of many external factors is strongly shifted by phallotoxins and virotoxins in the direction of F-actin, because these peptides bind with high affinity exclusively to the polymer ($K_{diss}$ about $10^{-8}$ M, that is: complete saturation of F-actin takes place in a solution of less than 1 mg peptide per 5 liters). In the presence of sufficient peptide F-actin can no longer depolymerize: it is more or less frozen, the filaments clot.

The molecular mechanism of the stabilization of F-actin filaments by the bound phallotoxins or virotoxins is not understood so far. Affinity and thereby toxicity are closely related to the structure of the toxins. We mention only a few points, relevant to the phallotoxins: cleavage of the thioether bridge leads to the complete loss of toxicity in the resulting monocyclic product, as does the absence of the 4-*cis* hydroxy group in the proline ring, or of the methyl group of alanine in position 5. The rigid structure of the bicyclic molecule, that can also be recognized in the characteristic CD spectrum, is a requirement for strong binding to the monomer units of F-actin. Interestingly, the monocyclic flexible virotoxins are bound in a similar way; it must be assumed that the additional hydroxyl groups (in hydroxyproline and D-serine) enhance the hydrogen bond mediated binding of the molecules docked, through induced fit, at the binding site.

Shortly after the structure elucidation of phalloidin in 1955 attempts at synthesis were started, but they led to success only in 1977 with the synthesis of phalloin by E. Munekata. Serious difficulties were caused by the γ-hydroxyleucine residue, which during the operations of peptide syntheses readily forms a γ-lactone yet this undesired side reaction cannot be prevented by the temporary blocking of the tertiary hydroxyl groups.

To generate the thioether bridge a suitable cysteine containing tetrapeptide was linked via the S-chloride to the indole ring of a tryptophan containing tripeptide that carried the lactone of γ-hydroxyleucine at its C-terminus. A shown in Fig. 22, the first cyclization led to the "secolactone" which, after the

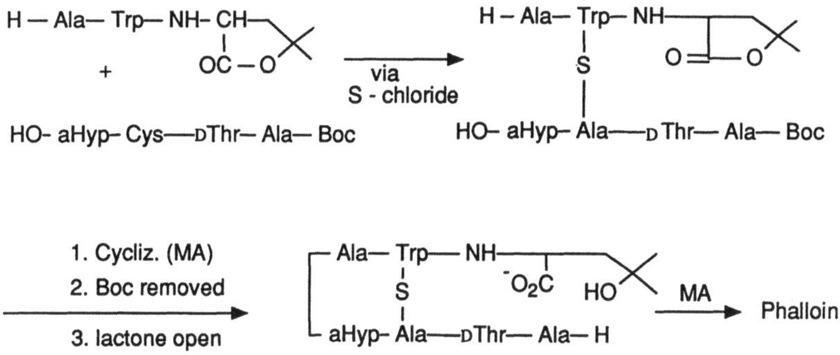


**Fig. 22.** Synthesis of phalloin

removal of the Boc group and opening of the lectone ring, contained the $H_2N$— and —COOH groups needed for the second ring closure. At this point all coupling methods led exclusively to lactone formation with the exception of the mixed anhydride procedure (p. 79) which works so fast that at least part of the compound to be cyclized escaped lactonization and closed the desired peptide bond.

Of the various aspects of the chemistry of phallotoxins and virotoxins only the numerous modifications of the hydroxyleucine side chain should be mentioned here; none of these affected the affinity to F-actin. Therefore, it was possible to introduce radioactive or fluorescent markers into the molecule, which then allowed the visualization of F-actin in histological preparations with great sensitivity.

## The Amatoxins

On resuming Amanita research in Heidelberg, after the Second World War in 1948, the original preparation of "amanitin" from Munich was analyzed by paper electrophoresis (p. 53). It turned out that the sample consisted of two individual substances yielding the same specific color reactions; a neutral one now called $\alpha$-amanitin and an acidic one, $\beta$-amanitin. Both amatoxins, and in addition $\gamma$- and $\varepsilon$-amanitin were isolated from methanolic extracts of several 10 kg batches of *A. phalloides* in the following years. These poisonous peptides were revealed on paper chromatograms by spraying with cinnamaldehyde and exposing the sheet to an atmosphere of HCl-gas: Almost immediately a deep violet-blue color develops. The same color is formed with HCl on an amanitin-impregnated pine splinter (a well known test for indole compounds which give a red color). As a consequence newsprint was recognized as a "reagent" for tracing amatoxins in mushroom sap. The color is generated by the acid-catalyzed reaction of the 6′-hydroxytryptophan moiety (Fig. 23) with unsaturated, cinnamaldehyde-like components of lignin.

An amatoxin present only in minimal amounts, "amanin", was found to yield no characteristic color in the systems just described. It was shown to be

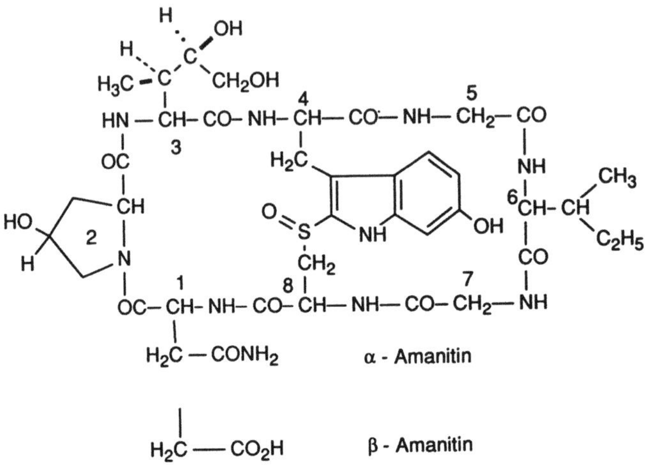


**Fig. 23.** Amatoxins

β-amanitin, lacking the hydroxyl group at the indole nucleus; its amide, corresponding to α-amanitin, amaninamide, is the deadly toxin of *Amanita virosa* [53].

Toxicity

The amatoxins, the real toadstool poisons, cause death several days after ingestion. The lethal dose for 50% of the animals differs widely among different species, the guinea pig being the most sensitive with a $LD_{50}$ of 0.1 mg per kg body weight. Assuming a comparable susceptibility of humans 5–10 mg of the toxin could be a fatal dose, an amount present in a single specimen of *A. phalloides*. The amatoxins are resorbed, at least in humans, from the gut into the blood and thus reach the liver, the target organ. Toxin, that is not held back immediately, will pass via bile flow back to the intestines and from there into the liver again, a so called enterohepatic circulation. Therefore as a therapeutic measure large amounts of charcoal should be administered together with injection of drugs inhibiting the uptake of the toxin by the liver, preferentially silymarin, a drug from the milk thistle (*Silybum marianum*). For diagnosis of toadstool intoxication radioimmunoassays are available. For details, see Refs. [41] and [54].

The recognition of the molecular mechanism of the toxicity of amatoxins began in 1965 when Luigi Fiume with R. Laschi in Bologna reported alterations of the structure of nuclei in hepatocytes by α-amanitin and, shortly after when F. Stirpe reported decreased RNA content in mouse liver nuclei after intoxication. The continued research in Italian, French, American, and German laboratories [41, 54], revealed the fact that amatoxins inhibit the transcription of DNA into pre-messenger RNA by binding to RNA polymerase form II (or B), the dissociation constant $K_{Diss}$ of the enzyme-amatoxin complex being extremely small, around $10^{-9}$–$10^{-10}$ M. A halt in m-RNA synthesis implies the stop of protein synthesis, the cause of death for cells. Its great specificity and sensitivity makes α-amanitin a frequently used tool in cell biology. Chemical investigations conducted mainly by Ulrich Gebert [55] ultimately led to the structure of α-amanitin (and β-amanitin) as shown in Fig. 23.

The 3-dimensional structure of β-amanitin was solved through X-ray diffraction analysis by Kostansek et al. [56] (Fig. 24).

The amatoxins are cyclic octapeptides of L-amino acids; the backbone is divided into two rings by a sulfur bridge analogously to the phallotoxins, but with an (*R*)-sulfoxide. The amanitins (α-, β-etc.) are derived from 6-hydroxytryptophan, amanin lacks the phenolic OH-group. As in collagen, 4-*trans*-hydroxyproline is the imino acid; the occurrence of γ, δ-dihydroxyisoleucine is uncommon.

Chemical modifications have shed light on the essential structural features required for toxicity (strong binding to RNA polymerase II). These are, besides the molecular shape, the presence of the OH-group at proline and a branched, hydroxylated side chain in position 3. From the spatial formula in Fig. 24 one can recognize that the indole ring projects over the folded plane; it is not involved in binding to the polymerase, its hydroxyl group can serve as a site for the addition of various groups without influencing toxicity.

Amatoxins, besides the already mentioned *Amanita* species, were found also in other mushrooms noted for their toxicity: the inconspicuous *Galerina* species, *G. marginata* and *G. venenata* and the *Lepiota* mushrooms as well, contain these peptides, synthesized not in the carpophore but in the mycelia, most probably through the non-ribosomal pathway. Several hundred more mushrooms have been examined, for instance with the help of the newsprint test, and it is unlikely that further amatoxin accumulating species will be found. Phallotoxins have been detected only in *Amanita* toadstools.

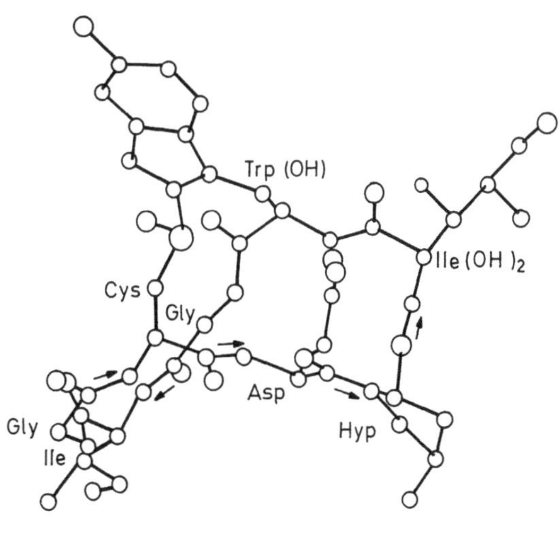


**Fig. 24.** Three-dimensional structure of β-amanitin [56]

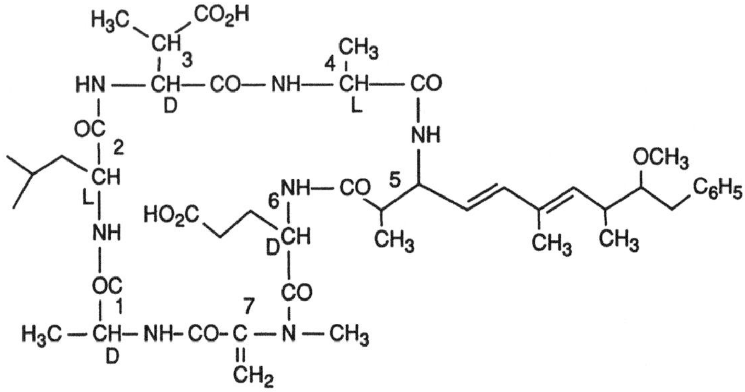


Fig. 25. Microcystin LA (Leu², Ala⁴)

## 9.4 Microcystins

In 1878 George Francis published the first written report of animal poisoning by a *Cyanobacterium* (blue-green alga). However, only in the last 30 years have significant amounts of information appeared on the structure and function of neurotoxin alkaloids and hepatotoxic peptides. Only the latter are discussed here. The peptide nature of the isolated "fast death factor" was recognized in 1959 by C.J. Bishop et al. [57] the structure of "cyanoginosin-LA" was revealed only in 1984 by Botes et al. (Fig. 25) [58].

Studies in several laboratories have shown that strongly hepatotoxic cyclic heptapeptides of a common type occur in different strains of Microcystis [59]. *Microcystis aeruginosa*, the species most frequently investigated contains the peptide formulated in Fig. 25. The two letter suffix LA designates the two L-amino acids in positions 2 and 4 (X, Y = Leu, Ala) which have been found variable in all of the toxins examined to date. X can also be arginine (R) or methionine (M). The 10-carbon chain $\beta$-amino acid is unique for this type of hepatotoxic compound. Another not quite common structural element is the unsaturated side chain of dehydroalanine (No. 7), the formation of which can readily be imagined by elimination of $H_2O$ from serine or $H_2S$ from cysteine.

## 9.5 Lantibiotics

Dehydroalanine (Dha) and dehydro-$\alpha$-aminobutyric acid, a putative dehydration product of threonine occur accumulated in a particular class of antibiotic fungal peptides recently named "lantibiotics" of which Nisin (Fig. 26) will be mentioned first. The structure of the 34-peptide as revealed by Erhard Gross and coworkers in the beginning of the seventies, contains three dehydro side chains, RCH=C(NH—)CO—, and five thioether amino acids, lanthionine, Ala-S-Ala and its homolog Abu-S-Ala. Total synthesis by the group of Tetsuo Shiba (Plate 43) [61].

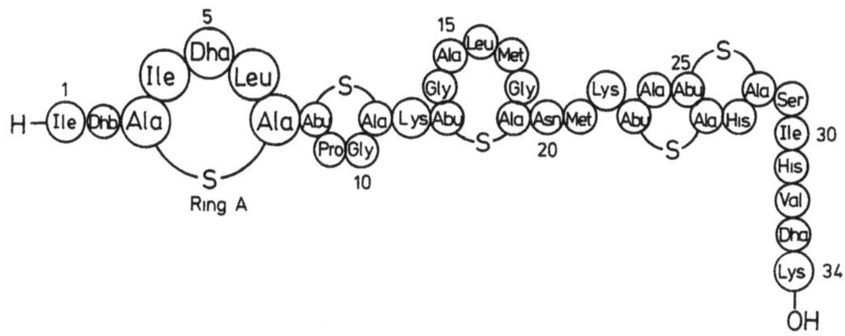


**Fig. 26.** The structure of Nisin. Ring A: residues 3–7; Abu = aminobutyric acid; Dha = dehydroalanine; Dhb = dehydrobutyrine ($\beta$-methyldehydroalanine); Ala-S-Ala = lanthionine; Abu-S-Ala = $\beta$-methyllanthionine

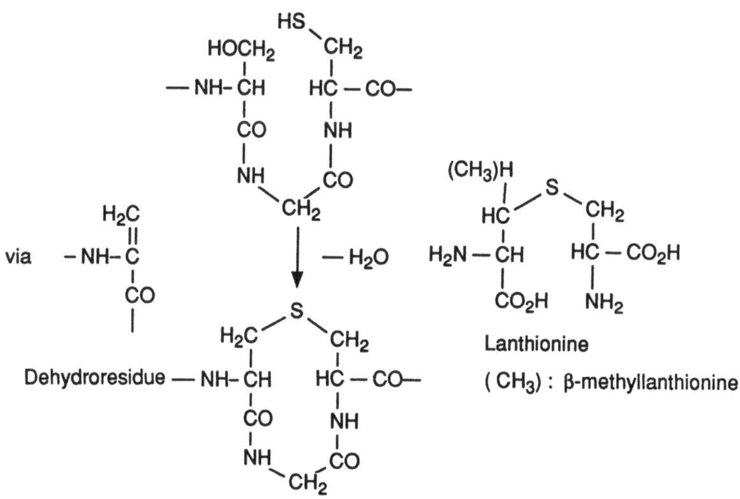


**Fig. 27.** Formation of lanthionine moiety by addition of cysteine to dehydroalanine

The thioether moieties are supposedly generated by addition of the thiol groups of cysteines to one of the unsaturated side chains formed by $H_2O$-elimination in the peptides after their biosynthesis (Fig. 27).

Quite recently it has been found that the sequence of 21 amino acids of a lantibiotic, epidermin from *Staphylococcus epidermidis* is encoded on a structural gene in the bacterium as a 52-amino-acid pre-peptide which is processed to the 21-peptide amide antibiotic. The actual sequence of epidermin contains the precursor amino acids Ser, Thr and Cys, from which the unusual amino acid constituents are derived [62]. This ribosomal mechanism is a very surprising observation since the accepted view so far has been that microbial peptides are usually biosynthesized by the thiotemplate pathway (p. 208).

## 9.6 Peptides Interacting with Nucleic Acids

At the end of this chapter, cyclic peptides will be mentioned one of them (a mixture) being indeed the first antibiotic peptide ever isolated: actinomycin from cultures of *Streptomyces antibioticus* by Waksman and Woodruff in 1940. The different actinomycins are orange-red peptides all containing the same chromophore, an aminophenoxazinone dicarboxylic acid, whose carboxyl groups are connected via the amino group of threonine with two identical cyclic depsipeptide rings consisting of five amino acids. In Fig. 28 the cyclic part of actinomycin $C_3$ is depicted as determined by H. Brockmann (Plate 13) and associates, who in the fifties contributed most to the analytical and synthetic chemistry of this class of compounds.

The antibiotic (and highly toxic) effect of the actinomycins is due to their strong binding by intercalation to the double helix of DNA so inhibiting transcription into RNA even with minimal concentrations [63].

Another group of cytostatic, antiviral and antibacterial yet highly toxic peptides are the "quinoxalines" [64] with two quinoxaline-carboxylic acids attached to eight membered, disulfide bridged symmetric peptide-lactone rings. Further inhibitors of protein biosynthesis isolated from Streptomyces species here mentioned only by name are streptogramin B and the tuberactinomycins (e.g. Viomycin) isolated, elucidated and synthesized mainly by T. Shiba's group in Osaka.

The chances of finding poisonous peptides not detected so far in higher plants, do not seem very great. Perhaps in studies of physiological reactions of plant cell cultures, small amounts of biologically active peptides will be recognized. From amino acids, many plants construct *peptide alkaloids*. The term first proposed by Goutarel in 1964 has been applied to a rapidly expanding group of closely related

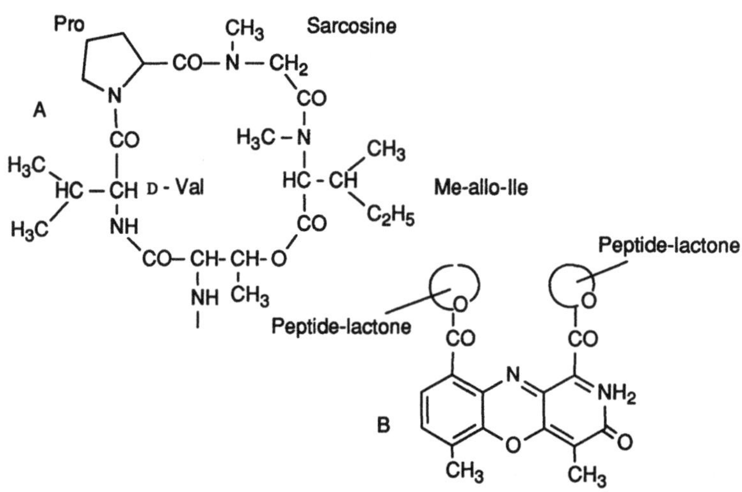


**Fig. 28.** Lactone moiety (A) and chromophore (B) of actinomycins

bases composed largely of simple amino acids in part connected by peptide bonds. Here only reference to a review article can be given [65]. Since then a great number of interesting natural cyclopeptides and cyclodepsipeptides have been added [66].

## References


1. A. Hofmann, H. Ott, R. Griot, P.A. Stadler, A.J. Frey, Synthese von Ergotamin, Helv. Chim. Acta 46: 2306–2336 (1963)
2. R.G. Griot, A.J. Frey, The formation of cyclols from N-hydroxyacyl lactames, Tetrahedron 19: 1661–1673 (1963)
3. M.M. Shemyakin, V.K. Antonov, A.M. Shkrob, Activation of the amide group by acylation, Peptides, Proc. 6th Europ. Pept. Symp., Athens 1963, 319–328
4. G. Lucente, A. Romeo, Synthesis of cyclols from small peptides via amide-amide reaction, Chem. Commun. 1605–1607 (1971)
5. M. Rothe, W. Schindler, R. Pudill, U. Kostrzewa, R. Theyson, R. Steinberger, Zum Problem der Cyclotripeptidsynthese, Peptides, Proc. 11th Europ. Pept. Symp. Wien 1971, 388–399
6. M. Rothe, K.-L. Roser, Conformational flexibility of cyclic tripeptides, Abstr. 20th Europ. Pept. Symp. Tübingen 1988, p. 36
7. Th. Wieland, H. Mohr, Diacylamide als energiereiche Verbindungen. Diglycylimid. Liebigs Ann. Chem. 599: 222–232 (1956); Th. Wieland, H. Urbach, Weitere Di-Aminoacylimide und ihre intramolekulare Umlagerung. Liebigs Ann. Chem. 613: 84–95 (1958)
8. M. Brenner, The aminoacyl insertion in: Ciba Foundation Sympos. on Amino acids and peptides with antimetabolic activity. G.E.W. Wolstenholme and C.M. O'Connor eds. Churchill, London 1958 pp 157–170
9. G. Zanotti, F. Pinnen, G. Lucente, S. Cerrini, W. Fedeli, F. Mazza. Peptide thiacyclols. Synthesis and structural studies, J. Chem. Soc. Perkin Trans. 1 1984, 1153–1157
10. H.T. Clarke, J.R. Johnson, R. Robinson, The Chemistry of Penicillin Princeton University Press, Princeton, N.J., 1949
11. H.R.V. Arnstein, D. Morris, The structure of a peptide containing α-amino-adipic acid in the mycelium of *Penicillium chyrsogenum.* Biochem. J. 76: 357–361 (1960)
12. D.J. Hook, R.P. Elander, R.B. Morin, Recent developments with cell free extracts on the enzymic biosynthesis of penicillins and cephalosporins in "Peptide-antibiotics". Biosynthesis and Function, H. Kleinkauf, H. von Döhren eds. de Gruyter, Berlin 1982, pp 84–100
13. E.P. Abraham, The cepholosporin C group. Quart. Rev. 21: 231–248 (1967)
14. R.B. Woodward, K. Heusler, J. Gosteli, P. Naegeli, W. Oppolzer, R. Ramage, S. Ranaganathan, H. Vorbrüggen, The total synthesis of cephalosporin C., J. Amer. Chem. Soc. 88: 852–853 (1966)
15. H. Brockmann, G. Schmidt-Kastner, Valinomycin I. XXVII. Mitt. über Antibiotica aus Actinomyceten, Chem. Ber. 88: 57–61 (1955)
16. C. Moore, B.C. Pressman, Mechanism of action of valinomycin on mitochondria, Biochem. Biophys. Res. Com. 15: 562–567 (1964)
17. Yu. A. Ovchinnikov, V.T. Ivanov, A.M. Shkorb, in Membrane-active complexones, BBA Library, Vol. 12 Elsevier, Amsterdam 1974
18. K. Neupert-Laves, M. Dobler, The crystal structure of a $K^+$-complex of valinomycin, Helv. Chim. Acta 58: 432–442 (1975)
19. M.M. Shemyakin, N.A. Aldanova, E.I. Vinogradova, M.Y. Feigina, The structure and total synthesis of valinomycin, Tetrahedron Lett. 1963: 1921–1925. See also V.T. Ivanov, Yu.A. Ovchinnikov, A.A. Kiryushkin, M.M. Shemyakin, Synthetic and natural desipeptides, Peptides Proc. 6th Eur. Pep. Symp. (L. Zervas ed) Pergamon 1963, pp 337–350
20. D. Baron, L.G. Pease, E.R. Blout, Cation binding of a cyclic dodecapeptide cyclo-(L-Val-Gly-Gly-L-Pro)$_3$ in an aprotic medium. J. Amer. Chem. Soc. 99: 8299–8306 (1977)
21. J. Dominguez, J.D. Demitz, H.H. Gerlach, V. Prelog, Stoffwechselprodukte von Aktinomyceten. Über die Konstitution von Nonactin. Helv. Chim. Acta 45: 129–138 (1962)
22. C.J. Pedersen, H.K. Frensdorff, Makrozyklische Polyäther und ihre Komplexe. Angew. Chem. 84: 16–26 (1972)
23. J.-M. Lehn, Supramolekulare Chemie—Moleküle, Übermoleküle und molekulare Funktionseinheiten (Nobel-Vortrag) Angew. Chem. 100: 92–116 (1988)

24. R. Consden, A.H. Gordon, A.J.P. Martin, R.L.M. Synge, Gramicidin S: the sequence of the amino acid residues. Biochem. J. XLIII-XLIV (1946)
25. R. Schwyzer, P. Sieber, Die Synthese von Gramicidin S, Helv. Chim. Acta 40: 624–639 (1957)
26 Th. Wieland, K.W. Ohly, Über Peptidsynthesen XVI., Peptidcyclisierungen mit Carbodiimiden, Liebigs Ann. Chem. 605: 179–182 (1957)
27. J.C. Sheehan, M. Goodman, W.L. Richardson, The product derived from the cyclization of triglycine azide. J. Amer. Chem. Soc. 77: 6391 (1955)
28. R.O. Studer, Synthesis and structure of fungisporin, Experimentia 25: 899 (1969)
29. R. Schwyzer, P. Sieber, Verdopplungsreaktionen beim Ringschluß von Peptiden, I. Synthese von Gramicidin S und von bis-homo-Gramicidin S aus den Pentapeptid-Einheiten. Helv. Chim. Acta 41: 2186–2189 (1958)
30. H.G. Zachau, G. Acs, F. Lipmann, Isolation of adenosine amino acid esters from a ribonuclease digest of soluble, liver ribonucleic acid. Proc. Natl. Acad. Sci. USA 44: 885–889 (1958)
31. See e.g. Th. Wieland, Sulfur in biomimetic peptide syntheses, In Roots of Biochemistry, Fritz Lipmann-Meeting, Berlin 1987. H. Kleinkauf, H.v. Döhren, L. Jaenicke eds., de Gruyter, Berlin 1988, pp 213–223. In this book one finds i.a. reviews on history of biological peptide syntheses
32. W. Gevers, H. Kleinkauf, F. Lipmann, Peptidyl transfers in gramicidin S biosynthesis from enzyme bound thiolester intermediates. Proc. Natl. Acad. Sci. USA 63: 1334–1342 (1969)
33. R. Sarges, B. Witkop, Gramicidin VIII. The structure of valine- and isoleucine-gramicidin C, Biochemistry 4: 2491–2494 (1965)
34. C.E. Meyer, F. Reusser, A polypeptide antibacterial agent isolated from Trichoderma viride. Experientia 23: 85–86 (1967)
35. G. Boheim, H.A. Kolb, Analysis of the multi-pore system of alamethicin in a lipid membrane. I. Voltage-jump current-relaxation measurements, J. Membrane Biol. 38: 99–150 (1978)
36. J.F. Borel, C. Feurer, H.V. Gubler, H. Stähelin. Biological effects of cyclosporin A: a new antilymphocytic agent. Agnets Actions 6: 468–475 (1976)
37. A. Rüegger, M. Kuhn, H. Lichti, H.-R. Loesli, R. Huguenin, Ch. Quiquerez, A.v. Wartburg, Cyclosporin A, ein immunsuppresiv wirksamer Peptidmetabolit aus *Trichoderma polysporum* (Link ex Pers.) Rifai, Helv. Chim. Acta 59: 1075–1092 (1976)
38. R.M. Wenger, Synthesis of cyclosporine. Total synthesis of "cyclosporin A" and "cyclosporin H", two fungal metabolites isolated from the species *Tolypocladium inflatum* Gams, Helv. Chim. Acta 67: 502–525 (1984)
39. H. Kobel, R. Traber, Directed biosynthesis of cyclosporins, Eur. J. Appl. Microbiol. Biotechnol 1. 14: 237–240 (1982)
40. A. Billich, R. Zocher, Enzymatic synthesis of cyclosporin A, J. Biol. Chem. 262: 17258–17259 (1987)
41. Th. Wieland, Peptides of poisonous Amanita mushrooms. A. Rich ed., Springer Verlag, Berlin Heidelberg New York 1986
42. Th. Wieland, G. Lüben, H.C.J. Ottenheym, J. Faesel, J.X. de Vries, W. Konz, A. Prox, J. Schmid, The discovery, isolation, elucidation of structure and synthesis of antamanide, Angew. Chem. Int. Ed. Engl. 7: 204–208 (1968)
43. Th. Wieland, M. Nassal, W. Kramer, G. Fricker, U. Bickel, G. Kurz, Identity of hepatic membrane transport systems for bile salts, phalloidin, and antamanide by photoaffinity labeling, Proc. Natl. Acad. Sci. USA 81: 5232–5236 (1984)
44. For details see e.g. W. Burgermeister, R. Winkler-Oswatitsch, Topics in current Chemistry, Complex formation of monovalent cations with bifunctional ligands. F.L. Boschke ed., Springer-Verlag Berlin Heidelberg New York 1977, pp. 91–196
45. I.L. Karle, J. Karle, Th. Wieland, W. Burgermeister, H. Faulstich, B. Witkop, Conformation of the Li-antamanide complex and Na [$Phe^4$, $Val^6$] antamanide complex in the crystalline state. Proc. Natl. Acad. Sci. USA 70: 1836–1840 (1973)
46. I.L. Karle, Th. Wieland, D. Schermer, H.C.J. Ottenheym, Conformation of uncomplexed, natural antamanide crystallized from $CH_3CN/H_2O$. Proc. Natl. Acad. Sci. USA 76: 1532–1536 (1979)
47a. W. Burgermeister, Th. Wieland, R. Winkler-Oswatitsch. Antamanide. Relaxation study of conformational equilibria, Eur. J. Biochem. 44: 311–316 (1974)
47b. H. Kessler, J.W. Bats, J. Lautz, A. Müller, Conformation of antamanide, Liebigs Ann. Chem. 1989: 913–928
48. H.P. Kaufmann, A. Tobschirbel, Über ein Oligopeptid aus Leinsamen, Chem. Ber. 92: 2805–2809 (1959)

49. H. Kessler, M. Gehrke, A. Haupt, M. Klein, A. Müller, K. Wagner, Common structural features of cytoprotection activities of somatostatin, antamanide and related peptides, Klin. Wochenschr. 64: 74–78 (1986)
50. H. Faulstich, A. Buku, H. Bodenmüller, Th. Wieland, Virotoxins: actin-binding cyclic peptides of *Amanita virosa* mushrooms, Biochemistry 19: 334–343 (1980)
51. E. Munekata, H. Faulstich, Th. Wieland, Totalsynthese des Phalloins und Leu$^3$-phalloins, Liebigs Ann. Chem. 1977: 1758–1765
52. A recent review: H. Faulstich, S. Zobeley, G. Rinnerthaler, J.V. Small, Fluorescent phallotoxins as probes for filamentous actin, J. Muscle Res. Cell Motility 9: 370–383 (1988)
53. A. Baku, Th. Wieland, H. Bodenmüller, H. Faulstich, Amaninamide, a new toxin of *Amanita virosa* mushrooms, Experientia 36: -36-34 (1980)
54. H. Faulstich, The amatoxins, Progr. Mol. Subcell. Biol. 7: 88–134 (1980)
55. Th. Wieland, U. Gebert, Strukturen der Amanitine, Liebigs Ann. Chem. 700: 157–173 (1966)
56. E.C. Kostansek, W.N. Lipscomb, R.R. Yocum, W.E. Thiessen, The crystal structure of the mushroom toxin β-amanitin. J. Amer. Chem. Soc. 99: 1273–1274 (1977)
57. C.T. Bishop, E.F.L.J. Anet, P.R. Gorham, Isolation and identification of the fast-death factor in *Microcystis aeruginosa* NRC-1. Can. J. Biochem. Physiol. 37: 453– (1959)
58. D.P. Botes, A.A. Tuimman, P.L. Wessels, C.C. Viljoen, H. Kruger, D.H. Williams, S. Santikarn, R.J. Smith, S.J. Hammond, The structure of cyanoginosin-LA, a cyclic heptapeptide toxin from the cyanobacterium *Microcystis aeruginosa*. J. Chem. Soc. Perkin Trans. 1984: 2311–2318
59. In order to unify the nomenclature 14 laboratories in a letter to the editor have suggested the name "microcystin" with the two letter suffic XY: Naming of cyclic heptapeptide toxins of cyanobacteria (blue-green algae), Toxicon 26: 971–973 (1988)
60. M.T.C. Runnegar, I.R. Falconer, Effect of toxin from the cyanobacterium *Microcystis aeruginosa* on ultrastructure morphology and actin polymerization in isolated hepatocytes Toxicon 24: 109– 115 (1986)
61. K. Fukase, M. Kitazawa, A. Sano, K. Shimbo, H. Fujita, S. Horimoto, T. Wakamiya, and T. Shiba, Total synthesis of peptide antibiotic Nisin, Tetrahedron Lett. 795–798 (1988).
62. N. Schnell, K.-D. Entian, U. Schneider, F. Götz, H. Zähner, R. Kellner, G. Jung, Prepeptide sequenz of epidermin, a ribosomally synthesized antibiotic with four sulfide-rings, Nature 333: 276–278 (1988)
63. H. Lackner, Die Raumstruktur der Actinomycine, Angew. Chem. 87: 400–411 (1975)
64. See Antibiotics III (J.W. Corwran and F.H. Hahn, eds.) Springer, Berlin Heidelberg New York 1975
65. For an overview see E.W. Warnhoff, Peptide alkaloids in Progress in the Chemistry of Organic Natural Products, W. Herz, H. Griesebach, A.I. Scott eds. Springer, Vienna New York Vol. 28, 162–203 (1970)
66. U. Schmidt, Natürliche Cyclopeptide und Cyclopeptolide, Nachr. aus Chemie und Technik 57: 1034–1042 (1989)

# 10 Peptide Research Around the World

As indicated in the first chapter of this book, for the origins of research in peptide chemistry one has to look back into the nineteenth century. At that time, with notable exceptions, such as the fundamental work of Jac. Berzelius (1779–1848), who lived and worked in Sweden, most of the important contributions came from France and from Germany. In the first half of this century the leading roles in peptide research were played mainly by German chemists, but during the Second World War this situation underwent a dramatic change. The spectacular progress of biochemistry stimulated worldwide interest in the field and many new investigators, trained as organic chemists, joined the biochemists and thus a new discipline, bio-organic chemistry, developed with peptide chemistry as one of its most productive branches.

Elucidation of the structure of insulin and synthesis of oxytocin became the starting points of an explosive growth in peptide research that soon transgressed national boundaries. It seems, therefore, justified to look around in the world and to identify some, certainly not all, research groups in individual countries. This brief survey also provides an opportunity to mention some of the best known investigators who initiated certain distinct line of study and the scientists who joined them in their efforts. Needless to say that no exhaustive listing of peptide chemists is meant, neither do we wish to suggest that the achievements briefly discussed here would be the only important contributions from a certain country. We hope, that in spite of such obvious limitations, this brief review of peptide research according to individual countries will be of some interest to the reader.

## 10.1 Australia

Peptide chemistry in Australia benefited from the experience of E.O.P. Thompson gained in the laboratory of Sanger (Plate 26) in Cambridge, England, where the determined the sequence of the A-chain of insulin, but after returning to the University of New South Wales in Sidney. Thompson dedicated his efforts to the study of proteins rather than peptides.

At Monash University in Melbourne, J.M. Swan, former coworker of du Vigneaud and known for his method of introducing glutamine residues through the incorporation of pyroglutamyl moieties, studied various important aspects of disulfide chemistry, a topic of general interest in Australia because of its relevance for wool. The same stimulus can be recognized in several of the research endeavors at the Australian national research organization (CSIRO) in Melbourne. Here the work of F.H.C. Stewart on the reactive esters and the observations of A.C. Beecham on the decomposition of tosylamino acid chlorides, both in the nineteen fifties, reveal the interest and participation of Australian investigators in peptide chemistry. At the St. Vincent's Hospital Medical School in Melbourne Pehr Edman (Plate 15) with G. Begg developed his automated peptide sequenator during 1957–1967 (p. 119).

## 10.2 Austria

The experiments of F. Wessely (Plate 47) with *N*-carboxyanhydrides in the nineteen twenties, have already been mentioned in Chap. 2 (cf. p. 37). Wessely continued to be interested in the development of the peptide field. His work at the university of Vienna was continued by his students H. Michl and K. Schlögl.

In 1953 a significant achievement was reported from the same university: the elucidation of the structure of oxytocin by Hans Tuppy. Tuppy had just returned from Cambridge, England, where, as one of the principal coworkers of Sanger, he determined the amino acid sequence of the B-chain of insulin. In Vienna, almost singlehandedly and with the simplest means, he solved the sequence of oxytocin and published it a few months ahead of V. du Vigneaud (cf. p. 140). From Tuppy's associates H. Nesvadba, first at the University later at the SANDOZ laboratories in Vienna, studied problems of peptide synthesis.

## 10.3 Belgium

From the contributions of A. Loffet, while still in Brussels, the acidic reagent $\beta$-mercaptoethanesulfonic acid, that can preferentially be used for the hydrolysis of tryptophan containing peptides is mentioned here. Currently he is active in France, engaged in the study of propeptides. In Brussels, G. van Binst, professor at the Vrieve Universiteit, is investigating the conformation of biologically active peptides, such as somatostatin and its analogs, with the help of 2D NMR spectroscopy.

## 10.4 Bulgaria

At the laboratories of the Bulgarian Academy of Sciences in Sophia, S.B. Stoev and G. Videnov, both former associates of H. Zahn (Plate 54) in Aachen, Germany, (Stoev also a former coworker of M. Manning in Toledo, USA) are the nucleus of a growing group of Bulgarian peptide chemists. Improvements in enzymatic synthesis is among the objectives of L. Mladenova-Orlinova, I. Mancheva and J. Calderon.

## 10.5 Canada

Isolation of insulin by F.G. Banting and C.H. Best, in 1922, is probably the earliest manifestation of peptide research in Canada. In the following years further significant contributions originated from the Connaught Laboratories at the University of Toronto, such as the discovery, by D.A. Scott, of zinc as the ligand in crystalline insulin and, later, in 1960, the ground-breaking experiment of G.H. Dixon and A.C. Wardlaw in which the two separated inactive chains of the hormone were combined to biologically active material. Today, at the same university, C. Deber (Ph.D. with E. Blout at Harvard) as professor of biochemistry is actively engaged in studies of problems of peptide-lipid interactions and of ion-transport across membranes.

At the University of Ottawa the work of L. Benoiton and his students, especially F.M.F. Chen, gained major importance in the study of racemization during peptide synthesis. The loss of chiral purity in activated *N*-methylamino acids is one of their numerous contributions. 1-Ethoxycarbonyl-2-ethoxy-1,2-dihydroquinoline (EEDQ), the coupling reagent mentioned on page 91 was introduced in 1968 by B. Belleau and Malek at Ottawa.

In Montreal, in 1957, C. Berse, R. Boucher and L. Piche introduced the *S-p*-nitrobenzyl group for the protection of the thiol function. In more recent years, the research of P.W. Schiller, Professor at the University of Montreal and director of peptide research the Clinical Research Institute of Montreal, is particularly noteworthy. Schiller received his training in Zurich and graduated at the ETH as a student of R. Schwyzer (Plate 39). His studies are dedicated to the exploration of relationships between the architecture and biological activities of hormones, such as corticotropin, mainly by means of the energy trasnsfer method, but they also include investigations of opioid peptides and their conformationally restricted cyclic analogs.

In 1973 an interesting method for the removal of the assembled peptide chain from the polymeric support through intramolecularly catalyzed transesterific-ation, was reported from the University of Alberta in Edmonton by M.A. Barton (Ph.D. with G.W. Kenner (Plate 27), Liverpool, England) and her associates. At the same university, R.S. Hodges, professor of biochemistry is known for his work on $Ca^{2+}$ binding peptides.

## 10.6 China

In the early nineteen sixties, simultaneously with efforts directed toward synthetic insulin in Germany and in the USA, a spectacularly successful recombination of the two insulin chains was achieved by Y.C. Du, R.Q. Jiang, and L.C. Tsou at the Institute of Biochemistry of Academia Sinica, in Shanghai. Soon after, this endeavor was concluded with the synthesis of the individual chains and their combination to the molecule of the hormone that was secured in crystalline form. This major achievement was due to the collaboration of three teams of peptide chemists: Y.T. Kung, Y.C. Du, W.T. Huang, C.C. Chen, L.T. Ke, S.C. Hu, R.Q. Jiang, S.Q. Chu and C.I. Niu of the Biochemistry Department of the Academy; J.Z. Hsu, W.C. Chang, L.L. Cheng, H.S. Li and Y. Wang (Plate 46) of the Institute of Organic Chemistry of the Academy, in Shanghai; and T.P. Loh, A.H. Chi, C.H. Li, P.T. Shi, Y.H. Yich, K.L. Tang and C.Y. Hsing of the Department of Chemistry of Peking University. An internationally renowned member of the Shanghai team in which he played a leading part is Y. Wang (Wang Yu). For a short biography see page 273.

## 10.7 Czechoslovakia

Stimulated by the 1953 synthesis of oxytocin by du Vigneaud and his associates, a research group led by Jose Rudinger (Plate 7) (cf. p. 155), the young head of the laboratory of organic chemistry of the Research Institute of the Czechoslovak Academy of Sciences in Prague, soon embarked on a new synthesis of the hormone and accomplished it with considerable improvements. The highly competent group of investigators around Rudinger continued to be active in peptide chemistry even after his emigration to Switzerland. The chemistry and synthesis of the neurohypophyseal hormones remained at the center of their interest and gave rise to a continuous flow of publications. The authors of these papers, K. Bláha 1926–1988, Plate 11), M. Zaoral, K. Jost (1932–1980), M. Pravda, K. Poduska, J. Honzl, V. Gut, I. Fric, J. Hlavacek, M. Lebl, established Prague as one of the important centers of peptide chemistry in Europe.

## 10.8 Denmark

Early production of insulin in this country stimulated the development of peptide research at two institutions, Nordisk Insulinlaboratorium and NOVO Pharmaceutical Company. At Nordisk, particularly after the formation of the Protein Chemistry Laboratory under the auspices of The Danish Academy of Technical Science, K. Brunfeldt initiated considerable effort toward improvements in solid phase peptide synthesis.

At NOVO by Jan Marcussen and his associates, at Nordisk by Bruno Hansen and his coworkers, significant advances in the production of human insulin were achieved, both through the conversion of porcine insulin and by the means of gene technology. At the Hagedorn Research Laboratory, insulin studies headed by B. Hansen were extended to include investigations on gastrointestinal hormones, for instance, together with H. Kofod, on the mechanism of action of secretin.

The Carlsberg laboratories have become, in recent years, a center of investigations in the area of peptide-bond formation with the help of proteolytic enzymes, especially carboxypeptidase Y (from yeast). In these studies the work of J.T. Johansen and his coworkers is remarkably productive: it has made possible the synthesis of smaller peptide hormones through the exclusive use of enzymes for coupling. (See on p. 60).

## 10.9 France

Around 1950 studies on the structure of oxytocin and vasopressin appeared from C. Fromageot's laboratory. (His son, P. Fromageot remains active in peptide research.) In 1956, L. Velluz and his associates, G. Amiard, J. Bartos, B. Goffinet and R. Heymes, published a refreshingly novel approach to the synthesis of oxytocin. It was based on the extensive use of the trityl group, including the protection of the SH function, a method of blocking that has gained new attention in recent years. Oxytocin provided the stimulus also for the studies of Roger Acher, a former coworker of du Vigneaud, professor at the University of Paris. In the nineteen-sixties he determined the structure of several of its natural analogs isolated from various fishes and amphibians. In that period E. Bricas (cf. also in the section on Greece), in Orsay, embarked on an extensive investigation of peptides from bacterial cell walls. A major figure in the chemistry of natural products, Edgar Lederer (1908–1988, Plate 28, p. 129) at the French national research organisation, Centre National de Recherche Scientifique (CNRS) in Gif sur Yvette dedicated a part of his efforts to peptides. With D.W. Thomas and B.C. Das, he pioneered the determination of the structure of peptidolipids and cyclic peptides through mass spectrometry. Later, in a collaborative study with the Swiss Nestle group, he established that the bitter principle of chocolate is a molecular adduct formed from a diketopiperazine and theobromine.

In the last decades several branches of CNRS became involved in peptide research; for instance, in Orleans (Y. Trudelle), in Marseille (J. van Rietschoten, synthesis of scorpion toxins), in Paris (C. Durieux, conformation of cholecystokinin); in Gif sur Ivette (S. Fermandjian, peptide architecture studies via CD spectra), and last, but not least, in Montpellier where Jean Martinez achieved considerable success in the study of gastrointestinal hormones. Martinez, a student of F. Winternitz, with whom at the University of Montpellier, he introduced a new method for peptide bond formation, and then a coworker of one of the present authors, (M.B.), recently proposed a novel hypothesis on the mechanism of action of gastrin involving an enzyme system with the receptor. In the same city, Montpellier, the pharmaceutical company Clin–Midy is an active participant in peptide research: a few years ago its investigators reported the large scale synthesis of somatostatin.

French universities have not been left behind in this development. At the Faculté de Pharmacie de Paris, B.P. Roques designed novel enkephalinase inhibitors, the group of A. Marquet is working on problems related to substance P. An important contribution originated from the University of Nancy, where B. Castro (now with CNRS in Montpellier) with J.R. Dormoy, developed the potent BOP coupling reagent (cf. p. 92) already in use in the practice of peptide synthesis.

## 10.10 Germany

By the end of the nineteen thirties the once flourishing peptide research came to a virtual halt in Germany, but it was reawakened after the War. Theodor Wieland's studies of the chromatography of amino acids represent a start that was followed by the development of the symmetrical and mixed anhydride methods of coupling in his laboratory. Through several decades the Wieland school (e.g. H. Determann, H. Faulstich, C. Birr, E. Munekata) remained engaged in the isolation, structure determination and synthesis of toxic peptides from Amanita mushrooms. Their work that included also the exploration of the biochemistry of these interesting substances remained productive after the transfer of Wieland from the University of Frankfurt to the Max Planck Institute for Medical Research in Heidelberg.

Munich became a similarly important center of peptide research. Fritz Weygand, (Plate 48) professor at the Technical University of Munich and his studies, A. Prox, W. Steglich (professor in Bonn), R. Geiger (Plate 19) and W. König (Plate 20) determined the extent of racemization through the separation of diastereoisomers by vapor phase chromatography. An important consequence of this work was the introduction of derivatives of trifluoroacetic acid in peptide synthesis and the proposal to use trifluoroacetic acid for the removal of blocking groups by acidolysis.

Weygand's successor, Ivar Ugi, invented the perhaps most surprising approach to peptide bond formation, the four-center-condensation (4CC) procedure. His principal coworker, D. Marquarding (1934–1982) is remembered here.

Also in Munich, the former Kaiser-Wilhelm-Institute für Lederforschung in Dresden, led by W. Grassmann after M. Bergmann's (Plate 3, p. 45) emigration in 1934 until its destruction by bombing in 1945 was reestablished in 1948 and renamed the Max Planck Institute for Protein and Leather Research. Here peptide studies were resumed by Grassmann and continued with increased vigor by his successor Erich Wünsch (Plate 51). His laboratory, now incorporated as the Department of Peptide Chemistry into the large Max Planck Institute for Biochemistry in Martinsried near Munich, was the first to accomplish the demanding synthesis of glucagon. The research efforts of Wünsch, E. Jaeger, L. Moroder and numerous other associates was later extended to the synthesis of several more gastrointestinal hormones. Wünsch and his coworkers deserve special credit for compiling and writing two impressive volumes of the Houben-Weyl handbook, Methoden der Organischen Chemie, that deals with the literature of the methodology of peptide synthesis until 1974 (see p. 62).

At the University of Mainz a remarkably successful attempt toward the synthesis of glycopeptides is being led by H. Kunz and his coworkers.

In the late fifties a daring endeavor, the synthesis of insulin, was undertaken by Helmut Zahn, (Plate 54) professor at the Technical University of Aachen and director of the Wool-Research Institute. His group, J. Meienhofer (Plate 54),

E. Schnabel, D. Brandenburg, H. Klostermeyer, T. Okuda, R. Zabel, W. Skroka, and H. Bremer, was the first to achieve the combination of the synthetic A and B chains of insulin to biologically active material. Their work was extended later to the preparation of numerous analogs of insulin and of proinsulin.

An exploration of the relationship between the structure of insulin and its hormonal activity was carried out also in Tübingen, where G. Weitzel and his associates (S. Hörnle, U. Weber) exploited the solid phase method for the preparation of the numerous insulin analogs needed for this purpose. Tübingen developed into center of peptide research through the activities of Ernst Bayer (Plate 10) and his school. With H. Hagenmayer, G. Jung and M. Mutter he developed the liquid phase technique in which peptide chains are built while attached to a soluble polymer, usually polyethyleneglycol. Mutter, now professor in Lausanne, belongs to the pioneers who attempt tailored synthesis of conformationally well-defined peptides. One of E. Bayer's students, W.A. König, who developed methods for the separation of enantiomeric amino acids by gas chromatography on chiral matrices (cf. p. 55) is now professor in Hamburg Günther Jung, working on several fields, is heavily involved in research on lanthionine containing antibiotics ("lantibiotics" cf. p. 222). The laboratory of W. Voelter, also in Tübingen, is engaged in a search for new blocking groups, analytical methods and immunologically active peptides. Not far from this city, in Ulm, the chemistry of cyclopeptides and of cyclols (cf. p. 196) is being studied in depth by Manfred Rothe.

In Stuttgart, Ulrich Schmidt with his associates have synthesized numerous cyclopeptides containing uncommon amino acids from plants (peptide alkaloids) and from moulds, (ansacyclopeptides with diphenylether units).

At the University of Frankfurt Horst Kessler, now at the Technical University of Munich, can list important achievements in the determination of the 3-dimensional structure of peptides with the help of 2D NMR spectra (cf. p. 127). At the nearby Hoechst pharmaceutical company Rolf Geiger (1932–1988) and W. König, both former students of F. Weygand, with R. Obermeier and H. Wissmann reported the synthesis of several complex, biologically active peptides, including secretin and insulin. At least equally important are their contributions to the methods of peptide bond formation and protection. The "additives" proposed by König and Geiger for the prevention of racemization (cf. p. 87) had a major impact on the practice of peptide synthesis.

In Berlin, in the laboratories of Schering A.G., E. Schröder and K. Lübke, made significant contributions to peptide chemistry.

Awakening of peptide research in Germany is not limited to the West. In the DDR important work has been carried out at various places. At the Institute of the Academy of Sciences in Berlin N. Niedrich pioneered the area of pseudopeptides through the incorporation of hydrazinoacids. The studies of G. Losse in Dresden have gained recognition. Fundamental investigations on novel active esters and on enzyme-catalyzed synthesis were carried out at the University of Leipzig, by H.D. Jakubke and his coworkers, of whom A. Baumert is mentioned here. At the

Friedrich Schiller University in Jena H. Arold and his group synthesized cystinepeptides and bradykinin analogs. Horst Hanson (1911–1978), a student of E. Abderhalden (Plate 8), and the successor of his teacher as the director of the Institute of Physiological Chemistry at the Martin Luther University in Halle/Saale, should certainly by mentioned. His work focused on intracellular proteolysis including synthesis of peptide substrates, in collaboration with Peter Hermann. In the Chemistry Institute at the same University, H. Jeschkeit is working on blocking groups and active esters.

## 10.11 Greece

The origins of peptide chemistry in Greece can be traced, without doubt, to Leonidas Zervas, (Plate 55) (1902–1980), one time principal coworker of Max Bergmann in Dresden (cf. p. 46). After returning to his native country Zervas, became profesor at the University of Athens, where he trained generations of peptide chemistry many of whom became major contributors to the field.

Of the Zervas-students, first and foremost, Iphigenia Photaki (Plate 35) (1921–1983), later professor at the University of Athens, should be remembered. Her collaboration with Zervas resulted in a whole array of sulfhydryl protecting groups. Later she turned here interest to peptides of lanthionine, including, with A.E. Yiotakis, the synthesis of lanthinonine containing cyclic peptides and, with C. Tzougraki, to chromogenic substrates.

Zervas himself remained interested in amine-protecting groups and made considerable progress in the application of the *O*-nitrophenylsulfenyl group and the trityl group. The latter represents a classical method of blocking, neglected for a long time, but brought back to life through the efforts of Zervas and his associates, particularly D. Theodoropoulos, now professor at the University of Patras. At the University of Ioannina scientific offsprings of the Athen school, A. Sakarellos, C. Sakarellos and M. Sakarellos-Daitsiotis are engaged in studies concerning the rules governing $\beta$-turns in peptides. Of the former Zervas coworkers G.C. Stelakatos remained in Athens.

Several students of Zervas followed distinguished careers abroad: E. Bricas became professor in Orsay, France. P.G. Katsoyannis (Plate 26) (cf. p. 161) in New York.

## 10.12 Hungary

In 1909 a young Ph.D., Géza Zemplén (1883–1956), traveled to Berlin to continue his studies in Emil Fischer's laboratory. Their joint publications deal with the synthesis of proline through ornithine. However, after his return to Hungary, where he became professor of organic chemistry at the Technical University of Budapest, instead of the peptide studies started with Fischer, Zemplén continued the second line of research in which he was involved while in Berlin: carbohydrates. Thus, interest in peptide chemistry was awakened only decades later in

Hungary and it happened in an unforeseen way. The professor of bacteriology at the University of Szeged, G. Ivánovics, separated from the anthrax bacillus a capsule that envelops the microorganism and asked Viktor (Gyözö) Bruckner (Plate 14) (1900–1980), professor of organic chemistry at the same university, to analyze the capsule material. Hydrolysis with hydrochloric acid yielded a single crystalline compound, the hydrochloride of D-glutamic acid. Further investigations revealed that the capsule consists of poly-$\gamma$-D-glutamic acid. This discovery was the start of a research effort that was continued through many years, first in Szeged, then after the transfer of Bruckner, in Budapest. The studies concerning the structure and biological properties of the capsule polypeptide were the first major undertaking in the peptide field in Hungary, the topic of many doctoral dissertations and training ground for a generation of peptide chemists of whom J. Kovács (1915–1985; later professor at St. John's University in Jamaica, Queens, New York), K. Medzihradszky, H. Medzihradszky-Schweiger, T. Vajda and M. Kajtár are mentioned here. Subsequently their research branched out to include the synthesis of biologically active peptides, such as corticotropin.

Independently from the work of the Bruckner school a second endeavor toward synthetic peptides was initiated, at the Institute for Drug Research in Budapest, by Miklos Bodanszky, whose interest in peptide synthesis was due to reading the 1953 communication of du Vigneaud and his coworkers on the synthesis of oxytocin. With a group of young chemists, M. Szelke, E. Tömörkény and E. Weisz, he first reproduced, then, by the application of active esters, improved the original procedure. Soon after S. Bajusz joined them to become later one of the most outstanding peptide chemists in the country. His work on ACTH, opioid peptides and synthetic thrombin inhibitors is indeed noteworthy. Subsequently, industrial laboratories took up peptide research; the investigators of the G. Richter Laboratories, L. Kisfaludy (1924–1988). M. Löw and I. Schön, through their studies of side reactions significantly contributed to the methodology of peptide synthesis.

When Bruckner accepted the invitation to work in Budapest, peptide chemistry did not come to a halt in Szeged. Led by his former associate K. Kovács, an active research group is engaged in synthesis and study of peptide hormones. The contributions of L. Baláspiri and B. Penke to the chemistry of gastrointestinal peptides are certainly noteworthy.

## 10.13 India

During the last three decades, at the Indian Institute of Science in Bangalore, a group of distinguished scientists, under the leadership of G.N. Ramachandran, established the fundamental principles of peptide conformation ("Ramachandran plots"). In the same city, at Central College, K.M. Sivanandaiah, with S.A. Khan, G.M. Anantharamaiah and others, improved certain methods of peptide synthesis, for instance introduced transfer hydrogenation for the removal of blocking groups. At Delhi University in New Delhi a former coworker of du

Vigneaud, V.V.S. Murti, trained numerous students in peptide chemistry, some of whom, e.g. S. Natarajan and N. Chandramouli, are now actively engaged in peptide studies in the USA.

## 10.14 Israel

The discovery of acidolytic cleavage of blocking groups initiated by A. Berger and D. Ben-Ishai was discussed in Chap. 3. This work, reported from the Weizmann Institute in Rehovoth, was followed by numerous significant contributions by members of this Institute. Outstanding research on poly-amino acids (see p. 36) by E. Katchalski (Plate 24) (who, as E. Katzir, was the fourth president of the State of Israel) and studies concerning acylation with polymer-bound reactive intermediates by A. Patchornik (Plate 25) and M. Fridkin are among their achievements. Also at the Weizman Institute, M. Sela and R. Arnon developed synthetic antigens based on poly-amino acids, while M. Wilchek, in collaboration with C.B. Anfinsen and P. Cuatrecasas of the National Institutes of Health of the USA, introduced affinity chromatography, a most valuable method for the purification of peptides and proteins.

At the Hebrew University of Jerusalem, M. Frankel, Y. Wolman, Y.S. Klausner, C. Gilon and M. Chorev were engaged in the development of synthetic methods, such as the design and preparation of polymeric carbodiimides and water-soluble active esters.

## 10.15 Italy

In the early sixties Ernesto Scoffone, (Plate 40) professor at the University of Padua, with his associates, F. Marchiori and R. Rocchi, embarked on a study of active esters and soon employed them for the synthesis of biologically important molecules such as the S-peptide. This 20-residue fragment of ribonuclease A is cleaved by subtilisin from the N-terminal part of the chain of the enzyme. The paduan school soon became a significant center of peptide research where a number of S-peptide analogs have been prepared and utilized in investigations of the ribonuclease S′ system. Several of Scoffone's coworkers, G. Borin, A. Fontana, C. Di Bello, L. Moroder, E. Peggion, A. Scatturin, A.M. Tamburro, C. Toniolo and G. Vidali turned into frequent contributors of the peptide field.

In Ferrara in the seventies C. Benassi, M. Guarneri and, for some years, R. Rocchi, carried out interesting work on novel types of reactive esters and also on protease inhibitors. Currently, R. Tomatis is interested in opioid peptides. In 1975 Rocchi was appointed professor in Padua where Di Bello, Peggion and Toniolo, all former associates of Scoffone, received similar appointments and are active in the study of structure–reactivity relationships via synthetic peptides. Fontana, also in Padua, is involved in protein chemistry, Moroder transferred his activities to the Max Planck Institute in Martinsried, Munich, Scatturin moved to Ferrara and Tamburro to Potenza.

In Milan, F. Chillemi and G. Pietta made significant improvements in solid phase synthesis. In the same city, at the research laboratory of the San Rafaele Hospital, R. Colombo investigated the synthesis of biologically active peptides by the liquid phase method and developed novel ways of protection of the histidine side chain. The laboratories of pharmaceutical company Farmitalia in Milan also became a major center of peptide research.

Following the discovery of potent peptides in the skin of amphibians by V. Erspamer (Plate 17), A. Anastasi and B. Bosisio, the Farmitalia group, L. Bernardi, R. de Castiglione, P. Montecucchi, G. Perseo and their colleagues developed procedures for the synthesis of a whole series of these compounds. One of them, caerulein (cf. p. 166) is prepared on a commercial scale.

In Rome peptide research is pursued both at the University "La Sapienza" (G. Lucente, F. Pinnen, G. Zanotti on cyclic and bicyclic, Amanita-peptides) and at Enricerche, the research laboratory of Enichem (S. Verdini). In Naples E. Benedetti, C. Pedone, P.A. Tamussi and L. Paolillo investigate the conformation of natural and synthetic peptides.

## 10.16 Japan

The famous Sakaguchi reaction for the detection of arginine indicates that amino acids and proteins had stimulated interest in Japan at an early stage of research in this field. In 1957, Shiro Akabori, (Plate 9) professor at Osaka University, with his students Kenji Okawa (now professor at Kwansei University) and Mikio Sato established a basic principle for the conversion of glycine to serine or threonine. Even more peptide related is Akabori's method, hydrazinolysis, for the determination of the C-terminal residue in peptides and proteins (p. 116). He launched numerous young researchers on a career in peptide research; several of them became distinguished contributors to various areas of this discipline. Thus Shumpei Sakakibara (Plate 37) introduced, with Yasutsugu Shimonishi, an important method for the removal of blocking groups: acidolysis with liquid hydrogen fluoride. As director of the Peptide Institute, in Osaka, Sakakibara leads a group of investigators in the demanding endeavor of synthesizing complex molecules, such as the 33-peptide cholecystokinin (cf. p. 166) or the parathyroid hormone, a 84-peptide, all in high purity. The Peptide Institute is now worldwide source of a large number of biologically active peptides synthesized for research purposes.

Also in Osaka, another former student of Akabori, professor Tetsuo Shiba (Plate 43) and his associates developed a procedure for the preparation of lanthionine-containing peptides and applied it, in 1988, to the synthesis of the antibiotic nisin (p. 223). In Tokyo, at Rikkyo University, professor Ichiro Muramatsu, also from the Akabori school, proposed a rapid method of synthesis and discovered novel side reactions, such as the formation of guanidine derivatives during coupling with carbodiimides.

Intensive studies in the synthesis of biologically active peptides followed the return of Haruaki Yajima (Plate 52), Chizuko Yanaihara and Noboru

Yanaihara from the laboratory of Klaus Hofmann (p. 267) in Pittsburgh. Yajima, as professor at the School of Pharmacy at the University of Kyoto, led a major research group and is the author of an impressive number of publications on the synthesis of peptide hormones. With Nobutaka Fujii as sole coworker he reported the synthesis of the enzyme ribonuclease A (124 residues), secured in crystalline form. The laboratory of professors C. and N. Yanaihara at the University of Shizuoka, is an important center of studies of the immuno-properties of peptides.

In the late sixties Teruaki Mukaiyama of the University of Tokyo, with his associates, R. Matsueda, H. Maruyama, M. Ueki and others, introduced a surprisingly novel method for the formation of the peptide bond, the oxidation-reduction condensation procedure in which the high affinity of trivalent phosphorus (triphenylphosphine) for oxygen is the driving force (p. 98). For a possible acyloxyphosphonium intermediate cf. the related coupling device of Mitin and Vlasov on p. 247.

In Fukuoka, professor Nobuo Izumiya and his numerous associates dedicated their efforts to the chemistry of cyclic peptides. They achieved considerable success in the preparation (e.g. with H. Aoyagi) and study of analogs of gramicidin S (cf. p. 205).

Eisuke Munekata, who was an associate of Theoder Wieland in Germany in the nineteen seventies (synthesis of phalloin, p. 219) now professor at the University of Tsukuba, is actively engaged in the study of neuropeptides.

Contributions from industrial laboratories are not less significant. For instance the work of Masumi Itoh on the prevention of racemization in coupling originates from the research laboratory of the Fugisawa company in Osaka. Masahiko Fujino of Takeda Chemical Industries, established the fundamental principles for the design of superactive LHRH analogs.

## 10.17 New Zealand

The efforts of D.A. Buckingham, in Duneden, to exploit the enhanced reactivity of peptides in Co-III complexes for rate enhancement in the hydrolysis of their esters and in the formation of a new peptide bonds brought interesting and potentially useful results. At Massey University the research of W.S. Hancock and his associates has led in the last decade to the design of new, selectively removable blocking groups.

## 10.18 Norway

In the 1970s, Kirsten Titlestad and J. Dale published studies on the conformational aspects of small cyclic peptides.

## 10.19 Poland

It was Emil Taschner's (Plate 45) unusually active mind that gave the first impetus to the development of peptide research in Poland. His wide interest encompassed such diverse areas as separation of diastereoisomers by paper-chromatography, preparation of **tert**-butyl esters of amino acids via transesterification, cleavage of blocking groups with pyridinium chloride or catalysis of the aminolysis of active esters by weak acids. His most important success lies, however, in the initiation of a generation of peptide chemists and thus the creation of a major center of research at the University of Gdansk. From the long list of his students A. Chimiak (protection of the SH group), B. Liberek (mechanisms of racemization, conformation of diketopiperazines, preparation of unusual amino acids), R. Rzezsotarska (modification of amino acids and analogs of biologically active peptides) and I.Z. Siemion (structure determination by NMR and CD, synthesis of bio-active peptides) are mentioned here. The University of Gdansk is the home of several more distinguished investigators, Z. Grzonka (synthesis of biologically active peptides), G. Kupryszewski (active esters, prevention of racemization) and F. Muzalewski (reactive intermediates).

At the University of Warshaw, S. Drabarek established a laboratory for peptide synthesis and her work is being vigorously continued by J. Izdebski. In Lodz the studies of M.T. Leplawi ($\alpha$-disubstituted amino acids and their peptides) are noteworthy.

## 10.20 Portugal

The sustained effort of M. Joaquina S.A. Amaral Trigo, onetime coworker of Rydon in Exeter, on selectively removable $\beta$-halogenethyl blocking groups, is frequently noted in the literature.

## 10.21 Spain

In recent years at the University of Barcelona E. Giralt with E. Pedroso, F. Albaricio and their associates have been working on improvements in the methodology of peptide synthesis. In collaboration with P. Van Rietschoten (Marseilles) they undertook the preparation of scorpion toxins by convergent solid phase peptide synthesis. This involves building of polymer bound segments of the target compound, cleavage of the intermediates in protected form from the resin followed by purification and finally their combination on a second insoluble polymeric support.

## 10.22 Sweden

The Karolinska Institute in Stockholm is one of the leading centers of peptide and protein research in the world. The definitive method of sequence determination, developed there by Pehr Edman (cf. p. 118), (Plate 15) is only one of the examples of early interest in peptide chemistry in Sweden. Peptide-related work can be found outside of the Karolinska as well. Thus, the studies of Georg Folch in Gøteborg, also in the nineteen fifties, on the synthesis of phosphoserine and phosphoserine containing peptides, should be remembered here.

In the years following the Second World War the pioneering work of J. Erik Jorpes (Plate 23), professor at the Karolinska Institute and of his young associate, Viktor Mutt (Plate 23), led to the isolation (from porcine intestines) of secretin and cholecystokinin-pancreozymin in the pure form. Determination of the amino acid sequence of the two hormones by Viktor Mutt soon followed and these sequences were confirmed by synthesis in numerous laboratories. Mutt's studies then broadened to include further gastrointestinal hormones (cf. p. 165) such as VIP, GIP, and several more biologically active peptides. During the last decade, Mutt has launched an entirely novel line of investigations, the discovery and isolation of biologically active peptides based on the identification of their characteristic C-terminal amino acid amide residues rather than on their biological activities. This ingenious method was tested in practice, in part in collaboration with H. Tatemoto, and has yielded several new gastrointestinal hormones.

In Uppsala, J. Porath and P. Flodin developed an extremely important method for the separation of peptides and proteins: gel permeation chromatography with crosslinked dextran (Sephadex). One of Porath's former associates, Ulf Ragnarsson contributed with innovative modifications to the solid phase method of peptide synthesis and in more recent years introduced the principle of complete protection of the amine function by diacylation. Another Porath student, Gunnar Lindeberg, explored the properties of vasopressin analogs.

Important contributions originated from the Swedish pharmaceutical industry as well. Ferring Pharmaceuticals in Malmø is a leader in the large scale manufacturing of vasopressin analogs for medical purposes. These were designed under the leadership of L. Carlsson, in collaboration with the Prague group (cf. the section on Czechoslovakia). At KABI, in Goteborg, based on the work of B. Blombäck and E. Blombäck, several enzyme substrates were developed. In the laboratories of ASTRA, B. Sjøberg and his coworkers designed novel approaches for the semisynthesis of new penicillins.

## 10.23 Switzerland

This small country occupies a leading position in the world of peptides. As early as 1949, publications on activation of N-blocked amino acids in the form of their reactive esters appeared from the laboratory of Robert Schwyzer (Plate 39) at CIBA in Basel. After his appointment to professor at the Eidgenössische Technische Hochschule (ETH) in Zürich, he established there, at the Institute of Molecular Biology and Biophysics, a distinguished school of peptide research. One of his students, P.W. Schiller, is professor at the University of Montreal. Another outstanding former student of Schwyzer, Alex Eberle, is a lecturer in Zurich, yet active in Basel, where he is engaged in affinity-labelling of hormone receptors.

From Schwyzer's major accomplishments, the introduction of active esters, the first synthesis of a cyclic peptide (the antibiotic gramicidin S, cf. p. 205) and the synthesis of the 39-peptide corticotropin (ACTH), in 1963, must be mentioned. His studies on the information content of biologically active peptides and on the mechanism of agonist-receptor interactions, including the membrane-compartment concept, remain in the forefront of peptide research.

K. Wüthrich, a leader in the field of determination of peptide-architecture by two-dimensional NMR spectroscopy, is also at the ETH in Zurich. At the University of Basel, professor Max Brenner (Plate 12) focussed his interest on several fundamental aspects of peptide synthesis, such as the problem of overactivation or peptide bond formation via rearrangements. One of his coworkers, Iphigenia Photaki (Plate 35) has already been mentioned in the section on Greece. Brenner's successor at the University of Basel, Manfred Mutter, excels with his work on liquid phase synthesis (cf. p. 111). Recently he accepted an invitation to the Free University of West Berlin and from there to Lausanne.

In Geneva, at the Centre Medicale Universitaire, R.E. Offord continues his pioneering work, started in Oxford, on protein semisynthesis, but extended his studies to the spontaneous coupling of segments as well.

The famous Swiss pharmaceutical industry is, of course, very much interested in biologically active peptides. When Schwyzer left CIBA, his former associates remained active in peptide research and several of them, W. Rittel, B. Iselin, P. Sieber, B. Riniker, H. Zuber (Ph.D. with H. Zahn, Aachen, now professor at ETH), H. Kappeler, B. Kamber, became well known for the excellence of their contributions, such as the development of highly acid-sensitive blocking groups or the first directed synthesis of insulin. With the merger of CIBA and GEIGY, A. Jöhl, a former du Vigneaud associate, joined the CIBA team.

At Hofmann–La Roche the work of K. Vogler and R.O. Studer on cyclic peptide antibiotics, for instance the polymyxins, is particularly noteworthy. At SANDOZ, R.A. Boissonnas was the primary mover of peptide studies. His work on mixed anhydrides (cf. p. 81) was the starting point of a continued research effort in which S. Guttmann, J. Pless. E. Sandrin, R. Huguenin, P.A. Jaquenoud, H. Bossert, W. Bauer and several younger associates participated. After early syntheses of the hormones of the neurohypophysis they became leaders in the synthesis of biologically active peptides on an industrial scale and first made oxytocin, then more complex molecules such as salmon calcitonin or more recently a cyclic peptide with somatostatin activity, commercially available.

This section on peptide research in Switzerland would be incomplete if we would leave without mention the laboratories (FLUKA, BACHEM, NOVABIOCHEM) that provide reliable intermediates for peptide synthesis in Switzerland and abroad.

## 10.24 The Netherlands

In 1955, J.F. Arens, at the University of Groningen, proposed a sophisticated and most interesting reagent for peptide bond formation, indeed the first coupling reagent, ethoxyacetylene (p. 88). With his associates, mainly with H.J. Panneman, he demonstrated the practical application of the new method in which ethyl acetate is set free as the only by-product.

E. Havinga, with K.E.T. Kerling, G. Heymens-Visser and C. Schattenkerk, at the University of Leiden, published an early synthesis of angiotensin and of many angiotensin analogs. He later devoted much of his peptide-work, in cooperation with Kerling, Schattenkerk, P. Hoogerhout and W. Bloemhoff, to the elucidation of the role of two histidyl residues which are essential for the exertion of enzymatic activity of ribonuclease-A.

An important research group emerged in Delft, at the Technical University, under the leadership of H.C. Beyerman. With his numerous coworkers, of whom W. Maassen van den Brink is mentioned here, Beyerman reported new syntheses of peptide hormones, such as oxytocin, secretin and VIP. The bifunctional catalysis proposed in his laboratory represent a noteworthy stage in the development of catalysis of active ester reactions.

In 1960 a novel and efficient method was invented by G.H.L. Nefkens, at the Catholic University in Nijmegen, for the phthaloylation of the amino group, then with G.I. Tesser and R.J.F. Nivard he applied the new reagent, *N*-carbethoxyphthalimide, in the synthesis of peptides. It also became the starting point for further improvements in peptide chemistry, such as protection of carboxyl groups in form of their *N*-hydroxymethylphthalimide esters and last but not least for the design of an entirely new class of active esters, *O*-acylaminoacyl *N*-hydroxyphthalimides. In 1975 a base sensitive amine-protecting group, the methylsulfonylethyloxycarbonyl (MSC) group, was proposed by Tesser, that enabled P.J. Boon, G.I. Tesser and R.J.F. Nivard to combine semisynthetically obtained fragments of cytochrome C. The alkali-lability of the 4-(methylsulfonyl)-phenylsulfonylethyloxycarbonyl (Mpc) group (C.G.J. Verhart and Tesser) allows its selective removal in the presence of Msc.

The pharmaceutical company Organon, in Oss, was one of the earliest sources of peptides manufactured on a large scale. Their research activities, for instance the publication of J.W. van Nispen or of H. Ottenheijm are certainly noteworthy.

## 10.25 United Kingdom

During the decade that followed the end of the Second World War Great Britain stepped into the forefront of peptide research. The insulin studies of Sanger (Plate 38) and his associates in Cambridge (cf. p. 158) had a major impact on the entire field of peptide and protein chemistry. In the same period elucidation of the structure of cyclic microbial peptides by A.J.P. Martin (Plate 31) and R.L.M. Synge (Plate 44) and the structural work of E. Abraham and C.G.F. Newton in Oxford on penicillins, and later on cephalosporins, provided a similar stimulus for the development of peptide research.

In the following years important research centers formed at several universities. In Exeter. N.C. Rydon (Plate 36) and his students, e.g. D. Jarvis, carried out fundamental studies on the rules governing cyclization via disulfide formation and on blocking groups that can be removed via base catalyzed $\beta$-elimination. The Exeter school was the starting point of the career of several distinguished peptide chemists, for instance of P.M. Hardy and B. Ridge. Some of Rydon's coworkers returned to their home countries to continue their work on peptides, thus L. Benoiton to Canada and M.J.S.A. Amaral to Portugal.

In Oxford, G.T. Young (Plate 53) and his coworkers, N.W. Williams, D.S. Jones, R.J. Knowles, G.A. Fletcher, initiated the first systematic studies on the problem of racemization in peptide synthesis. A novel and valuable procedure for the facilitation of chain building, the "handle method" also originates from Young's laboratory. Oxford University is the home of several more groups engaged in peptide research. Dorothy Hodgkin's (Plate 21) studies include the determination of the three dimensional structure of peptides (insulin, gramicidin S, thiostrepton, etc.) by X-ray crystallography. D.B. Hope, former coworker of du Vigneaud, made significant contributions to the chemistry of the neurohypophyseal hormones and the pioneering work of E.R. Offord on semisynthesis is especially noteworthy.

London itself became a center of peptide studies. At the National Institute for Medical Research (Mill Hill) the investigations of D.F. Elliot on hypertensin and on bradykinin are mentioned here. At the Postgraduate School of Medicine of the University of London M. Szelke made major contributions in the area of peptide hormones and their antagonists, particularly in the development of novel inhibitors of the angiotensin-renin system. The two dimensional NMR work of W. Gibbons and the X-ray structure studies of T.L. Blundell of biologically active peptides are similarly significant.

In Liverpool the fundamental contribution of G.W. Kenner (Plate 27) (1922–1978) are indeed memorable. He established an original and effective method of coupling via sulfuric acid mixed anhydrides and, with J.A. Farrington and J.M. Turner, designed reactive intermediates, such as the *p*-nitrothiophenyl esters of acylamino acids. With his research associates, among whom R.C. Sheppard played a key role, Kenner determined the structure of gastrin (cf. p. 165) and carried out its first synthesis. Sheppard himself became the head of an important research group at the laboratory of the Medical Research Council in Cambridge, where he is engaged in the further development of the methodology of solid phase peptide synthesis. In this endeavor the contributions of E. Atherton must be mentioned. From Kenner's former associates I.J. Galpin remained at the University of Liverpool, R. Ramage moved to Edinburgh, where he works on the development of novel base sensitive protecting groups. C.N.C. Drey is active in Aberdeen, Scotland.

Industrial laboratories are not less interested in peptide research. Thus the contributions of J. Morley, at Imperial Chemical Industries (ICI) in the gastrin field were unusually extensive. C. Hassal conducts interesting studies at Roche, G.W. Hardy at Welcome Research Laboratories, R. Wade at CIBA-GIEGY.

## 10.26 USA

Early studies in peptide chemistry by Max Bergmann (Plate 3) and his associates at the Rockefeller Institute in New York City were discussed in Chap. 3 where the important contributions made by W.H. Stein (Plate 4),

S. Moore (Plate 4) and L.C. Craig (Plate 16) after the Second World War have also been mentioned. In the early 1960s new and extensive research was started in the Institute (by then The Rockefeller University) by R.B. Merrifield. His discovery of solid phase peptide synthesis (Chap. 5) stimulated the minds of many young investigators. One of his first associates, J.M. Stewart, like Merrifield himself, a former coworker of D.W. Wooley at Rockefeller, became instrumental in the automation of the new process. The presence of Merrifield (Plate 5) at the Rockefeller Institute acts as a magnet for scientists from all over the world to go, to New York to participate in the further development of the method; to mention just one: B. Gutte, from Germany. Of course Merrifield attracted numerous American scientists as well. Several of them, for instance B.W. Erickson, A. Marglin, G.R. Marshall, J.P. Tam, S.B.H. Kent, A.R. Mitchell, distinguished themselves in the rapidly progressing field of solid phase peptide synthesis. A few years ago, E.T. Kaiser, professor of biochemistry at the University of Chicago, joined the Rockefeller Institute and established there a laboratory for the study of the mechanism and architecture of enzymes and their inhibitors. His vigorous activities were cut short by his untimely death.

In the immediate vicinity of the Rockefeller Institute is the New York Hospital which houses the Medical College of Cornell University. Its department of biochemistry, headed by V. du Vigneaud (Plate 6), after the War, became one of focal points of peptide research in the United States. The list of investigators who worked with du Vigneaud, members of the "VduV Club", is indeed impressive. It includes names of Nobel laureates (R.W. Holley, F. Lipmann) and of a long series of associates who later became known for their contributions. To mention a few: R. Acher (who returned to France as professor at the University of Paris), O.K. Behrens, R.A. Boissonnas (cf. Chap. 4), F.H. Carpenter (professor at the University of California in Berkeley), W.D. Cash, S. Drabarek, D.F. Dyckes (professor, University of Houston, Texas), M. Ferger, P.S. Fitt (professor, University of Ottawa), G. Flouret, D. Gillessen, J. Glass, F.R.N. Gurd (professor at the University of Indiana in Bloomington), K. Hofmann (professor, University of Pittsburgh), H. Hagenmaier, G.P. Hess (cf. Chap. 4), D.B. Hope, V.J. Hruby (professor, University of Arizona in Tucson), D. Jarvis, A. Johl, P.G. Katsoyannis (Plate 26) (professor, Mount Sinai University, New York City), A. Light, E.O. Lundell, M. Manning (professor, Ohio State University, Toledo), J. Meienhofer, A. Meister, who became du Vigneaud's successor at Cornell, V.V.S. Murti (professor, University of Delhi, India), J.J. Nestor, I. Photaki (cf. section on Greece), J.G. Pierce, E.A. Popenoe, J.R. Rachele, C. Ressler, R.W. Roeske (professor, University of Indiana, Indianapolis), S. Simmonds, J. Stedman, F.H.C. Stewart, J.E. Stouffer (professor, Baylor University, Houston, Texas), R.O. Studer, J.M. Swan, R.J. Vavrek, R.W. Walter (professor, University of Illinois, Chicago), D. Yamashiro.

Research groups took up peptide chemistry at other American Universities as well, for instance in the laboratories of E.R. Blout and of R.B. Woodward (Plate 50), both at Harvard, of J.C. Sheehan (Plate 41) and D.S. Kemp at MIT. The studies on peptide conformation carried out at Cornell in Ithaca,

New York, by H. Sheraga, by K.D. Kopple at the Illinois Institute of Technology and by L. Gierasch at the University of Texas in Dallas were indeed fruitful. The sustained investigations of biologically active peptides in the Hormone Research Laboratory of the University of California in San Francisco, headed by C.H. Li (1912–1988) (Plate 30), are quite impressive. Several members of this group, J. Ramachandran, D. Yamashiro, J. Blake, R.L. Noble, D. Chung, R.A. Houghten deserve considerable credit for important contributions. Intensive hormone studies were pursued in Pittsburgh by Klaus Hofmann, later in collaboration with his wife, F. Finn, at Mount Sinai Medical School in New York City; by P.G. Katsoyannis (Plate 26) and his associates; and also by J. Glass and Angeliki Buku. Fundamental work in peptide chemistry was carried out by J.S. Fruton (Plate 18) at Yale University and by M. Goodman and his many associates at the University of California in San Diego. L.A. Carpino at the University of Massachusetts at Amherst is a most productive innovator in the chemistry of blocking groups. On sulfur containing peptides, major results were achieved at Chapel Hill, University of North Carolina by R.G. Hiskey and his students, for instance J.T. Sparrow. In Tucson, Arizona, V. Hruby is continuing studies on the conformation of biologically active peptides. Of his students, A.F. Spatola should be mentioned. Dehydropeptides receive special attention by J. Stammer at the University of Georgia in Athens.

The solid phase method found dedicated followers who advanced its methodology: J.M. Stewart in Denver, G.R. Marshall in St. Louis, G. Barany in Minneapolis, and D.H. Coy in New Orleans come first to mind.

In addition to universities several research institutes participate in peptide research in the U.S., e.g. the National Institutes of Health in Bethesda, Maryland with C.B. Anfinsen (Nobel laureate), B. Witkop (Plate 49) and E. Gross (1928–1981) among its researchers. At the Laboratory of Naval Research, I. Karle made basic contributions to the study of peptide conformation by X-ray crystallography. In the Salk Institute in La Jolla, California, extensive work on peptide hormones and releasing factors is being pursued by R. Guillemin and J. Rivier. A search for hypertension controlling peptides at the Cleveland Clinic Research Foundation, initiated by I.H. Page and F.M. Bumpus is being kept active by M. Khosla and R.R. Smeby. Also in Cleveland, at the Veterans Administration Hospital, the antihypertension problem is being attacked via inhibitors of renin, with L. Skeggs and K. Lentz in leading roles.

Some of the most successful ventures in peptide research took place in laboratories of the pharmaceutical industry. The Lederle Laboratories, with G.W. Anderson, J.R. Vaughan, Jr. and F.M. Callahan as principal investigators, made early contributions to the methodology of synthesis and to the synthesis of hormones such as oxytocin and calcitonin. The Squibb Institute for Medical Research was the first to report the synthesis of secretin, an effort led by Miklos Bodanszky. More recently the Institute scored a major success with the development of the first medically useful inhibitor of the angiotensin converting enzyme, an intellectually most appealing discovery of M.A. Ondetti (Plate 32). Another Squibb researcher, J.T. Sheehan (1913–1985), should be remembered for his contributions to synthetic methods. A proficient peptide group was assembled at the Merck Sharp and Dohme Research Laboratories by R. Hirschmann. Several members of this group, e.g. D.F. Veber and R.F. Nutt, are noted for the excellence of their contributions. At Searl the work of R.H. Mazur and his coworkers led to the discovery of the sweetener ASPARTAME, a dipeptide ester. In Philadelphia, at the Smith Kline French laboratories, G.W. Huffmann undertook research in the field of peptides, while Abbott Laboratories in North Chicago, under the direction of K.W. Funk and J.C. Tolle, built a unique facility for the production of biologically active peptides on an industrial scale.

An important role is played by some research supply houses that provide synthetic peptides for biological biochemical and pharmacological studies. The lists of peptides offered by Penninsula and by Bachem, both in California, are indeed impressive.

## 10.27 USSR

In the early nineteen sixties substantial contributions to peptide chemistry were made in the Institute of Natural Products of the USSR National Academy of Sciences in Moscow. V.K. Antonov, V.T. Ivanov, A.K. Kiryushkin and Yu.A. Ovchinnikov (Plate 33), coworkers of M.M. Shemyakin (Plate 42), reported novel synthetic routes to cyclic depsipeptides and simultaneously clarified the structures of the enniatins (cf. p. 20). Their studies also led to the establishment of the structure and geometry of valinomycin and to the elucidation of the role of architecture in ionophores. Following the first publication by Merrifield on solid phase peptide synthesis, the Moscow group introduced a new approach to facile chain building by attaching the C-terminal residue of the peptide to be constructed to a soluble macromolecule (liquid phase). They also developed improved methods for the determination of the amino acid sequence of larger peptides via mass spectrometry. In the following years the investigators of the Institute (now the Shemyakin Institute of Bioorganic Chemistry, USSR Academy of Sciences) extended the area of their studies to include the synthesis of biologically active peptides. For instance Ivanov and his associates reported the synthesis of the $\delta$-sleep inducing peptide and its analogs. V.F. Bystrov established high resolution NMR spectroscopy for studies on peptide and protein conformation. He died in Sŭmmer 1990.

In Leningrad, G.P. Vlasov and his coworkers published the preparation of various opioid peptides. In 1967, Vlasov together with Yu. Mitin (now at Pushchino) developed a "redox-method" for preparing amino acid thioaryl esters from anions and sulfenyl chlorides in the presence of triethyl phosphite, which is thereby oxydized to triethyl phosphate. This procedure, further worked out by Mitin for peptide synthesis, possibly involves acyloxyphosphonium compounds as reactive intermediates and thus is reminiscent of the Mukayama method mentioned on pages 98 and 239.

After the decease of Professor Shemyakin (1908–1970) (Plate 42) Ovchinnikov (Plate 33) succeeded him as director of the Institute. After the untimely death of Ovchinnikov (1934–1988), The Institute is now headed by V.T. Ivanov.

G.I. Chipens and his associates at the Institute of Organic Chemistry of the Academy of Sciences of the Latvian SSR in Riga, investigate the functional organization of biologically active peptides. In more recent years their work led to the discovery of the immunopeptide *rigin*.

# Appendix

In this Appendix biographical sketches are compiled of many scientists who have made notable contributions to the development of peptide chemistry up to its present state. We have tried to consider names mainly connected with important events during the earlier periods of peptide history, but could not include all authors mentioned in the text of this book. This is particularly true for the more recent decades when the number of peptide chemists and biologists increased to such an extent that their enumeration would have gone beyond the scope of this Appendix.

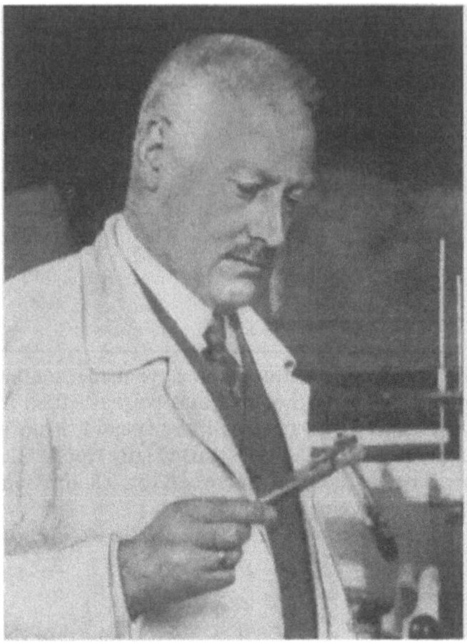

**Plate 8.** Emil Abderhalden (1877–1950), Photo Leopoldina, Halle

**Plate 9.** S. Akabori

**Plate 10.** Ernst Bayer

**Plate 11.** Karel Blaha (1926–1988)

**Plate 12.** Max Brenner

**Plate 13.** Hans Brockmann (1903–1988)

**Plate 14.** Victor Bruckner (1900–1980)

**Plate 15.** Pehr V. Edman (1916–1977)

**Plate 16.** Lyman C. Craig (1906–1974)

**Plate 17.** Vittorio Erspamer

**Plate 18.** Joseph S. Fruton, Biochemist and Historian

**Plate 19.** Rolf Geiger (1923–1988)

**Plate 20.** Wolfgang König

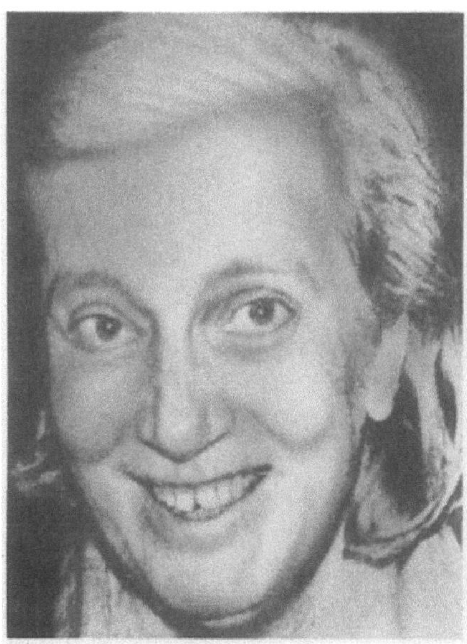

**Plate 21.** Dorothy Hodgkins

**Plate. 22.** Franz Hofmeister (1850–1922), (Fischer, biograph. Lexikon)

**Plate 23.** The picture shows the late Professor J.E. Jorpes (r.) and Professor V. Mutt during their favorite pastime in the archipelago on the Baltic near Stockholm

**Plate 24.** Ephraim Katchalski (Katzir)

**Plate 25.** Abraham Patchornik

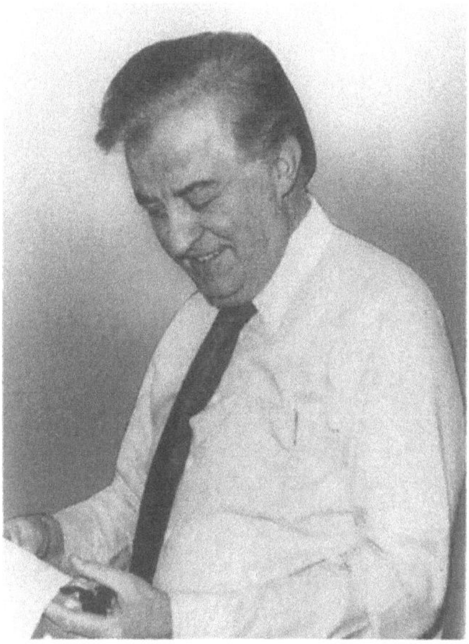

**Plate 26.** P.G. Katsoyannis

**Plate 27.** George W. Kenner (1922–1978)

**Plate 28.** Edger Lederer (1908–1988)

**Plate 29.** Hermann Leuchs (1879–1945)

**Plate 30.** Choh Hao Li (1913–1987)

**Plate 31.** A.J.P. Martin

**Plate 32.** Miguel A. Ondetti

**Plate 33.** Yuri A. Ovchinnikov (1934–1988)

**Plate 34.** Linus Pauling, 40 years α-Helix
"We have now used this information (about interatomic distances, bond angles and other... [the author]) to construct two reasonable hydrogen-bonded helical configurations for the polypeptide chain; we think that it is likely that these configurations constitute an important part of the structure of both fibrous and globular proteins, as well as of synthetic polypeptides".
(From Linus Pauling, Robert B. Corey and H.R. Branson, The structure of proteins: two hydrogen-bonded helical configurations of the polypeptide chain, Proc. Natl. Acad. Sci. USA 37: 205–211, 1951)

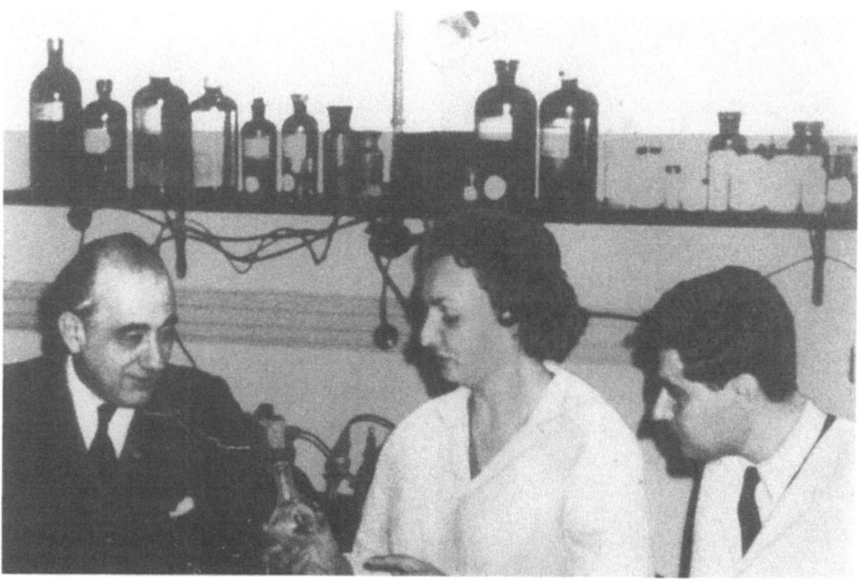

**Plate 35.** Iphigenia Photaki (1921–1983) with Leonidas Zervas

**Plate 36.** H. Norman Rydon

**Plate 37.** Shŭmpei Sakakibara

**Plate 38.** Frederick Sanger

**Plate 39.** Robert Schwyzer

**Plate 40.** Ernesto Scoffone

**Plate 41.** John C. Sheehan

**Plate 42.** M.M. Shemyakin

**Plate 43.** Tetsuo Shiba

**Plate 44.** R.L.M. Synge

**Plate 45.** Emil Taschner (1900–1982)

**Plate 46.** Wang Yu

**Plate 47.** Friedrich Wessely von Karnegg (1897–1967) (Courtesy Prof. K. Schlögl)

**Plate 48.** Friedrich Weygand (1911–1969)

**Plate 49.** Bernhard Witkop

**Plate 50.** Robert B. Wordward

**Plate 51.** Erich Wünsch

**Plate 52.** Haruaki Yajima

**Plae 53.** Geoffrey T. Young

**Plate 54.** The insulin group of Helmut Zahn (Photo M. Forschelen, Aachen) from left: H. Bremer, O. Brinkhoff, H. Zahn, R. Zabel, E. Schnabel, J. Meienhofer

**Plate 55.** Leonidas Zervas (1902–1980) at the age of about 60

*Abderhalden, Emil*, 1877–1950 (p. 34, Plate 8) born in Oberuzwil, Switzerland, studied medicine in Basel where he graduated as M.D. in 1902. In the same year he moved to Berlin to join Emil Fischer, where he combined chemistry with enzymology (p. 35). In 1908, he was appointed Professor of Physiological Chemistry at the School of Veterinary Medicine in Berlin. In 1911 he became Professor and Head of the Institute of Physiology of the University of Halle/Saale where he worked on the physiology and biochemistry of peptides, proteins and enzymes. Abderhalden was President of the "Deutsche Akademie der Naturforscher Leopoldina" from 1931 until 1945, the year in which he and his family were moved to West Germany by American troops. He died in 1950 in Zürich, Switzerland.

*Akabori, Shiro* (p. 116, Plate 9) born in Shizuoka Prefecture, Japan, in 1900, graduated from Tohoku Imperial University in 1925, research associate in the laboratory of Prof. Toshiyuki Majima and received his Doctor of Science degree from the same university in 1931. He studied in the laboratory of Prof. E. Waldschmidt-Leitz in the German University at Prague, Czechoslovakia from 1932 to 1934, appointed as Assistant Professor at Osaka Imperial University in 1935 where he advanced to full Professor (1939), Dean of the Faculty of Science (1947), and the President (1969). He established the Insitute for Protein Research at Osaka University and served as the first Director of the Insitute (1957). Since retiring from Osaka University (1965), he has been the President and a trustee of the Protein Research Foundation to which the Peptide Institute belongs. He is a member of the Japanese Academy, and a member of numerous foreign Academies.

*Bayer, Ernst* (pp. 54, 234, Plate 10) born in 1927 in Ludwigshafen/Rhine, studied chemistry in Heidelberg, Freiburg/Br. Dr. rer. nat. 1954, from 1958 at the Technical University, Karlsruhe and University of Tübingen, there associate professor in 1962, full professor in 1965.

*Bergmann, Max*, 1886–1944 (p. 45, Plate 3) born in 1886 in Fürth, Bavaria, begun studies of botany at the Technical Highschool in Munich, transferred to chemistry in E. Fischers laboratory in Berlin, where he worked on his doctoral thesis, 1911. Fischer engaged him as assistant in his personal laboratory to work in the field of, among other topics, amino acids and peptides. In 1920 he was appointed Acting Director at the Kaiser-Wilhelm-Institute for Fiber Research in Berlin-Dahlem. In 1922 he moved to Dresden as Director of the Kaiser-Wilhelm-Institute for Leather Research, where he, with Leonidas Zervas, invented the benzyloxycarbonyl group for the reversible protection of amino groups (p. 46). In 1934, Bergmann had to leave Nazi Germany. He was kindly received by the Rockefeller Institute for Medical Research in New York City, where he successfully continued his investigations of amino acids and peptides with excellent collaborators (S. Moore, E. Stein et al.) (see p. 50) Bergmann died in 1944, 58 years old, after a long illness in New York.

*Biemann, Klaus* (p. 128) born in Innsbruck, Austria in 1926. He studied chemistry at his home Univ., and received his PhD in 1951. He was instructor of chemistry there until 1955 when he became Professor of Chemistry at the Massachusetts Institute of Technology, Cambridge USA and has applied mass spectrometry in his investigations since then.

*Bláha, Karel*, 1926–1987 (pp. 122, 231, Plate 11) was born in 1926 near Pilsen, Bohemia. In 1949, after graduating in chemistry at the Technical University in Prague, he joined the Laboratory of Heterocyclic compounds led by Rudolf Lukes where he worked and gained experience of organic chemistry in depth. In 1960, Blaha went to the Institute of Organic Chemistry and Biochemistry of the Czechoslovak Academy of Sciences, and started his successful research in peptide chemistry, at first as a coworker of Joseph Rudinger, and after Rudinger's emigration (1968) his successor as Head of the Department.

*Bodanszky, Miklos* born in 1915 in Budapest, Hungary, received his doctorate at the Technical University of Budapest. After a few years in the pharmaceutical industry he was appointed as Head of the natural products department of the Institute for Medical Research and at the time became a lecturer in Medicinal Chemistry at the Technical University of Budapest. His work on active esters started in this period (p. 85). In 1956, he left Hungary and joined V. du Vigneaud, at Cornell University Medical School in New York City. Here he demonstrated the stepwise strategy in a novel synthesis of oxytocin. In 1959 he joined the Squibb Institute for Medical Research, where with his coworkers (among them Miguel A. Ondetti) he reported the first synthesis of the gastrointestinal hormone secretin (p. 167). From 1966 until his retirement in 1983 he was professor of Chemistry and Biochemistry at Case Western Reserve University in Cleveland, Ohio.

*Brenner, Max* (pp. 58, 197, 242, Plate 12) born in 1915 in Chur, Switzerland. Studied chemistry at the Technical University, Zürich, Diploma Chemical Ing. in 1937, Dr. Sc. techn. after thesis with L. Ruzicka. Thereafter one year working at Rockefeller Institute in New York with Max Bergmann on synthetic substrates for proteases, 1941–1947 research in the pharmaceutical-chemical industry, 1947 assistant to T. Reichstein, 1949 Lecturer, 1954 Professor until 1980 at the University of Basel.

*Brockmann, Hans*, 1902–1988 (p. 118, Plate 13) born in 1903 in Altkloster near Hamburg, received his Dr. degree at the Univ. of Halle/Saale (with E. Abderhalden, in 1930, went to Richard Kuhn at the Kaiser-Wilhelm (later Max-Planck)-Institute for Medical Research in Heidelberg where he did pioneering work on chromatography, e.g. of carotenoids. In 1935 he became Head of the biochemical department of the Chemistry Institute of the Univ. of Göttingen (Director Adolf Windaus), in 1941 Prof. at the Univ. Posen, and after the war, in 1945, successor of A. Windaus in Göttingen until retirement in 1972. Antibiotica (actinomycin, p. 224 etc) ionophores (valinomycin, p. 201).

*Bruckner, Viktor*, 1900–1980 (pp. 41, 236, Plate 14). Born in 1900 in Késmárk a small Carpathian town, graduated from the Technical University in Budapest, thereafter studied with Alexander Schönberg (Berlin) and Fritz Pregl (Graz), then with Albert von Szent-Györgyi in Szeged (Hungary). Engaged in peptide chemistry from 1937 (bacterial capsular substances), in 1950 appointed as Director of the Institute of Organic Chemistry of the Eötvös Lorand University in Budapest. Died in Budapest in 1980.

*Craig, Lyman C.*, 1906–1974 (p. 55, Plate 16) son of a farmer, was born in Palmyra, Iowa. In 1932, with a fresh Ph.D. degree from the University of Iowa at Ames, he moved to Baltimore to work at the Johns Hopkins University as a National Research Council fellow. Two years later Craig joined the Rockefeller Institute in New York City to collaborate with A. Jacobs on the structure-elucidation of ergot alkaloids. Here he developed the fractionation of mixtures by repeated extraction procedures. Countercurrent distribution (CCD) turned out to be of great value in the isolation of natural and synthetic products. The definitive determination of the molecular weight of insulin, for instance, became possible through the isolation of the least substituted derivative by CCD. For the concentration of solutions recovered from CCD, he invented the rotary evaporator.

Lyman C. Craig remained at the Rockefeller Institute during his entire career and was appointed Professor when the Institute became a University. His work was recognized by numerous awards. He died in 1974.

*Curtius, Theodor*, 1857–1928 (p. 25, Plate 2) born in Duisburg (Rhineland). He first studied music and natural sciences in Leipzig until 1876, and went after his military service in 1878 to Heidelberg to study chemistry with Robert Bunsen. He returned to Leipzig where he received his Ph.D. degree after a thesis guided by Hermann Kolbe in 1882 (first synthesis of a peptide, p. 26), moved to the University of Munich (A. Baeyer), where he was assistant from 1884–1886. In 1886 he graduated Dr. habil. at the University of Erlangen, in 1889 he moved as full professor to the University of Kiel. A call in 1892 to the University of Würzburg as successor to Emil Fischer was not accepted, but he succeeded A.V. Kékule in Bonn in 1897, although for only one year until 1898 when he succeeded Victor Meyer at the University of Heidelberg as Head of the Chemical Institute. There he worked until the end of his life in 1928.

*du Vigneaud, Vincent.* See biography on pp. 153–155.

*Edman, Pehr Victor*, 1916–1977 (pp. 118, 240, Plate 15) born in Stockholm in 1916, matriculation examination in 1935, studied medicine at the Karolinska Institute Medical School in Stockholm from 1935, Bachelor of Medicine in 1938, graduation as a physician in 1946. Concurrently with his studies in medicine he started his training in biochemistry with Erik Jorpes, for a short time also with Hugo Theorell, and soon started a project on angiotensin that led to a MD thesis. Then he widened his experience in protein chemistry during one year at the Rockefeller Institute in Princeton with Northrop and Kunitz (crystallization of proteolytic enzymes). On his return to Sweden, Edman was awarded an associate professorship in Lund in 1947 where he conducted his stepwise peptide degradation work (p. 118) between 1950 and 1956. In 1957 Pehr Edman accepted an offer to be Director of Research at St. Vincent's School of Medical Research in Melbourne, Australia, where he remained for 15 years, during which the work on an automated sequence analyzer was finished in 1967. From 1972 until his death from a brain tumor in 1977 he was Director of the Department of Protein Chemistry of the Max-Planck-Institute for Biochemistry in Martinsried near Munich.

*Erspamer, Vittorio* (pp. 134, 238, Plate 17), born in Malosco (Trento, Italy) in 1909, studied Medicine at the University of Padua (M.D. in 1939), was there assistant at the Institute of Comparative Anatomy and Physiology from 1935–1938, then postdoctoral fellow at the Universities of Berlin and Bonn, at the Institute of Pharmacology University of Rome, 1940–1947, Professor of Pharmacology University of Bari Medical School until 1955, University of Parma Medical School until 1967 and from then to 1984 Director of the Institute of Pharmacology, University of Rome. Ersparmer's work comprised comparative pharmacology and biochemistry of biogenic amines (17 new compounds) and, particularly, bioactive peptides (40 compounds) from invertebrates and lower vertebrates (amphibian skin) (see p. 184). The opioid peptides (demorphis and deltorphins) are the most potent among all natural peptides.

*Fischer, Emil*, 1852–1919 (p. 24, Plate 1), son of a merchant was born in 1852 in Euskirchen (Rhineland) studied chemistry in Bonn (Kekulé), Straßburg (A. Baeyer, physics; A. Kundt) where he graduated Dr. phil. in 1874. Assistant with Baeyer, from 1874 in Munich. Professor in Erlangen (1882–1885), then in Würzburg until his call to the famous Chemical Institute of the University of Berlin as successor to A.W.v. Hofmann, in 1892. Amino acid and peptide research from 1900, after extremely successful research in the field of carbohydrates and purines. Fisher died in the summer of 1919 in Berlin. He was the second winner of the Nobel Prize for Chemistry in 1902, after van t'Hoff in 1901.

*Fruton, Joseph S.* (pp. 56, 246, Plate 18), born in 1911 in Czestochowa, Poland Ph.D. at Columbia University, New York, in 1934, at the Rockefeller Institute with Max Bergmann 1934–1935 (p. 47). In 1945 he became Associate Professor of Physiological Chemistry at Yale University, New Haven, Conn., Chair of the Department of Biochemistry 1951–1967, Eugene Higgins Professor of Biochemistry 1957–1982, Professor of the History of Medicine 1980–1982. Published, among others, General Biochemistry (with S. Simmons, 1953), Molecules and Life (1972), and selected articles on the History of Biochemistry and Chemistry since 1800.

*Geiger, Rolf*, 1923–1988 (pp. 93, 234, Plate 19), born in Bodman on Lake Constance, studied chemistry at the University of Tübingen (1949) and completed his training in 1956, at the Technical University in Berlin, with a doctoral thesis under Friedrich Weygand. In the following year he initiated peptide chemistry in the Pharma Synthese Laboratories at Hoechst A.G. In 1969, Geiger became the leader of the peptide group. First a part time lecturer at the University of Frankfurt a.M., in 1979 he was named Honorary Professor. With his group he worked successfully on the synthesis of hormones, such as adrenocorticotropin or insulin, and of releasing factors.
Among his many excellent coworkers Wolfgang *König* (Plate 20) also a student of F. Weygand, is mentioned here because of his large share in the development of racemization-free coupling methods.

*Goodman, Murray* (p. 246) born in New York N.Y. in 1928, studied chemistry at Brooklyn College and the Univ. of California and received his Ph.D. in 1952. He was a Research associate at MIT 1952–55, was Prof. of Chemistry at Polytech Institute Brooklyn 1956–71 and is now Prof. of Chemistry, Univ. of California, San Diego.

*Gross, Erhard*, 1928–1981 (p. 62) was born in 1928 in Wenings near Frankfurt a. M., studied chemistry at the Universities of Mainz and Frankfurt. After his Dr. Degree with Th. Wieland he went to Bernhard Witkop at the National Institute of Arthritis and Metabolic Diseases, NIH in Bethesda, Maryland, USA. In 1968 Gross was appointed Chief of the Section on Molecular Structure, Laboratory of Biomedical Sciences, and in 1973 Section Reproduction Research Branch at the National Institute of Child Health and Human Development, NIH. He died tragically in a road accident in 1981. His name is connected with the cyanogen bromide cleavage of peptides (p. 117), the structure of nisin (p. 222), channel forming peptides, and outstanding literary contributions to the peptide and protein field.

*Guillemin, Roger* (p. 246) born in Dijon, France, in 1924 studying medicine at the Univ. of Dijon, received his M.D. in 1949 and his Ph.D. at the Univ. of Montreal in 1952. There he became assistant professor and assistant director of the Institute for Experim. Medicine and Surgery (1951–53), then at Baylor College of Medicine, Houston Tex. until 1963. He is now Chairman of the Laboratories of Neuroendocrinology, Salk Institute, La Jolla, Cal. He was awarded the Nobel Prize for Medicine and Physiology (shared with A. Schally) in 1977.

*Hirschmann, Ralph Franz* (pp. 41, 247) born in 1922 in Bavaria, Ph.D. in organic chemistry in 1950 at the University of Wisconsin. Started professional experience in Research Laboratories at Merck Sharp u. Dohme in 1950 and advanced to become director of peptide research, 1968, then of protein research, director medicinal chemistry of the company in West Point, Pa. 1974–76 Vice President of basic research in the Company at Rahway, N.J.

*Hodgkin, Crowfoot Dorothy* (p. 132, Plate 21) born in 1910 in Cairo, she studied chemistry at Oxford from 1928–32 and concentrated on X-ray crystallography with J.D. Bernal at Cambridge; in 1934 she returned to Oxford where she has remained, except for brief intervals, ever since. She became University lecturer and demonstrator in 1946, University Reader in X-ray crystallography in 1956 and Wolfson Research Professor of the Royal Society in 1960–1983. Apart from X-ray studies of insulin (p. 159) she also did research on penicillin (in 1942) and on vitamin $B_{12}$ (in 1948). She received the Nobel Prize in chemistry 1964. In 1937 she married Thomas Hodgkin, historian of Africa and the Arab world.

*Hofmann, Klaus Heinrich* (pp. 38, 41, 163) born in Karlsruhe, Germany, in 1911, studied chemistry at the Swiss Federal Institute of Technology (ETH, Zürich), was awarded his Ph.D. in 1936. He became fellow of the Rockefeller Foundation 1938–40; assistant at the Medical College, Cornell Univ. 1940–42; assistant and then associate prof. 1944–47. He has been research prof. of biochemistry and Prof. of the Biochemistry School of Medicine, Univ. of Pittsburgh, Director of Protein Research Laboratories since 1964.

*Hofmeister, Franz*, 1850–1922 (p. 2, Plate 22) born in Prague, son of a physician, studied medicine at the ancient University, worked on "peptones" (Habilitation). Thereafter, six months with the pharmacologist O. Schmiedeberg in Straßburg (then German Alsace), from 1885–1896 director of the Institute of Pharmacology in Prague. In 1896 professor of Physiological Chemistry, successor of Felix Hoppe-Seyler in Straßburg. After the return of Strasbourg to French rule (1919) he left and found a haven in Würzburg, where he was named Honorary Professor and worked on vitamins until shortly before his death in 1922.

*Hruby, Victor J.* (p. 246), born in Valley City, N. Dakota USA in 1938, Ph.D. Cornell Univ. 1965, Prof. of Chem. Univ. of Arizona since 1977. He became interested in peptide hormones as a postdoctoral fellow with Nobel prize-winner Vincent du Vigneaud at Cornell University in the late 1960s just before he joined the University of Arizona faculty. In the 20 years since, he has become a world leader in this important research field (synthesis, isolation, conformations, dynamics, mechanisms of action, and structure-activity relationships of peptide hormones and neurotransmitters and their analogs). He is the editor of the International Journal of Peptide and Protein Research as successor to Choh Hao Li who died in 1988.

*Jorpes, Erik J.*, 1894–1973 (pp. 166, 241, Plate 23) was born on the Finish island Kokar in the Aland archipelago, when Finnland was part of the Russian Empire. He worked together with V. Mutt with great success on the chemistry and physiology of the gastro-intestinal hormones at the Karolinska Institute in Stockholm, Sweden, where he died in 1973.

*Karle, Isabella L.* (p. 133) born in Detroit, Mich., in 1921, studied chemistry at the Univ. of Michigan, Ph.D. 1944. Associate chemist at Univ. Chicago, instructor there and physicist 1946–59, now Head of X-ray analyt. section of U.S. Naval Res. Lab. Washington D.C. Married to Jerome Karle (born in 1918 in New York City, Nobel Prize for Chemistry 1987)

*Katchalski, Ephraim* (*Katzir*) (pp. 40, 237, Plate 24) born 1916 in Kiev (Ukraine); Hebrew University Jerusalem, 1951–73; Professor and Head of the Dept. Biophys. Weizmann Institute of Sciences, Rehovot; President of Israel 1973–78; 1978 Professor at Tel-Aviv University. Syntheses and properties of polyamino acids are linked with his name.

*Katsoyannis, Panayotis* (p. 160, Plate 26) born in Greece in 1924, studied chemistry at the National Univ. Athens with L. Zervas; Ph.D. in 1952. Research associate at the Medical College of Cornell Univ. Ithaka N.Y. (1952–56), associate prof. School of Medicine, Univ. of Pittsburgh (1958–64), at Medical Res. Center Natl. Lab. Brookhaven until 1968 now Professor of Biochem. and Chairman Mount Sinai School of Medicine, New York City. The synthesis of insulin was his field.

*Kenner, George Wallace* (pp. 24, 81, Plate 27). Born 1922. Graduated in Chemistry, University of Manchester, 1939. His early research was with A.R. Todd at Manchester and then at Cambridge, on purine and pyrimidine chemistry and the synthesis of nucleosides. In 1957 appointed Head of the Department of Organic Chemistry, University of Liverpool. Elected Fellow of the Royal Society in 1964. Just before his death in 1978 he was appointed a Royal Society Research Professor, holding his appointment at Liverpool.

*Lederer, Edgar*, 1908–1988 (pp. 129, 232, Plate 28). Born in 1908 in Vienna, Austria, studied Chemistry at University of Vienna (Dr. phil., 1929); postdoctoral research at Kaiser-Wilhelm-Institut Heidelberg, 1930–1933, with Richard Kuhn, where he revived in 1931, Mikhail Tswett's chromatography with carotenoide mixtures. He was from 1933–35 Research fellow at the Inst. de Biologie Physico-Chimique, Paris, director of Dept. of Organic Synthesis, Vitamin Inst. Leningrad 1936–37. After Army service, he carried out research at Laboratoire de Biochimie, Lyon, from 1940 to 1947, then Maître des récherches with Inst. de Biologie Physico-chimique until 1960. Director Dept. de Chimie Biologique, Inst. de Chimie des Substances Naturélles CNRS 1960–1978. Died in 1988 in Paris. Lipopeptides, immuno adjuvants, mass spectrometry were among his studies.

*Leuchs, Hermann* (1879–1945, p. 35, Plate 29) born in 1879 in Nuremberg, studied chemistry in Munich and graduated in 1902, Ph.D. with a thesis guided by E. Fischer in Berlin. He then became associate professor there and, in 1926 Associate Director of the Institute of Chemistry of the University. After his discovery of the "Leuchs' substances" (p. 35) in 1906 and some of their reactions he turned to strychnine chemistry on which he worked until his death in the destroyed Berlin in 1945.

*Li, Choh Hao*, 1913–1987 (p. 246, Plate 30) was born in Canton, China, in 1913 and had his early schooling there, graduating from Pui Ying High School in Canton in 1929, and receiving his B.Sc. in Chemistry from the University of Nanking in 1933. In 1938 he was granted provisional admission to the University of California, Berkeley, and went on there to his doctorate in physical-organic chemistry. It was in the laboratory of Herbert Evans at Berkeley that Li began to develop the techniques that would eventually lead to the isolation and structure determination of several peptide and protein hormones found in the brain including growth hormone and beta-endorphin. From 1950 he headed the Hormone Research Laboratory as Professor of Biochemistry and of Experimental Endocrinology first at Berkely then at San Francisco. After his retirement in 1983 he continued his research and his numerous editorial activities until his death in 1987.

*Lipmann, Fritz*, 1899–1988 (p. 208) born in Königsberg, East Prussia in 1899, he became Dr. med. at the Univ. of Berlin in 1924. Assistant at Kaiser-Wilhelm-Institutes for Biology in Berlin and Medical Research in Heidelberg 1927–31, Carlsberg Institute Copenhagen 1932–39, research associate Medical College of Cornell Univ. 1939–41, Head of Biochem. Res. Lab. and Prof. at Harvard Univ. and Mass. General Hospital 1941–47, then Professor at Rockefeller Univ., New York. Nobel Prize for Medicine and Physiology 1953. Coenzyme A, peptide biosynthesis.

*Martin, Archer John Porter* (pp. 52, 54, Plate 31) born in 1910 in London, studied chemistry in Cambridge (Ph.D.) worked there in the physical chemical laboratory on nutritional research, then in the Wool Institute Leeds until 1946. Member of staff of the Medical Research Council, director Natl. Inst. Medical Res., Mill Hill (until 1956), later various professorships, e.g. École polytechnique, Lausanne 1980–1983. Nobel Prize in chemistry 1952, shared with R.L.M. Synge with whom he worked in Leeds on chromatographic methods.

*Merrifield, Bruce R.* (p. 103, Plate 5) was born in Fort Worth, Texas, in 1921. His parents moved to California when he was two years old and it was in California where he received all his education and also his training in chemistry. After two years in Pasadena Junior College he transferred to the University of California in Los Angeles (UCLA). Following graduation he worked for a year at the P.R. Park Research Foundation but returned to UCLA for graduate training in biochemistry. His studies, guided by Professor M.S. Dunn, were concluded with the Ph.D. degree, that was awarded on June 19, 1949. The next day he married Elizabeth Furlong (later his coworker) and the day after they moved to New York City to joint Dr. D.W. Woolley at the Rockefeller Institute. Since 1959 he has dedicated his research to the solid phase idea and developed his famous method of peptide synthesis.

Merrifield advanced in rank at the Rockefeller University to Professor and received numerous awards and honorary degrees elsewhere. In 1984 he was named John D. Rockefeller professor and in the same year received the Nobel Prize in Chemistry.

*Moore, Stanford*, 1913–1982 (p. 50, Plate 4). Born in Chicago, Illinois, in 1913. On graduation from Vanderbilt University in Nashville, Tennessee, with summa cum laude and as a Founder's Medalist, he entered the University of Wisconsin for graduate studies. There he studied organic chemistry under Homer Adkins and worked on his Ph.D. thesis in biochemistry under the guidance of Karl Paul Link, a friend of Max Bergmann whom he joined at the Rockefeller Institute in 1939. The work involved the determination of amino acids by precipitation with selective reagents used at two different concentrations ("solubility product method") in collaboration with W.H. Stein. During the war Dr. Moore served from 1942 to 1945 in the Office of Scientific Research in Washington, D.C. On returning to the Rockefeller Institute, after Bergmann's death in 1944, Moore continued the effort in collaboration with William Stein applying chromatography and the important "amino acid analyzer" (p. 51). The new method was applied by Moore, Stein and Hirs, in the same laboratory for the elucidation of the primary structure of ribonuclease A with 124 amino acid residues. Stanford Moore, in recognition for these invaluable achievements, shared the Nobel Prize in Chemistry in 1972 with W.H. Stein. He held a chair as a visiting professor at the University of Bruxelles, Belgium, and received an honorary doctorate from the same university and was a chair-professor at Rockefeller University. The names of Stanford Moore and William H. Stein are linked together forever. Yet, fate added still another link to their association, a tragic one. In 1982, threatened by paralysis, similar to that of Stein's, Professor Moore took his own life.

*Mutt, Viktor* (pp. 167, 241, Plate 23) was born in 1923 in Tartu, Estonia, moved to Sweden in 1944 after one year in Finnland, worked with E.J. Jorpes with great success on the chemistry and physiology of gastrointestinal peptides at the Karolinska Institute in Stockholm, Sweden.

*Ondetti, Miguel Angel* (Plate 32) born in 1930 in Buenos Aires, Argentina, completed his studies in chemistry at the University of Buenos Aires in 1957 with a Doctor of Natural Sciences degree received for his work on alkaloid chemistry under V. Deulefeu. Soon after, he joined the Squibb Institute for Medical Research and continued his research at the Institute, also after his transfer to the United States in 1960. There, for several years, he participated in an effort, led by M. Bodanszky, toward the synthesis of peptide hormones, including the first synthesis of secretin and became coauthor of one of the earliest monographs on peptide synthesis. Subsequently Ondetti initiated and led a major endeavor in the area of anti-hypertensive agents that resulted in the discovery of ACE inhibitors (p. 183), a new class of compounds that has gained considerable importance in the treatment of hypertension and other cardiovascular disorders.

*Ovchinnikov, Yuri A.*, 1934–1988 (p. 202, Plate 33) born in Moscow, studied chemistry at the Lomonosov Moscow State University and entered in 1960 the laboratory of M.M. Shemyakin in the Institute of Chemistry of Natural Products (later "Shemyakin Institute of Bioorganic Chemistry") of the USSR Academy of Sciences in Moscow. Research on depsipeptides, in 1964 in Zürich with V. Prelog stereochemistry of cyclic peptides; back to Moscow—membrane active peptides, proteins. After Shemyakin's death in 1970 he became Academician, Head of the institute, and at the age of 40, Vice President of the Academy. He died in 1988 after a two years struggle with advancing disease.

*Patchornik, Abraham* (p. 237, Plate 25) born in 1926 in Nes-Ziona, Israel, studied chemistry at the Hebrew University Jerusalem, where he obtained his Ph.D. with Prof. E. Katchalski. In this time he was research assistant at the Weizmann Institute in Rehovot. There he became Professor at the Department of Biophysics in 1968, and Head Department of Organic Chemistry in 1972. His carrier in Israel was supplemented by a Fellowship with B. Witkop in Bethesda, Md. (1957/58) and Research visits with N.O. Kaplan (Brandeis, 1961) and R.B. Woodward at Harvard, 1965/66. Patchornik developed the field of non-enzymatic degradation of peptides and proteins, e.g. by bromination at the indole and imidazol nucleus, studied photosensitive blocking groups and polymeric reagents.

*Pauling, Linus* (p. 257, Plate 34) born in 1901, educated at Oregon State College, Calif. Inst. of Technol. (Caltec) and Universities of Munich, Copenhagen and Zürich. After being assistant at Oregon State Coll. and Caltec, he was there Asst. Prof. 1927–29, Assoc. Prof. 1929–31, Prof. 1931–64; Prof. of Chemistry Univ. of Calif., San Diego 1967–69, Stanford Univ. 1969–74, since then Prof. emer. Member of numerous Academies, among a multitude of degrees and prizes he won the Nobel Prize for Chemistry in 1954, for Peace in 1962. Among numerous fundamental contributions to general chemistry he suggested secondary structure of peptide chains (with R.B. Corey α-helix, pleated sheets, p. 13), Vitamin C and longevity.

*Photaki, Iphigenia*, 1921–1983 (p. 257, Plate 35) born in Corinth, Greece, studied chemistry at the University of Athens. Her research, guided by Leonidas Zervas, led, in 1950, to the degree of Doctor of Natural Sciences. After postdoctoral studies in Switzerland, with Max Brenner and H. Erlenmeyer, and later also with V. du Vigneaud in New York, she rejoined the Department of Chemistry at the University of Athens, advanced in rank and, in 1977, was promoted to full Professor. Her outstanding work, particularly her studies of cysteine- and serine-containing peptides, gained worldwide recognition, but was sadly terminated by her untimely death in 1983.

*Rothe, Manfred* (pp. 40, 196) born in 1927 in Halle/Saale studied chemistry in Rostock. Habilitation in Halle, transferred to the Univ. of Mainz in 1961 where he became associate Professor in 1966, since 1974 full Professor at Univ. of Ulm (peptides of proline, cyclols).

*Rudinger, Joseph.* See biography on pp. 155–157.

*Rydon, Norman* (p. 243, Plate 36) born in 1912 studied chemistry at London (B.Sc. 1931, Ph.D. 1933, D.Sc. 1938), D. Phil., Oxford 1939. Chemical Defence Experimental Section, 1940–1945. From 1945–1947 Member of Scientific Staff of the Lister Institute, then Reader at Birkbeck College, London, until 1949 and from 1949–1952 Reader at Imperial College, London. Professor at Manchester College of Science and Technology 1952–1957, Professor and Head of Department, Exeter University 1957–1977. Rydon's work outside of peptide research was on natural substances; peptide topics at Exeter included cysteine peptides, sequential polypeptides, synthetic studies of ferredoxins. Chlorination followed by starchh-potassium iodide for the detection of peptides is known as Rydon–Smith reagent.

*Sakakibara, Shumpei* (p. 67, Plate 37) born in Kobe, Japan in 1926, graduated from Osaka University in 1951, research associate in the laboratory of Professor Shiro Akabori at Osaka University and received his doctor of science degree from the same University. From 1960 to 1962, he studied in the laboratory of Professor George Hess in Cornell University, Ithaca, where he had an opportunity of handling liquid HF as a solvent of peptides and proteins. After returning from the Cornell University, he was appointed Head of the Peptide Center, Institute for Protein Research, Osaka University as an Associate Professor, he demonstrated usefulness of HF as a deprotecting reagent. In 1971, he resigned from Osaka University, organized the Peptide Institute in the Protein Research Foundation serving as the research director. In 1977, he organized the Peptide Institute, Inc. outside of the Foundation, served as the president, headed a research group for peptide synthesis and made a success of the Foundation's business of distributing biologically active peptides.

*Sanger, Frederick* (Plate 38) born in 1918 in Rendcombe (Gloucestershire), studied at St. John's College, Cambridge. Biochemical research at Cambridge from 1940, Ph.D. in 1943, 1951–1983 Member of the Scientific Staff of the Medical Research Council, Nobel prizes in Chemistry 1958 and 1980 for sequencing polypeptides (insulin) and nucleic acids, resp. "Sangers reagent", 2,4-dinitrofluorobenzene (p. 115).

*Schally, Andrew V.* (p. 172) born in Wilno, Poland in 1926, studied biochemistry at McGill Univ. Montreal. Ph.D. 1957. Assistant at Natl. Institute for Medical Res. England (1949–52), associate endocrinologist at Allan Memorial Institute of Psychiatry, Canada (1952–57), assist. prof. protein chemistry and endocrinology at Baylor College of Medicine, Houston, Tex. until 1962, associate prof. until 1966. Prof. School of Medicine, Tulane Univ. New Orleans, Chief Endocrinol. and polypeptide Laboratories at Veterans Administration Hospital. Nobel Prize for Medicine and Physiology 1977 (shared with R. Guillemin).

*Schwyzer, Robert* (pp. 84, 205, Plate 39) was born in Zurich, Switzerland, in 1920, but he received his primary and secondary education in the United States. The family later returned to Switzerland and he studied chemistry in Zurich, completing his experimental work for his Ph.D. degree under Paul Karrer. After some years in industry (CIBA), he accepted an invitation to join the faculty of the Federal Institute of Technology (ETH) in Zurich, as professor. Schwyzer first synthesized angiotensin (p. 205), gramicidin S (where he detected cyclodimerization) and, last but not least, the 39-residue sequence of ACTH. He proposed cyanomethyl esters for the formation of the peptide bond.

Schwyzer introduced the concepts of "homodetic" and "heterodetic" cyclopeptides. Similar conceptualization of architectural aspects of biologically active peptides and of the relation of sequence and geometry to biological activity was expressed in the nomenclature proposed by him with terms such as "address sequences", "sychnologic organization" (continuate words) and "rhegnylogic organization" (discontinuate words). These thoughts and suggestions, for instance his more recent membrane-compartment model, were not merely intellectually stimulating but also pointed out new directions in hormone research.

*Scoffone, Ernesto*, 1923–1973 (p. 237, Plate 40) born in 1923 in Udine, Italy, studied pharmacy at Univ. of Bologna (Ph.D. 1946) and chemistry at Univ. of Padua (Ph.D. 1950). Postdoctoral work at Rockefeller Univ. New York City in 1959/60. From 1960 associate Prof. Univ. Padua, in 1965, full Prof. of organic chemistry at the same Univ., and director of the Centre for Macromol. Chemistry of C.N.R.

Scoffone established synthetic peptide chemistry in Italy by work of his group on S-peptide of ribonuclease A, new protecting groups and methods of modification and fragmentation of proteins.

*Sheehan, John C.* (pp. 88, 199, Plate 41) born in Battlecreek, Michigan, 1915, received his Ph.D. degree at the University of Michigan in 1941. After a few years in the pharmaceutical industry (Merck) he joined the faculty of the Massachusetts Institute of Technology where he advanced to full Professor and where he remains actively engaged in research even after his retirement in 1976. He also served as Scientific Liaison Officer of Naval Research and as Editor in Chief of the Journal of Organic

Chemistry. For his important contribution to the synthesis of penicillins, amino acids, peptides, alkaloids and explosives Sheehan has received numerous awards.

*Shemyakin, Michail M.*, 1908–1970 (pp. 20, 202, Plate 42). Laid the basis of bioorganic chemistry in the USSR. Academician (1958). Graduated from Moscow State University (1930). Founder and first Director of the Institute for Natural Products, USSR Academy of Sciences. (In 1974 the Institute was given the name "The Shemyakin Institute of Bioorganic Chemistry"). Chemistry of naturally occurring biologically active compounds. With A. Braunstein he developed a general theory of pyridoxal enzyme action. Synthesis of antibiotics, vitamins, amino acids, quinones, total tetracycline synthesis (1966). He published many papers on the structure and function of proteins, peptides, and biologically membranes.

*Shiba, Tetsuo* (p. 222, Plate 43) born in Hiroshima Prefecture, Japan in 1924, received his Ph.D. degree at Osaka University in 1959. After having studied at the National Institute of Health, Bethesda, USA, as Visiting Scientist for two and a half years, he returned to Osaka University as Associate Professor in 1962. In 1971 he was appointed Professor, Laboratory of Natural Product Chemistry, Department of Chemistry, Faculty of Science, Osaka University. Here he developed his structural and synthetic studies on biologically active substances (peptides and others). In the meantime, he served as Vice President of the Chemical Society of Japan, and now as a Member of the Science Council of Japan. He successfully planned the Japan Symposia of Peptide Chemistry and the first international Symposium of Peptide Chemistry in Japan as President of the Organizing Committee in 1987.

He retired from Osaka University in 1988, however, he is still continuing his scientific activity as the Director of the Peptide Institute, Protein Research Foundation, Japan, which was founded by S. Akabori.

*Stein, William H.* 1911–1980 (p. 50, Plate 4) was born in New York City, in 1911. He graduated from Harvard University in 1933 with a B.Sc degree, after one year he transferred to Columbia University. Research guided by Hans Clark, on the protein elastin had a lasting influence on the direction of his scientific interests. Another Columbia professor, Erwin Brand, introduced him to the recently arrived Max Bergmann. Thus, after graduation from Columbia with a Ph.D. degree in 1937 he joined Bergmann at the Rockefeller Institute where he remained throughout his career. At Rockefeller the young researcher was surrounded by other young scientists of real excellence, to name a few: Stanford Moore, Joseph S. Fruton, Emil Smith, Klaus Hofmann, Paul Zamecnik. W.H. Stein participated in an effort aiming at the quantitative determination of the amino acid composition of proteins by the "solubility product" method of Bergmann (p. 48). In the years following Bergmann's death in 1944, Stein and Stanford Moore applied chromatographic procedures for the separation of amino acids in the hydrolysates of proteins, at first partition chromatography, later by application of ion-exchange chromatography (p. 50). A fully automated instrument, the "amino acid analyzer" was developed in collaboration with D.H. Spackman in 1958. Stein and Moore together with C.H.W. Hirs, addressed themselves to the elucidation of the structure of the enzyme ribonuclease A, and they solved the entire covalent structure of the chain of 124 amino acid residues (p. 51). Dr. Stein's work received wide recognition. When the Rockefeller Institute was reorganized to Rockefeller University he was appointed Professor. In 1972 he received the Nobel Prize in Chemistry (shared with Stanford Moore). In 1980 he died, paralyzed by a rare disease, in New York.

*Synge, Richard Laurence Millington* (p. 52, Plate 44). Born in 1914 in Liverpool, he studied Biochemistry at Cambridge, Int. Wool Secr. 1938, Biochemist at Wool Industries Research Association, Leeds 1941–1943, Staff Biochemist Lister Inst. of Preventive Medicine, London, until 1948, then Head of Dept. of Protein Chemistry, Riwett Research Inst. Bucksburn, Aberdeen (1948–67), Biochemist, Food Research Institute, Norwich, 1967–1976. Shared Nobel Prize for Chemistry with A.J.P. Martin, 1952 for invention of partition chromatography.

*Taschner, Emil*, 1900–1982 (p. 239, Plate 5) was born in Cracow, studied chemistry there at the Jagiellonian University in 1920, later in Vienna where he received his Ph.D. degree in 1927. After postgraduate studies in Paris (Pasteur Institute, University) he returned to Cracow to work in the Pharmacological Department and in pharmaceutical industries. The years of World War II he spent in Lwow as a workman in hiding. After the War he moved to the Chemistry Department of the University in Wroclaw, in 1953 he was appointed Head of the Department of General Chemistry at the Technical University of Gdansk, later of Peptide Chemistry Department until his retirement in 1971. Taschner was the founder of peptide chemistry in Poland in the early fifties with studies of protection, deprotection and racemization of amino acids.

*Ugi, Ivar* (pp. 99, 233) born in 1930 in Estonia studied chemistry at the Univ. of Munich, became lecturer there in 1960. Honorary Professor Univ. Cologne in 1967, 1968–1971 Professor of Chemistry Univ. of Southern California, Los Angeles, at present Professor at the Technical University Munich. 4-Component peptide synthesis.

*Wang, Yu* (pp. 161, 231, Plate 46) born 1910 in Hangshou, China, he studied in Nanking, Peking and graduated in 1937 in Chemistry at the University of Munich, Germany. After one-year's postdoctoral work with Richard Kuhn at the Kaiser–Wilhelm–Institute for Medical Research in Heidelberg (1938/39) he advanced via Peking to Shanghai where he, as director of the Institute of Organic Chemistry of the Academia Sinica, has been working in different fields of bioorganic chemistry since 1950. Essentially involved in the synthesis of insulin.

*Wessely von Karnegg, Friedrich*, 1897–1967 (pp. 29, 37, Plate 47). After being severely wounded in the First World War, he studied in Vienna from 1919 and graduated there under Franke in 1922. Among his fellow students, we find names such as Hermann Mark and Richard Kuhn. His interest in amino acids and peptides was aroused during the years 1923–1924 spent in Berlin–Dahlem at the Kaiser Wilhelm Institute for Fiber Research where he worked on silk-fibroin. Although he had other interests, he remained faithful to this area of research for over 30 years. At the 2nd Institute of Chemistry of the University of Vienna he was engaged in studies on carbonyl-bisamino acids and collaborated with E. Späth in work on complex natural products of the terpene series. He was named "Privatdozent" in the early thirties. "Extraordinarius" in 1937 and in 1946 invited to be Professor and Chairman of the Institute for Medicinal Chemistry of the University. After the sudden death of E. Späth in 1948, Wessely was appointed as his successor. His last publication from the peptide field appeared in 1957; he died suddenly in 1967. His life outside the laboratory was dedicated to music, he was himself an accomplished pianist, and to the mountains. An artificial leg notwithstanding he skied in the Alps and was an excellent swimmer. An entry in the books of the Polar Institute of Norway in Oslo shows a Wessely Peak on the Magdalena Fjord in the Spitzbergen.

*Weygand, Friedrich*, 1911–1969 (pp. 70, 129, Plate 48). Born in 1911 in Reichelshausen (Upper Hassia) studied chemistry in Frankfurt a. M., moved in 1934 to the Kaiser–Wilhelm–Institut für medizinische Forschung in Heidelberg where he took his doctorate with Professor Richard Kuhn (1935) and worked in the Institute until 1940, interrupted by a one-year's stay at Oxford with Sir Robert Robinson (1935/36). He worked in Heidelberg until 1943 when he moved to the University of Strasbourg, then as associate professor to Tübingen, in 1955 as Full Professor to the Technical University of Berlin, and finally in 1958 as director of the Organic Chemistry Department of the Institute of Technology (now Technical University) in Munich. Beside peptides (trifluoroacetyl derivatives, gaschromatography, mass spectroscopy, p. 129) his interest was also in carbohydrates and heterocyclic compounds.

*Wieland Theodor* born in 1913 in Munich, he studied chemistry in Freiburg Brsg. and Munich where he received his Dr. phil. in 1937. From 1937 until 1946 he was assistant, (Lecturer 1941), with Richard Kuhn at the Kaiser–Wilhelm/Max–Planck–Institute in Heidelberg. In 1946 he became Prof. at the Univ. of Mainz. Between 1951 and 1968 he was Prof. and Director of the Institute of Organic Chemistry at the Univ. of Frankfurt. After R. Kuhn's death in 1967, he was appointed Director of the Department of Chemistry, later Natural Products, of the Heidelberg Institute until 1981. He is hon. prof. of the Universities of Frankfurt and Heidelberg. His contributions include electrophoresis on paper of amino acids, peptides and proteins, isoenzymes of lactate dehydrogenase (with G. Pfleiderer), mixed anhydride method of peptide synthesis (p. 79) peptides of *Amanita* mushrooms (p. 211).

*Witkop, Bernhard* (pp. 117, 209, Plate 49) was born 1917 in Freiburg/Breisgau. He studied chemistry at University of Munich and there received his Dr. phil. in 1940 (Heinrich Wieland) and became Lecturer in 1946. He became Mellon Fellow at Harvard University in 1947. Instructor and Lecturer 1948–1950; Special Fellow U.S. Public Health Service 1950–1953. He was at Natl. Institute of Health. Bethesda, Maryland USA. from 1953, there the Chief of the Laboratory of Chemistry of the Institute of Arthritis, Metabolic and Digestive Diseases from 1957 until 1987. Witkop, a very all-round scientist (alkaloids, oxidation mechanisms, pharmacodynamic amines, amphibian venoms etc.), in peptide chemistry, studied hydroxyamino acids, non-enzymatic cleavage of proteins (with E. Gross cyanogen bromide cleavage, p. 117), gramicidins.

*Woodward, Robert Burns*, 1917–1979 (Plate 50). Born in Boston in 1917, he entered Massachusetts Instutute of Technology in 1933 and graduated as Ph.D. in 1937. Postdoctoral fellow at Harvard (1937–40). Instructor in Chemistry (1941–1944). Assistant Professor (1944–46), Associate Professor (1946–50), Professor from 1950 until his death in 1979. Nobel Prize in Chemistry 1965, "for his outstanding achievements in the art of organic synthesis". Woodwards reagent (p. 90), synthesis of cephalosporin C (p. 200), poly-$\alpha$-amino-acids (p. 37).

*Wünsch, Erich* (p. 168, Plate 51) born in 1923 in Reichenberg, Bohemia, began studying chemistry in 1941 at the Karls–University, Prague, and after Army service and being a prisoner of war, resumed his studies in Regensburg, Bavaria, in 1946. Graduated Dr. rer. nat. at the Ludwig–Maximilian University in Munich in 1956 and Dr. habil at the Technical University in Munich in 1956. There he was nominated professor in 1973 at the same time he was scientific member of the Max–Planck–Gesellschaft, and director of the department of peptide chemistry at the Max–Planck–Institute for Biochemistry in Martinsried near Munich. Methods of protection in peptide synthesis, synthesis of hormones, author of 2 volumes, Peptides in Houben–Weyl–Müller's handbook (see p. 62) are among his achievements.

*Yajima, Haruaki* (p. 238, Plate 52) born in Takakarazuka, Japan, in 1925, received his Ph.D. degree as an alkaloid chemist at the Kyoto University in 1956. He then studied peptide synthesis with Klaus Hofmann in the Department of Biochemistry at the University of Pittsburgh until 1962, mainly contributing to the first synthesis of a tricosapeptide with full ACTH activity (p. 164). He returned to Japan in 1962 as Associate Professor at Kyoto University and was promoted to full Professor in 1973. After his retirement in 1989, he moved to Niigata College of Pharmacy as president. Synthesis of a great number of biologically active peptides, also ribonuclease. A are some of his accomplishments.

*Young, Geoffrey Tyndale* (pp. 95, 244, Plate 53). Born 1915. Graduated in Chemistry at the University of Birmingham in 1936. Then moved to Bristol University, working on carbohydrates with E.L. Hirst. In 1938, he received his Ph.D. and was appointed Assistant Lecturer. In 1939–1943, he did war research on explosives at Bristol University. From 1943 to 1945 he was with the British Commonwealth Scientific Office at Washington, U.S.A., exchanging information on war research. In 1947, he was elected Fellow and Tutor in Chemistry, Jesus College, Oxford; 1952, University Lecturer in Organic Chemistry; 1970 Aldrichian Praelector in Chemistry. 1973–1977, Acting Principal, Jesus College, Oxford. 1982, Emeritus Fellow, Jesus College. 1983, O.B.E. (Order of the British Empire). One of the founders of the European Peptide Symposia in 1958.

*Zahn, Helmut* (p. 161, Plate 54) was born 1916 in Erlangen and studied chemistry and graduated at the Technical University in Karlsruhe (Dr. Ing. in 1940). From 1940 to 1949 he was assistant to Professor Elöd at the Institute for Textile Chemistry in Badenweiler, Baden, from 1949–1957 Assistant, Lecturer and Associate Professor at the University of Heidelberg. From 1952–1985 he was Director of the German Institute for Wool Research and in 1960 was appointed full Professor at the Technical University in Aachen. With his team (Plate 54) he was the first to obtain insulin by chemical synthesis.

*Zervas, Leonidas*, 1902–1980 (p. 46, Plate 55) was born in Megapolis in the Peloponese, Greece. He studied chemistry at the University of Athens and then in Berlin where, in 1926, he received his Ph.D.

degree. The experimental work leading to this degree was carried out in Dresden at the Kaiser Wilhelm Institute and was gruided by its director, Max Bergmann (see p. 44). After graduation, Zervas remained in Dresden as research associate; a few years later he was promoted to Head of the Organic Chemistry Section and then to Associate Director of the Institute. After the resignation of Bergmann in 1933, Zervas stayed in Dresden to complete the ongoing studies but a year later he followed Bergmann to New York City to participate in his research at the Rockefeller Institute.

In 1937 Zervas returned to Greece where he was appointed Professor at the University of Thessaloniki. Two years later he accepted an invitation to the University of Athens, where he remained, even after his retirement in 1968. He became a member of the Academy of Athens, was named, in 1964, Secretary of Industry and was one of the initiators of The National Hellenic Research Foundation. In spite of such new responsibilities he continued to be active in the laboratory until the end of his life. His carbobenzoxy group and the consequences for peptide chemistry are discussed in chapter 3.

# Author Index

Boldface page numbers refer to illustrations

# Subject Index

Page numbers in italics refer to figures and formulas